THE BENEVOLENCE OF ROGUES

by

JOHN RIGHTEN

Thank you to the Rogues, alive and dead, who saved my life and the lives of others.

John

Contents

Chapter 1 – Catholic Tastes 1

Chapter 2 – Growing Pains 16

Chapter 3 – School for Scandal. 28

Chapter 4 – The Joker is Wild. 42

Chapter 5 – Nurse! 55

Chapter 6 – Teenage Kicks 63

Chapter 7 – A Grave Matter 68

Chapter 8 – Please, There's No Need for Tanks 81

Chapter 9 – "There are No Homosexuals in the Soviet Union" 86

Chapter 10 – Foreign Affairs 98

Chapter 11 – The Special Relationship 106

Chapter 12 – The Greenhouse Effect 118

Chapter 13 – A Flock of Cuckoos Flying Over the Nest 123

Chapter 14 – I Think I'm Turning Japanese 128

Chapter 15 – The End of an Error. 137

Chapter 16 – The Simple Bear Necessity 146

Chapter 17 – "When I Grow Up, I Want to Join a Boy Band" 155

Chapter 18 – This is How Wars Begin 163

Chapter 19 – "Why Are There Not More Lesbians in the World?" 178

Chapter 20 – Death by Stiletto 195

Chapter 21 – "It's Alright For You, I Have To Live In Here!" 221

Chapter 22 – The University of Hard Knocks 228

Chapter 23 – A Game and Too Many Halves 246

Chapter 24 – Magpies and Custard..276

Chapter 25 – That's a Relief...293

Chapter 26 – Know Your Left From Your Right...310

Chapter 27 – "Englander!"..323

Chapter 28 – If Only I Had Learnt Latin at School..333

Chapter 29 – The Demise of Rogues..355

Chapter 30 – Trumpton-The Lunatics Have Taken Over the Asylum..........................362

Chapter 31 – If Darwin Had Found Jockey on the Galapagos Islands?..................381

Chapter 32 – The League of Extraordinary Lunatics..413

Chapter 33 – The Empire Strikes Back..427

Chapter 34 – Was Ever a Man More Misunderstood?...457

Chapter 35 – Rule Britannia...464

Chapter 36 – It's Not the Years, it's the Mileage..474

Chapter 37 – With Friends Like These Who Needs Enemies?.......................................492

Chapter 38 – Twilight of the Godless...497

CHILDHOOD ROGUES

"Why does everyone take an instant dislike to you?"

"Saves time."

Spike Milligan

AID WORKER'S MISSIONS FIND UNLIKELY SUPPORT FROM PRISON FORGERS, GANGSTER's HENCHMEN AND SYMPATHETIC POLICE
Hampstead & Highgate Express – Interview 21st June 2012

'John Righten has been in the wrong place at the right time since the 1980s. Then, he was in Romania, delivering medical supplies to orphans suffering from Aids. Subsequently he was in Bosnia in the 90s, sneaking in medical supplies and in South America – Brazil, Chile and Peru – during the 2000s. Righten is now back and has put together his experiences in his autobiography, The Benevolence Of Rogues. "The title of the book is really referring to people who many would class as rogues,' 'but then who do these things to help other people that most wouldn't expect them to do," says Righten. To decipher what Righten means by "a rogue" you need only take a look at the book. "Looking back over the years," he writes, "I've formed strange alliances. These ranged from friends in military intelligence and sympathetic police officers who carried out criminal checks on prospective co-drivers joining me on humanitarian convoys; to forgers in Pentonville prison who assisted me in drawing up papers to get medical supplies across hostile borders." He was even abducted by the henchmen of a notorious gangster, who, to his surprise, offered to help him get diabetic drugs over the border to help sick children. Righten has been criticised by some for his treatment of such serious topics with humour. "Some people think it is in bad taste, but I think that when you are in such a situation'…'you have to have humour, otherwise you wouldn't get through it."

Kirkus Independent review (US) 'The World's Toughest Book Critics.

'Debut memoirist Righten describes his working-class childhood and erstwhile young adulthood amid a motley crew of relatives and friends in Ireland and England. Righten's memoir will remind readers of a drunken evening in a pub spent listening to tales of a sordid and colorful life. He takes turns as a rollicking fighter, pub denizen and gambler before, eventually, becoming a humanitarian. With a sly, ironic voice, Righten avoids sentimentalizing his life by undercutting every harsh observation with humor; ever the "Fenian bastard," Righten has a gift for rendering the eccentricities of his friends and relatives in a comedic way. Scenes of brawls, workplace shenanigans, football matches gone awry and wacky travel anecdotes are lively and engrossing. However, this is not a memoir for the straight-laced, politically correct or faint of heart: Massive quantities of alcohol are consumed, many teeth are knocked out and sarcasm is in generous supply. Righten's life philosophy, represented in the title, plays a large part in the narrative. He believes that even the most unlikely of rogues has a moral compass and is capable of unexpected acts of kindness. Of course, the author employs a uniquely flexible definition of good, which may include violence against cheating husbands and abusive fathers. But readers will find it difficult to disagree with the thrust of Righten's arguments in the second half of this memoir, which focuses on the author's career as a humanitarian worker in Bosnia and Latin America. His run-in with a sociopathic mercenary is particularly chilling, as are his descriptions of delivering medicine to hospitals filled with traumatized, needy children during the Bosnian War. Righten's own near-death experiences will convince readers that the author is one of these benevolent scoundrels he so admires. A roguishly charming memoir. -

Chapter 1 – *Catholic Tastes*

I am nothing special and I have achieved little. I have not averted a major disaster or caused one. I have no discernible artistic or scientific talents. I am not a leading politico, which is the normal route for a talentless individual. I have not achieved ephemeral fame by washing my backside live on terrestrial television or delivering my 'gift', as Chelsea Mark calls it, to a famous pop star, actress or obscure member of the Royal Family. I have met some outrageous characters, rogues as I call them, but then who hasn't? But these men and women would, on occasion, use their wit and guile to perform acts of extraordinary kindness. I call these characters benevolent rogues and it's their stories I now wish to tell.

My story is the spine that brings together their tales. Please don't feel duped when I make reference to myself as the narrative is based on my experiences, after all. This is the only method I can think of to bring coherence to chaos.

I was five years old when my father died. I can't get sentimental about him, since I have hardly any memories of him. Grieving, I believe, should be the rightful domain of those who were close to the deceased. He counted for fifty percent of my Irish genes, my mother Biddy accounts for the other half. Biddy's real name is Bernadette, but I called her Biddy after a donkey I saw on a beach in Clacton when I was a kid. The donkey held me with a fixed permanent stare which reminded me of my mother's look of disapproval.

Soon after my birth, there was confusion over my parentage. Let me explain. On the day of my birth, my father, a forward thinking man in some respects, had already begun the wetting the baby's head. So by the time the main event occurred he could not find the hospital. It's possible that he didn't even try; it was probable that he had forgotten why the celebrations had been initiated in the first place. Meanwhile, my Uncle Paddy had convinced the maternity nurse that he was the father of Little Johnny, so he could congratulate the mother, confirm the sex of the child and then claim on the bets he'd taken on the outcome. Paddy and his partner Julia, Biddy's sister, were admitted, only for my real father to appear minutes later. This convinced the pious nurse that my mother was a bigamist or worse "Swedish" as she called it – perhaps having two lovers was seen as typical Scandinavian behaviour at the time.

There was little love that I could evidence in my parents' relationship. One day, I approached the difficult topic of parental departure.

"Biddy, when you kick the bucket, shall I drop you down the chute with dad?"

"You stick me with him and I will haunt you till the day you die," she replied.

Despite the absence of love, and perhaps even of passion, Irish women seem to have a gene that ensures they are able to produce a football team (including substitute), irrespective of such obstacles as contraceptives, flaccid members, zero sperm counts, or no physical contact with their spouse whatsoever. My father's death was the only absolute guarantee that I would remain an only child.

Biddy and my father first met in The Buffalo, a notorious Irish club in the north London area of Camden Town. The chief bouncer had lost an arm whilst trying to dislodge a branch from a combine harvester. Prosthetics were presumably rather limited at the time and the National Health Service replaced his arm with what looked like the base of a metal lamppost. In the event of trouble, his technique was to stand just behind the main antagonist and swing his leaden arm, with the aid of his good arm, in a pendulum motion. After a few swings, his metal arm would reach its highest point and then descend onto the skull of his intended target. The risk was that while the bouncer was getting his mechanical arm into position, his target would have gone home and would be drinking a mug of cocoa in bed, before the metal arm descended. However, the bouncer was aware that most Irishmen, once they are in position to fight, hardly ever move, as it takes both combatants a good ten minutes to focus on each other due to the amount of alcohol they've consumed and anyway. Irishmen also rarely drink cocoa.

At the time of Biddy's marriage, she was thirty years old. Her best friend had recently married and moved up to the North of England – Biddy was left all alone. My father had no recognisable social skills, was drunk most of the time and was, surprisingly for a young Irishman with such attributes, unmarried. They wed for no other reason than they were expected to, as both were Irish and still alone.

They married in a bitterly prejudiced London in the fifties. There were notices on pub doors stating 'No Dogs, No Irish'. Despite this and the 'Troubles' in Northern Ireland, the vast majority of Irish men and women who came to England at that time were peaceful. Most wanted to work, to raise a family and, when possible, send money back to those who remained in the 'old country'. Despite the occasional sensationalist headline, we are fortunate to live in one of the most liberal countries in the world. When we talk of what it means to be English, 'tolerant' should be up there alongside other descriptive labels such as 'bulldog', 'St George' and 'lager-induced violence'. If you are reading this in a cafe, bookshop, or jail-cell, look up and observe the various races and creeds that surround you. Look at the melting pot we live in and the cultures, languages, foods and perspectives which all add their own distinct flavour to the mixture. However, the melting pot does need constant attention so that all the constituents mix, or you'll end up with isolated lumps and risk the pot overheating. Of course, I am assuming you are not reading this in the London suburb of Warshalton, where the turban and Burka are donned purely for fancy dress or as a disguise to cover visits to the sexual health-clinic.

Like many Irish people, my father was one of those who had come to England to find work. When he landed in England, aged seventeen, he received the first of many letters from Her Majesty the Queen. However, on further reading, it was not an invitation to the Palace to have tea with the corgis, but a note to inform him that he was being called up for National Service. My father, a forward thinker as I mentioned, pondered his responsibilities as an immigrant to his new country for a good three seconds and concluded that 'bollocks' was an acceptable response. A few weeks later, two military policemen tracked him down to the building site where he was working, to challenge his rationale. The two military-carthorses, as

I later heard them described by my Uncle Paddy, highlighted the benefits of becoming a member of Her Majesty's Armed Services, by sticking their foot-long truncheons under his chin. My father immediately saw the logic of their argument and agreed to go quietly. First though, he asked if they would accompany him to collect his gear from the work hut. As soon as he was away from public view, he made his escape by kicking both carthorses in the privates and diving out the window.

My father was in the minority amongst his fellow Irishmen as he found it hard to completely assimilate. Many Irishmen did indeed enlist for National Service and went willingly. It was not that the male members of the Righten clan were averse to conflict, in fact far from it – many of my ancestors fought in the engagement known to the Irish as the 'English problem' and to the rest of the world as the 'Troubles.' For my father, assimilation was, on the whole, not a problem. He bore no animosity to his new country but he had no wish to be enlisted in anyone's army. He was just here to have fun.

As I grew up, my family rarely spoke of my father. I recently discovered a rare wedding photo of him taken from the left-hand side. It showed a black eye that would be a credit to any panda. Unsurprisingly, on his stag night he had had an encounter with a nightclub bouncer. Every other wedding photo on view showed his profile from the right. I still, to this day, have not seen a single shot of him on his wedding day taken head-on. Yet the wedding photos were the only pictures of him in the houses I grew up in. Until the discovery of the photo, I had the impression that he had some deformity. I was even of the view that he had an embarrassing two-foot-wide head, like Steven Spielberg's alien E.T. He would have been less of an embarrassment to the family if that was the case.

The Rightens, as a proud auntie once informed me, are descended from the Irish aristocracy. As with many an Irish 'factual' account, the truth is slightly different. The Rightens are descended from French mercenaries who aided William the Conqueror's Norman armies when he invaded England. The Normans slaughtered the indigenous Anglo-Saxons, in addition to sticking an arrow in their King's eye. This resulted in the defeated locals fleeing towards the hills to Wales. Despite the invasion being over, and the locals subjugated, the Rightens kept fighting. The Normans and the defeated Anglo-Saxons had, by this time, settled down for dinner. William wanted to consolidate his victory, but now had an embarrassing problem in trying to accommodate the violent tendencies of the Rightens tearing up the kitchen. I guess we were the Gordon Ramseys of our time, full of rage, rampaging around, though without the slightest interest in food preparation. William enticed the Rightens to leave England and head to Ireland, no doubt with promises of even greater carnage. My patriotic auntie was correct in linking us to the Kings of Ireland, as the Rightens then proceeded to massacre the lot of them. Amends were made later as many Rightens died in the establishment of the Irish Republic. Whether that was patriotism or hereditary blood lust is still up for debate.

A work colleague once tried to goad me into a fight by pulling out a book entitled *The Origin of Surnames*. He proudly read out to all that Kennedy originated from the medieval name *Ugly-head*. I responded that this was fascinating, but that my name wasn't Kennedy but Righten. After reading that the medieval French origin of my surname was *Attacker*, he returned to his book. Today, a growing number of people desperately try to bring colour to their lives by creating tenuous links to a famous, even infamous, ancestor. This trait is often exhibited by hanging a coat of arms on the toilet door or by pointing at some supposedly famous rollicking ancestor's picture super glued on to a personalised tankard containing low-alcohol bitter.

Recently, my doctor checked me over with the sole aim as she put it "of trying to discover how I am still alive," rather than to find out if there was anything wrong with me.

"Any history of family illness?" she asked.

"No," I replied.

"No?" she responded, incredulously.

"No, all the women in my family live to their nineties, while generations of young male Rightens have died due to wars or accidents; none made it past forty. I am the oldest male Righten for generations and have no idea who or what will punch my clock."

At that point, I conjured up an image of my demise because of Bear falling on me, but more of that fine example of 'the best of British' later.

My father died at the age of thirty-four. His was a long painful death from injury having falling sixty-feet from a scaffold and caving in his skull. Despite his failings, he did keep up the family commitment to the 'Darwin Awards.' This is the term given to the method of improving the human gene pool, by removing oneself from it. A recent award winner thought it was a good idea to header a bowling ball. Later, the story went that my dad had heard some old cockney ex-RAF pilot working on the building site, talk about how he had "Flown in Wellingtons during the war" and my father took it literally.

An Irish woman with a teapot in her hand is the nearest physical science has produced to perpetual motion. In addition to this, elderly Irish women never listen to anything, particularly when in conversation. Biddy, God bless her, is a typical yet fine, exemplar of this phenomenon. A rare example of her listening to what someone had said, was when as a teenager, I told her that two friends of mine H and Bourgeois were popping around for five minutes to pick me up for the Chelsea game at Arsenal's Highbury Stadium (more commonly known as the 'Library' due to its lack of atmosphere). On receiving news of impending guests, she dropkicked a joint of meat into the oven. This in itself did not imply that she had heard what I had said; her violent bombardment of the cooker with lumps of meat was a daily occurrence. Without any expectation of a reply, I told her that Bourgeois was vegetarian and as I expected I didn't receive one.

When H and Bourgeois turned up, Biddy corralled them to the dinner table. As we sat down for lunch, Bourgeois prepared himself for a small rabbit portion of lettuce leaf. To his horror, three thick slices of beef were heaped on to his plate with an implement that was similar to, although smaller than, a coal shovel. In abject fear of this shovel laden Irish woman, he bravely stuttered.

"I am ... sorry, Mrs Righten, but ... Johnny should have told you I am a ... vegetarian."

"He did," replied Biddy, "that is why I only gave you a little bit of meat."

Biddy and I do not have what you would call a typical mother and son relationship. Although we love each other, it is not laced with any 'blood is thicker than water' sentiment. Our relationship is based on merits and God help the other if they do not keep their wits about them.

When I was a cocky ten-year-old boy, Biddy took me to Hampstead Heath funfair in north London. Biddy had been pontificating all day about her great love of Irish culture, in particular its playwrights, so I drew her attention to one of the fair's exhibits. "Biddy, there is an exhibition here of great Irish painters such as Wilde, Joyce and the excellent landscape artist Brendan Behan."

"Lovely," she exclaimed, "they're my favourites."

Off we went to the exhibit where such great works of 'art' were hung in a wooden cylindrical chamber. After the door was locked behind us, the room began to spin and the floor disappeared. This, and the fact that there were no pictures hanging on the revolving wall, made Biddy slightly suspicious. The funfair's 'Rotating Chamber' was not what Biddy expected. She was pressed to the wall sideways as the chamber rotated at a speed of twenty miles per hour. The language emanating from the old woman stuck to the wooden panels was shocking. Believe me, the children, who were stuck to the panels like butterflies in a lepidopterist scrapbook, were more petrified by Biddy than the floor receding below them. When the ten-minute cycle was over, Biddy fell to the ground like a cat emerging from a tumble dryer. Naturally, I was through the door quicker than shit off a shovel; the words of my dear mother "I'll fucken kill ya, ya little fecker", ringing in my ears.

There was one member of my family whom I adored – my Auntie Julia, Paddy's 'wife'. She was not the brightest of Biddy's nine sisters or the richest, but she was the most loving and the funniest; two traits that kids love even more than presents. She always took an interest in me and never judged me. Despite my insurmountable shortcomings, her love for me was unqualified. If there is a heaven, my lovely aunt will be there, though I imagine that she would have accidentally trodden on God's toe quite a few times by now, for as you will discover she was slightly accident-prone.

Biddy was the first of the family to cross the water to England. Julia followed a few months later aged just sixteen. Biddy or Julia hardly talked about their early years, but one day Biddy described the first family she worked for.

"Lazy bastards – I even had to turn the lights on for them at the weekends."

I thought it was best to leave her to her angry innocence rather explaining that it was the Sabbath and they were Jewish. The pay for this and similar unskilled work was poor, so such jobs would be left to the latest influx of immigrants, on this occasion the Irish, West Indians and Philippinos. While immigrants shared menial jobs, it did not mean they shared a bond and that all was harmonious. Left-wing activists romanticise about such relationships but poverty is not an indicator of man's greater humanity. The norm for your average worker is to hang on to a job, provide security for a family and secure any available positions for those of a similar race and creed. This strengthens your standing in the community, as well as providing greater protection through safety in numbers for you and your family.

These days it isn't unusual to hear members of immigrant families, and not just white immigrants, talk of 'Pakis' or 'Gypos' flooding the country. Some, without thinking, adopt the bigotry of their peers to raise themselves up the social hierarchy at the expense of the latest influx of immigrants. Maybe it's part of the human condition, a tribal trait; as we move up the pecking order as soon as another group or individual can be identified to take our place.

Julia's first job was cleaning at the local hospital in London and she spoke of being welcomed into work each morning by her West Indian workmates singing Tommy Steel's hit record of the time *The Little White Bull*. Julia did not see the harm in it and said it was one of her favourites. I loved her innocence and warmth. I am sure this story was part of the camaraderie at her workplace yet it made it clear to Julia that she was in the minority and it taught me racism can be found anywhere and it is indiscriminate of race or colour.

My Uncle Paddy, Julia's 'husband', was a porter at the same hospital and was always around in my childhood, with the exception of my time in Ireland. Uncle Paddy was not really my 'Uncle', as I found out years later. When he met Julia, Paddy was married with four kids but as soon as number five popped out his wife eloped with the manager of the local Odeon cinema. I do not know, or care, who was at fault since Paddy always treated Julia with kindness and that was good enough for me. However, perhaps it did explain why as a kid he never took me to the pictures.

Paddy was the first rogue I encountered. He was known as 'the Bookie' on the hospital wards; taking patients' bets only to pocket the money and take the next day off sick if they won. He would only return to work after the winning patient had died or been discharged. He was also always up for a scam. Once he tried to claim insurance on an Afghan carpet stating on his claim form that it had caught fire in his living room. The insurance assessor challenged Paddy about the lack of scorch marks on the floor although the carpet remnants were now ashes. Having been rumbled, Paddy once again went missing for a few days.

Paddy, like Julia, could make me laugh and had some tales to tell. As a teenager in the Republic of Ireland, he could not find work. Paddy was older than my parents and as this was in the early nineteen forties, rather than heading to England, he headed to Northern Ireland to take the 'King's shilling'. This was the term for the payment made to those willing to enlist in His Majesty's Armed Services. Like my father, he cared little for the military and confessed to me years later that he had never heard of the word fascism, let alone known of its threat. He just wanted to experience the world and have a job to put money in his pocket. Like so many others, he lied about his real age; he was too young to enlist. Such stories are so common that I think we fought the Axis powers with an army in schoolboy shorts armed with lollipops. In Paddy's case it was true; following his death, I came across his birth certificate; he was sixteen when he was sent to Burma.

Within five days of leaving his parents' home in Cork; he was fighting the Japanese in Burma's jungles. His life was immediately in peril, but not in the way he had imagined. His first commission was to clear away a troop of baboons, so that his platoon could make camp for the night. He walked over to the largest baboon, known as the Chief by his comrades, with a coconut in his hand. Paddy's platoon, aware of the fault in his plan, ran in all directions, diving over the nearest bushes for safety. The Chief looked up nonchalantly as Paddy raised his arm and brought the large coconut down on top of the spiky-haired baboon's head. The enraged baboons, led by the now flat-topped haired Chief, went ballistic and attacked the entire platoon, inflicting pain and humiliation on all and sundry. Paddy spent a week in hospital next to his lieutenant and others in his platoon who had incurred injuries in the conflict. Next stop on Paddy's world tour was the Japanese army, though he later told me.

"They were easier to fight than those hairy, purple-arsed bastards."

As a kid, I wanted to hear heroic war stories but Paddy rarely talked about the war. I do know that even till the day he died he never forgave the Japanese, when he saw the condition of prisoners of war released from camps in 1945. He was an emotional man and, years later, Julia told me that if a war film was on the television his eyes would well up with tears; no doubt he was thinking of comrades lost. It may also be due to memories of the winnings he had to pay out to any survivors, as he was also known as the Bookie back then. As a little boy, I understood none of this and I continually pressed him for more stories of times he was trying to forget. Kids like me would pay money to read about war exploits in comics and go to the cinema to watch war films. Thinking about it now, knowing Paddy, I'm sure we could have reached a deal where I could have simply paid him for his stories by running his bets back and forth from the 'terminal' ward.

I kept asking him about a long scar he had down his back, but he always refused to say, until he got a little bit tipsy one Christmas night and relented. He related a tale of one night in the jungle when he went to take a leak against a tree and as he did so blood appeared in the growing circle around his feet.

"What happened?" I asked, completely wide-eyed in anticipation of the heroic tale that was about to unfold.

"At first I thought my dick was rotting again from one of the women from the brothel in town."

My mouth fell open, as I did not remember this storyline in my any of my war-comics.

"But as I looked behind," he continued, "I found this little Japanese bastard sitting dazed on the ground with a machete in his hand. I was lucky; he must have been hiding in the branches above me and jumped down on top of me to stab me while I was having a leak. Fortunately, their daggers were short and curved at the end; it only sliced through my skin. If he had used a British Army Surplus dagger, it would have cut right through into my entrails."

"What did you do then?" I eagerly enquired, but he refused to go into any detail, as he was not the kind of man who revelled in acts of violence.

His only comment on the matter was: "Johnny that was the last time he would disturb a man when he was pointing his turkey at the porcelain."

This was obviously another member of the animal community he had upset. At that point, I decided that travel does indeed broaden the mind and I too wanted to experience the world, though without aggravating too many baboons or turkeys.

Apart from stirring a desire to discover the world for myself, his story also troubled me. It was not the kind of heroic tale I had imagined from watching old Hollywood stars like

Errol Flynn who, on screen, had defeated the Japanese single-handed in Burma, with hardly a reference to the British who fought there. Paddy, I knew, resented how Hollywood and the press sanitised the war in the Pacific and he described Burma, as the 'dirty war' or the 'forgotten war'. I later read how as many of Paddy's comrades had died from disease as from a bullet. The press and the state withheld news of the true horrors of the war in Asia from the civil population at home, in the belief that it would undermine morale.

I was troubled by Paddy's tale. As a young boy I found it hard to believe that Julia's partner, this very funny, gentle rogue, could be involved in a fight to the death. Perhaps the greatest shock was that you could catch a painful cock-rotting disease from making love to a woman. I did learn one important lesson from all this, though it was many years before I understood what Paddy meant when he told me, "Never get into the saddle without wearing an Englishman's riding jacket."

Despite his shortcomings, Julia fell in love with Paddy, and aged just nineteen; she helped him raise his five children.

Julia went through life in her own inimitable manner, which regularly resulted in Paddy and me convulsing in hysterics. She was a very nervous person, which led to all sorts of problems. One day, Julia decided she would cook a surprise Christmas dinner. The surprise being that the date selected was mid-November.

I happily offered to help Julia prepare our festive treat. Firstly, we went to the supermarket, where Julia bought a bottle of brandy to pour a few drops over the Christmas pudding. As we left the supermarket, a young student carrying out a market research exercise outside Unwins off-licence, asked what her favourite drink was. Julia hesitantly replied, "Nivea Cream," which was a face moisturiser. She meant *Enva* Cream, a kind of dry Sherry. The bewildered student was left with the vision of a mad Irish woman licking out jars of face cream to induce some kind of alcoholic state. Julia and I then set off to the butcher to buy the festive bird.

The butcher asked, "Madam, do you want its insides out?"

A bemused Julia replied, "Err... no thank you."

Later, back at her flat, she looked at the turkey and admitted that she did not have a clue about how to gut it. When I explained that they had offered to in the shop, she sheepishly replied, "I thought he said do you want it inside out."

God knows what disturbing image was in her mind as we walked home.

Julia had also treated herself to a Christmas present of an economy pack of twelve pairs of tights. Julia opened them at the dinner table only to find that they were for pregnant women. Paddy and I exchanged glances, knowing this would really get the family talking. After devouring an excellent Christmas dinner, apart from the plastic wrapped around the giblets still inside, Paddy and I sat back, as Julia set fire to the Christmas pudding, along with the plastic holly and tablecloth after using the entire contents of the bottle of brandy.

A few weeks before he died, Paddy told me that Julia was the love of his life but in all their fifty years together they had never made love. Julia's strong Catholic views forbade her from having sex outside marriage. He was divorced and though he told me he would have happily married Julia the Church forbade second marriages, so the term 'Julia's husband' was used to avoid awkward questions. As a sexually active young man, I found it hard to believe that Paddy, or any man or woman could forego sexual intimacy when they were in love. I also heard that her priest scolded her when she asked him for guidance on marrying Paddy when his divorce came through. It seemed perverse that the institution that professed to love them denied them the opportunity of having a family of their own. I know they would have been the best of parents and their children would never have been short of love or laughter.

It was not only Julia's situation that gave me a jaundiced view of the Catholic religion. I found some of its practices hard to accept. The Confessional absolves you of personal responsibility. All you have to do is go into the darkened 'Tardis' once a week, or in my case daily, repeat chants as directed by a man wearing a collarless frock behind a curtain and round this off with an appearance at church on a Sunday. Monday morning or Sunday night, dependent on your vices, you can start all over again.

Despite support from my mother, the loss of Paddy led Julia to withdraw from the world, mentally and physically. Though her Church had offered little solace at her time of need, I admired Julia for retaining her religious beliefs. Her devotion to her beliefs was unshakable. I have never had any sympathy for those who imply they have some mutual contract with their God, one where their God should deliver eternal happiness, a healthy family and good fortune, and if he doesn't they 'lose their faith'. Julia had no such contract with her maker. She was devoted to her God and accepted with good grace all the pain that came with it. If her God exists, at least he was kind enough to give her an Irish rogue called Paddy.

Recently, leading atheists have gone on the attack and some ridicule the religious beliefs of others. Some rationalist thinkers even imply that believers are somehow less intelligent than them. To attack religious intolerance is one thing, but to attack someone's religious belief in an equally intolerant and superior manner is not only hypocritical, it's irrational. I have often witnessed the comfort religion can provide to those grieving or looking for answers. Paddy on his deathbed wanted to give his last confession, even though he was not a devoted churchgoer. Mind you, he always liked to 'lay-off' his bets. Each to their own, I say. However, when others try to force religious views on me, I give a reading from another little book of mine, '*A Reading from the Book of Intolerance*' and quote a commonly used passage, "Away with ye God botherer," or words to that effect which usually endeth the lesson for the day.

I've experienced how the church can entice you into its arms; while at the same time delivering severe retribution if you stray from the faith. It was the day I received my second induction ceremony into Christ's church, the first being baptism. This was my Holy Communion at the age of seven, a big event in a Roman Catholic child's calendar. I started the day very excited at being part of this huge ceremony, but like the rest of the kids there, I did not really understand what it was about.

When it was my turn to walk up to the altar to receive the 'Body of Christ', in the form of a small circular sliver of bread, I was overwhelmed by the spectacle of it all. There I was, centre-stage, surrounded by all the splendour in the "Temple of our Lord", as the priest called it. It is an excellent device to indoctrinate impressionable children into an archaic institution and I can see why the church invests so much in gold, gilt, candles and incense. If anyone was to invent such a ceremony today, they would be locked up for life for 'grooming' young children en masse.

Yet for all this and though I was enthralled by it, I was by nature a mischievous child. So when the priest blessed himself, I stood up and recited in a low voice; "Get your tent down" (right hand moves from forehead to chest) "and get the fuck out of my garden" (hand shoots from left to right) mimicking a joke I had heard about the Pope on his balcony, clearing campers who had sneaked into the Vatican's garden. At that moment, Sister Winifred, the most violent of all the 'penguins', appeared behind me. She had spotted my blasphemous act and sprinted over from the church organ.

"Why do you think I have a shaving mirror hanging on my church organ you evil boy?" she screamed.

"I thought it was because you have a moustache," I replied

In return for my astute observation, I got a slap on the back of my head and a month's detention. My short dalliance with the Lord was over.

Chapter 2 – *Growing Pains*

While my dad was dying, even Julia could not save me from being sent to Ireland. This was to protect me, for after my father's accident the pain was physically unbearable for him and mentally unbearable for Biddy. He resorted to more drink, which led to further violence until it got to the point where Biddy was worried for my safety, though she too was a risk. Before his accident, I heard he doted on me. However, my vague memories are of a different kind. He was not a bad man and it was not his fault, but because of his severe head injuries, he was in constant pain and prone to lashing out in random fits of rage. As he was working cash-in-hand at the time of the accident, he received no state support. Biddy had to take on more cleaning jobs to support the increasing cost of his medication; his growing reliance on drink, pay the rent and provide food for the three of us. Inevitably, the money ran out earlier each week, as his drinking increased. This is not an unusual story for anyone married to an addict, in this case an alcoholic, with an additional need to dull his increasing pain.

Now that Biddy, as the sole breadwinner, had to work longer hours, my dad and I were often left alone. Biddy asked the family if they could take her and Little Johnny in for a short while, but they all refused, as no one wanted to condone splitting up an Irish Catholic family, irrespective of the reason. They were also terrified of my father. My Auntie Julia was the only one who offered to take us, but she had no room in her bed-sit. Biddy feared that my chances of survival, if I was alone with him when he launched one of his violent attacks were less promising than those of a one-legged man in an arse-kicking contest. As Biddy admitted to me recently, she had to get me away or he would have killed us both. It still left her at risk.

We lived in a small flat in Kings Cross and we all suffered, but none more so than Billy, my pet budgie. I believe he was a pious bird; as my father's bad language seemed to play on his sensitivities. Often, when my Dad was having one of his swearing attacks, he would submerge his little feathered head in his water trough. Poor Billy, those were futile attempts, as his water trough was an impractical instrument for suicide, being only half the size of his head. I think it was a welcome relief when one day Billy suffered a stroke and died just as my dad was informing him that he going to use him to wipe his backside. It didn't help matters that Billy's cage had been punched off its stand and was lying at the other side of the living room floor.

You probably think at this point that I am callous and unfeeling, injecting humour into such a tragic tale, but extracting humour from the bleakest of situations, helped me deal with it. Gallows humour will be found throughout this book, it is not an indication of heartlessness, but a defence mechanism. It's only when I look back at this time, I realise the impact it had on me. It helped me cope with the tough times ahead, in particular, having to live in the same age as the television show *Britain's Got No Talent*.

So, at the age of four, I was smuggled out of England to Ireland, to avoid suffering a fate similar to that of Billy. It was an interesting time; I was dumped on a succession of relatives, each with an armada of kids of their own. I was an unwelcome addition to families already busting at the seams, having a horrible cocky little English kid was the last thing they needed.

Even before I was sent to Ireland, I was a handful. The week before I was "extradited", as my Uncle Mick called it, the family gathered in Biddy's living room to discuss what was to be done with me. At that point, I climbed onto the kitchen table and ran along it to gate-crash the family meeting in the front room. Unfortunately, as I jumped I hit my head on the door arch and my flight continued on a horizontal trajectory until I landed unconscious on the coffee table. The meeting abruptly moved on to 'any other business' and the decision was unanimous; I was to be sent to the 'motherland'.

To some my impending travels were not only for survival, but on the premise; "the move will do Little Johnny good," as my Auntie Edna was heard to say. This explained why, when I was coming to my senses before I was taken to hospital to get my head stitched, I swung a kick at her. I guess my reaction fully justified her reasoning. As a consequence of the impact of my shoe with her knee, the standard issue compass in the heel of my Tuft shot out of the open window. My Uncle Paddy was dispatched to telephone for two ambulances; it was agreed by everyone in the room that it was best if my aunt and I travelled separately. However, in his rush to place a bet in the betting shop up the road from the telephone-box, Paddy forgot to mention my aunt's incapacity when he called the hospital. Mind you, as he told me later, he didn't like her much either.

As with all deportations, the authorities ensured my humiliation was a very public one. It was slightly different to terror-suspects on their way to Guantanamo Bay, as I was not handcuffed and forced to wear an orange boiler suit. No, for mine I was made to wear my school-uniform and my hands were constrained by flowers and a box of Quality Street chocolates for any of Biddy's nine sisters and one brother who were stupid enough to take me in.

Once in Ireland, I again appeared to bring out the worst tendencies of reasonable Irishmen who had never raised a hand to their own children. If my father was the first to take a dislike to me, it wasn't long before a disorderly queue of angry Irishmen formed behind him. I continued to make further violent enemies in life, but I convinced myself to look upon this as 'a gift', rather than this being due to my lack of social skills.

As soon as my size-four shoe set foot on Irish soil, the 'Little' was dropped from my name and I was called Johnny. Later in life, if I was not known by one of my drop-down menu of nicknames, most people referred to me only by my surname like TV's Inspector *Morse* and Hitler. I was full of rage, based on my belief that I had been simply abandoned amongst a bunch of male Irish lunatics. It was only years later that I realised Biddy had to make a decision that no parent should ever have to make, to send their child away for its own protection.

I've also learnt recently, the reason that I was passed from family to family, like a package during a pass-the-parcel game at an al-Qaeda annual general meeting. My father had crossed the Irish Sea, and when he wasn't unconscious in a pub, he was smashing down the doors of my mother's sisters desperately trying to find me. In hindsight, I realise that I've done my male relatives in Ireland a disservice over the years, as I did not appreciate that part of their anger towards me was because my presence put their own families in danger.

Though Biddy's sisters protected me, smuggling me from one to the other when my father was getting too close, I think one or two of their husbands could have been a bit more hospitable. At my Uncle Thomas's house one morning, I was blowing porridge through a straw back into my little cousin Dermot's bowl when his father walked in and exploded: "Out of my house you little English ferker."

He had his faults, but he was a man of his word and threw me out of the first floor window and I fell through the roof of their greenhouse. Amazingly, I survived, as the greenhouse roof was covered in polythene. I was saved from being forced to move on to the next family by my cousin Marisa. Every day after school, she would put me on the back of her bike and cycling around the town until her dad had collapsed into a drunken stupor and it was safe to smuggle me back into the house.

Though I had only just turned five, it was then that I fell in love with women. I noticed how fantastically smooth her legs were when she was pedalling that bike. I could smell her freshly washed hair and felt the softness of her body as I wrapped my arms around her to stop myself falling off. For the rest of my life, women have always saved me, usually from myself. When I think of that part of childhood in Ireland, it is still vague, but I remember Marisa vividly; I remember her kindness and the risks she took to protect me. She was my first love. So I have mainly good memories of my time in the 'motherland', but if it wasn't for Marisa's bike rides perhaps I could have grown up differently. If she'd worn a gas mask and pissed her knickers while pedalling, I might now be a Tory MP.

Apart from Marisa, I had two elderly fans amongst my family in Ireland. These were my wonderful aunties Mary and Ellen. They were very much in my Auntie Julia's mould, as they spoiled me with what little they had. Despite my being a surly little bastard, they did everything they could to make me happy. Once, I complained my sausages were too greasy, the next day I found them to be mouth-wateringly crispy and dry. A few days later, I woke up early one morning, crept downstairs and discovered my Auntie Ellen preparing my breakfast. I was amazed to find that she had been getting up early each morning to cook and then squeeze the grease out of each of my sausages into a handkerchief. My ungracious comment had resulted in this act of kindness, which again is one of those wonderful memories I pray I will never forget. Most of the bad memories have faded with time. I know I have been very lucky in life and have no grounds for self-pity or any wish to torture the reader by adding to the anthology of published works such as *Tears Before my Daily Beating* and the heart wrenching *My Childhood, as a Make-do Brush in my School's Shithouse.*

My humorous memories of my time in Ireland centred on my visits to Mary and Ellen. My two aunties were always a joy to a mischievous child who loved the surreal world that elderly Irish women inhabit. A typical welcome would be:

"Hello, Johnny. Great to see you, when are you going?"

Or at dinner, a typical exchange such as:

"Are you hungry Johnny?" "No," I would reply, over my big sulky lip.

But my protest went unheeded, as a mountain of food would be slapped on the plate in front of me. This was then followed by the question, "You're not going to eat all of that are you?"

The skill of communicating through absolute nonsensical ramblings is an art passed from each generation of Irish mothers to their daughters. It is rarely mastered by outsiders, with perhaps the exception of the Mayor of London, Boris Johnson. I would strongly advise those in the criminal fraternity to learn this skill, as nothing you would ever say could be used in court against you.

When I returned to England, little had changed. One of the first family functions I attended was a sacrifice to the Black Widow Spider, or as it was officially known, the marriage of Biddy's other sister, Mad Kathrina. The sacrificial mate was a lovely guy called Mick. During the ceremony, the priest reached the stage where he asked:

"Kathrina Josephine Toomey, do you take Michael Richard Fitzpatrick to be your lawfully wedded husband?"

Initially the congregation was completely bewildered, as it was the first time anyone had ever addressed her as Kathrina, without the prefix 'Mad'. In the most insincere tone she replied softly, "I do."

The priest then turned to her meal.

"Do you, Michael Richard Fitzpatrick, take Kathrina Josephine Toomey to be your lawfully wedded wife?"

There was no reply.

I looked at Mick and saw that he was looking at a stained glass window. He appeared to be transfixed by its beauty. I thought it would have been better if he had made his escape through it. I looked at Mad Kathrina; her face now as red as the claret-coloured pieces of stained glass when the sun rays pierced them. Everyone in the church shared the same horrific thought, that the spider would not wait until after mating, but would instead devour her mate right in front of the terrified congregation. I also knew that she would never get convicted for the slaughter; everyone would be too intimidated to give evidence. Even the priest looked terrified now fearing for his life; his sermon had a professed a belief in a better life hereafter, but he was in no rush to get there.

I ran up to Mick and tugged his arm to bring him back to reality. He looked down at me and smiled. The priest repeated the question, this time louder, his voice trembling with fear. Then to the great relief of all, Mick replied in a full and confident voice; "Sure, I might as well while I am here."

For Mick, these were his last freely spoken words. For the remainder of his life he became the object of Mad Kathrina's wrath. Death, when it ended his short life, came as a welcome release. Mick had not meant to embarrass his bride-to-be. It was just that he was simply an easy-going young Irishman who was relaxed about the occasion, indeed he seemed to be happily content. After that day, Mick sadly turned to drink, which gave Mad Kathrina an even firmer footing to kick the living crap out of him. Every day her chisel-like tongue chipped away a little more of his self-respect, from the moment he awoke until he collapsed into a drunken stupor after the nine o'clock news.

The Irishmen I know have the utmost respect for their mothers and this was transposed onto their wives and partners. My dad only turned on Biddy and me after his brain injury. Uncle Mick was a big man, similar in looks and build to the late laconic actor Robert Mitchum. He was typical in that he saw his wife through the same eyes as he saw his mother, the exception being for the carnal three minutes on Tuesday evenings during the break between *Neighbours* and the *6 O'clock News*. When the image of female perfection crumbled in front of him, he turned on himself rather than her. It was the most extreme example of a failed marriage, with respect on both sides disappearing ... followed by his clothing out of the window, so he could not go out to the pub.

On these occasions, housebound through want of clothes, Uncle Mick was able to adapt to his confinement. His back-up plan was the 'Woolworth's homemade beer kit'. I was out on the town one Saturday afternoon with a girl friend, Georgina, we decided to call on Uncle Mick, without a thought that Mad Kathrina might be prowling nearby. As expected, we found Mick smashed out of his head on the cheapest vodka, a vicious brand called 'Popov Vodka.' I believe it had no name in its country of origin, with only a government health warning slapped on the bottle. The warning had been poorly translated into English and used as the product's name. It should have been called 'Pop-Off'. At a cost of two pounds a litre, considering that Uncle Mick was not long for this world, it would have been aptly named. This was an aperitif, as he waited for his homemade brew to mature. I am sure his home brew-kit would have produced the finest of homemade beers, providing it had been allowed to ferment for the time stated on the box. Unfortunately, Uncle Mick did not have the patience required of a master brewer; he was an alcoholic after all.

Mick offered us a sample of his latest brew. We both looked at our two pints of green liquid, which had a layer of hops on top.

"Jesus, when did you make this stuff?" I asked.

"I added the water to it fresh from the tap this morning," he replied.

Georgina, who could charm the frock off the Pope, captivated him. He leant towards her and said, "I know everything," then, after a silence, he leant back into his armchair and sighed. "But then again I know nothing."

Like any beautiful woman; Georgina could manipulate men with a glance, but even she melted at the warmth and vulnerability of the persecuted man in front of her.

The moment was lost, as the door flew open and there was Mad Kathrina screaming. I hate it when people put on airs and graces when they welcome guests, so it was good to see Mad Kathrina playing it natural.

"You useless fecking eejet," she spat from her lips as easily as one would say: "How was your day, dear?" I stood up and took an ironic bow.

"Thank you Auntie, I love you too," in a futile attempt to distract her impending onslaught on Uncle Mick.

However, her talons were already unsheathed and she brushed me aside as she leap towards her prey. As Uncle Mick looked up dejectedly, she hit him right across the temple with a teapot. That was the first time that Mick had been in contact with tea in any shape or form since his marriage. Georgina and I fled, leaving him flat out on the matchstick remnants of the coffee table. I did not fear for him at this stage as he was in a blissful coma, but I did fear for

Georgina. I gave one last look back, to witness Mad Kathrina standing over Uncle Mick like a victorious adrenaline-pumped prize-fighter standing over his blooded victim. In any other situation, I would have felt sorry for the wife of an alcoholic, but I believe she broke him as a man. Before he met my aunt, Mick had signed 'the pledge' in Ireland; this was an oath of abstinence given to the church. Until then alcohol had never passed his lips.

I had no father figure to look up to and Uncle Mick did not fill that role. I saw him more as someone who needed a father, which perversely, was the role I assumed; I did all I could to protect him. He taught me to play chess and there were rare moments when the weight appeared to lift from his shoulders. Unfortunately, the games took a while, as he tried to focus on the sixty-four pieces, double the standard number, his vision blurred by the two-litre bottle of cider he needed to help him concentrate. The other male figure in the family was Paddy, but he was more a member of the audience while Julia spread humour and confusion all around her. Biddy was not exactly 'motherly' in a typical sense, but she had tremendous strength coping with my father as his violent rages grew worse during his final months. Julia was the nearest I had to 'family', as she was always willing to listen and to overindulge me. This was still the case later in her life, until the onset of Alzheimer's disease which meant she could not comprehend what was being said, or even who I was. I believe the disease was triggered when Paddy died; her loss tipping her into a frightening world without reason, structure or hope.

My childhood was filled with every kind of noise. Usually some Irish ballads on the record player and along with this incessant wailing, the TV was always on. I think the broadcasters' plan was to produce safe, respectable programmes, expounding middle-class values in children's programmes like *Blue Peter*, sitcoms such as *Crossroads* and mainstream comedies such as *The Good Life* and *Terry and June*. However, with no real male mentors, I found anti-heroes amongst the TV dross. These characters leapt forth in the form of comedy creations such as Ronnie Barker's prison lag Norman Stanley Fletcher in *Porridge*, and Jimmy

Nail's bricklayer Oz in *Auf Wiedersehen Pet*. These were edgy characters, inhabiting darker comedies, who did what they must to survive, using wit and street guile learnt from growing up in tough working-class neighbourhoods. These characters cared little for the establishment or its laws, but held 'higher' values based on their own 'moral code', Fletcher at any rate. Acting without malice, they always triumphed in their own way, sustaining a sharp mental edge in times of adversity. Fletcher was a north Londoner and Oz a Geordie, I could relate to both as working-class wide-boys. Irrespective of their different backgrounds, regional dialect and slang, they were of the world that me and my friends knew.

The comedic trilogy was later completed by the genius of Rowan Atkinson's *Blackadder*, whose eponymous hero, the ultimate screen rogue, once described himself as having "three gold stars from the kindergarten of having the shit kicked out of me." These characters exhibited the blackest of humour in the direst of situations. I had finally discovered characters I could relate to.

Though all three were rogues, from an early age I could see a distinction between them, which I also make between the various rogues whose stories I now tell. Oz and Blackadder were outright 'rogues'. They really only cared about themselves. When asked who was the most important person in his life, Blackadder replied "Well, me really." Fletcher, on the other hand, was a 'benevolent rogue', though he would fiercely deny it. He took on the establishment in the form of Mr Mackay, the chief prison warden, and helped his fresh faced cell-mate Godber, cope with being in jail for the first time. He taught Godber not only how to survive, but to make sure he never returned to Slade Prison. I have known many rogues but my story is of that rare breed who, because of their roguish traits (rather than despite them), committed acts of extraordinary kindness.

There were many hard-hitting TV dramas, such as Ken Loach's, *Cathy Come Home*, but these dramas were just too bleak and depressing with no hope of salvation. As a little boy, I needed humour with an edge to entice me into a story; where mischievous characters could wring a joke out of the darkest moments. At such times, humour is hope; it is a signal that all is not lost and you have the will to challenge what is happening. The one drama series that delivered this, as well as opening my eyes to fractured eighties' Britain, was Alan Bleasdale's TV play, the *Boys from the Black Stuff*, which showed the deprivation of Liverpool brought about by the onslaught of Thatcherism. It brought a smile to my face, even in its darkest moments. If you have never seen it, the last episode dealt with the death of one of the lead characters. The final act showed the remaining protagonists attend his wake in the local pub, while at the same time a group of dockers arrive waving their redundancy wages. Pure mayhem ensues. It's done with pathos, mixed with wit, and is pure genius, if I may say so, Mr Bleasdale.

When it came to films, my rogues were of a more violent and brutal nature; on celluloid there was more freedom to cross boundaries of what is acceptable. Clint Eastwood's *Dirty Harry*, James Cann's 'Jonathan E' from *Rollerball*, John Belushi's Bluto Blutosky from *Animal House* and Donald Sutherland's Hawkeye from *MASH*, were the antiheroes of my youth.

For light reading, there were literary rogues, such as Captain Yossarian from Joseph Heller's *Catch 22* and Ken Kesey's Mr McMurphy from *One Flew over the Cuckoo's Nest*. All were films based on the books, but it was the films that made me want to pick up the books and learn more of these characters.

These self-sufficient rogues were my idols, getting through life despite the rules of the establishment, rather than because of them. This is not to say that I went out shooting my schoolmates in the arse with a spud-gun while asking them to "make my day" like some kind of Dirty Harry Junior. But, after the death of my father, my exile to Ireland and then my return to England, these idols mirrored my view of myself as a loner, never comfortable fitting in anywhere and never thinking that anything would last.

During my time in Ireland, I was commonly known as that 'fecking English boy'.

When I returned to England, I was known as 'that fucking Irish boy'. I was a foreigner, an alien in both countries. I remember getting into a fight with a couple of local English boys for coming over and taking 'our' jobs. As we were all about ten years old, I thought paper rounds must have been in high demand.

No doubt, this goes some way to explaining what a dysfunctional kid I was. As I went home that day, with my new batch of fresh cuts and bruises, I remember thinking 'Does life have to be this hard?' Throughout my teenage years, I held a lot of anger and had an over-developed sense of mistrust. While I acknowledge that I am emotionally crippled, and rarely put my fate in the hands of others, I have learnt to trust a few people, and to give people the benefit of the doubt. It took me a long time to evolve to the stage that most normal people reach watching TV's *Sooty* after school, whilst eating their jam sandwiches.

I developed an independent streak, which remains with me today. I try to help others, but I lack the confidence to accept help in return. H, one my best friends, once said that I was "The loneliest wolf he had ever known." This was condemnation indeed from someone who would not even speak to my Trumpton friends. I should explain that Trumpton, based on the TV children's series, is a term I coined for the area known as Highbury Quadrant, where all the residents, to Chelsea supporting H's irritation, were Arsenal fans. Actually, I must qualify this last statement, as there was one Tottenham fan, but more of this unique rogue later.

Looking back on my family, all of them were good people and, although they seemed unable to mould or control me, they all contributed to the man I became and for that I count myself lucky, even if society is not. As a teenager, my anger subsided and I grew in confidence - or arrogance if you prefer. I began directing my own path in life, a trait similar to the benevolent rogues you will learn more of in this book. Most of these characters came from dysfunctional backgrounds, but they refused to become victims, indeed each in their own way helped protect others who were vulnerable.

Chapter 3 – *School for Scandal*

St Vesuvius was a tough school characterised by drink and violence, but perhaps surprisingly, no drugs or truancy; that came later. I think we needed to get the hang of alcohol first. When I was twelve, for the first and only time, Biddy attended an end-of-term parents' evening; not because she had any interest in my mathematics or English studies, but to investigate the progress of my religious education.

"Well how is Johnny doing?" she asked my teacher of Religious Education, Mr Burns.

"Well he is one of the best in the school," replied the proud teacher.

"One of the best, you mean at religious studies?" asked a startled Biddy.

"God, no," said Mr Burns, chuckling. "In that he has no interest whatsoever, no I mean the other day I watched him in the playground standing his ground against three lads. I tell you, Mrs Righten, he has one of the best right hooks I have ever seen."

Mr Burns also taught boxing and as any aficionados of Jimmy Cagney films will tell you, religion and pugilism were taught from the pulpit. At St Vesuvius, turning the other cheek was an opportunity to have your jaw detached from your cranium. Biddy was no novice in this field either and delivered a verbal uppercut to the enthused tutor.

"I don't give a damn about his fighting skills. How did you get a teaching diploma, saving up bottle-tops?"

Mr Burns was on the ropes attempting to steady himself against Biddy's counterpunch, but he was finished. Irish women don't have to do much when they want to beat an opponent; a remark is often enough. She turned on her heel without a backward glance and never set foot in the school again.

What made these parents' evenings such a gladiatorial event was that teachers would highlight failings in their pupils, yet naively failed to recognise that these traits originated in the parents. Having more experience, they were in many ways far better masters of their vices than their offspring. The only subject that teachers were given carte blanche to criticise was truancy, for the parents of these kids rarely turned up. It was no surprise then, that when Mr Burns highlighted my belligerent ways; Biddy's natural instinct was to go on the attack, even though she frowned on my aggressive tendencies.

Over the years, various teachers tried to instil order at St Vesuvius. Mr Topped opted for the cane for the slightest misdemeanour; for his troubles a rubbish bag filled with water was dropped on top of him from the roof of the school building, resulting in a broken collarbone. Mr McGreevy, the Deputy Head, was hard but fair. He had a glass eye and no sense of direction, which undermined his position of authority. Once, at morning assembly he cried out to a talkative student, "You boy, to the front now!"

His finger was pointing one way and both eyes staring in random directions. Three boys then headed to the front, resulting in absolute pandemonium for the next ten minutes, as Mr McGreevy kept shouting "Not you, you!" the true culprit disappeared out of the back of the hall for a cigarette.

I continued lashing out, as I had in Ireland. At junior school, I nearly strangled an older boy who tried to beat me up for my lunch money; I was dragged off him after one of his eyes popped. The next day I was exhibited at the front of school assembly by the headmaster who informed the school "this is an evil boy, whom you should all shun." Unfortunately, his message was lost as Eamon, a sarcastic little bastard like me, who shouted from the back, "How's Popeye?"

This, however, was one of those events that can scar a kid for life and not just the victim. I was horrified that I had nearly killed that boy. After that, I entered what I would now call a state of 'pugilist paralysis', often seen in older boxers terrified of severely hurting their opponent, or least on celluloid such as in John Wayne's film, *The Quiet Man*. The incident left me with a complete terror of hurting someone else. The reaction of my family and teachers was that I was a psychopath. Only my Auntie Julia refused to condemn me, telling me that "I should always stand up to bullies". The next few years were hell. In our neighbourhood, your social standing was based on whom you had beaten in a fight.

Over those years, I was beaten, bullied and worst of all, spat on, yet I did nothing. My arms and legs became lead, I froze and I was unable to lash out; my mind filled with images of the eyes popping out of my assailant's head. After each incident, the humiliation made my stomach tighten. I was a regular target, especially as I was not part of a gang and had no back up. I had no dad or brothers to protect me, and no sister that I could cry my eyes out to and who would then tell me to grow a pair.

My period of paralysis ended when two gang members, went too far and seriously tried to hurt me. I returned to type and went on the offensive. Each punch to my face seemed like a key gradually turning to releasing my former savage self. I could just about take being verbally or physically abused, but with my own blood welling up in my mouth, I reverted back to my nature. My rule since is to avoid fighting and I am often the peacemaker when trouble starts. I care little for what is termed 'respect' of one's image and I will walk away, but I do have an aversion to pain. I believe my choice of education saved me, as I went to a comprehensive rather than a public school; I have instinctive sadistic rather than masochistic tendencies.

As a young man, my life revolved around drink and girls. The inevitable mistakes that you make as a stumbling teenager in these pursuits are valuable lessons, providing you learn from them. It's a bit like essays at school, where poor grades were only useful when the teacher took the time to make notes in the margin on where you went wrong. Thankfully, I never received bland markings, my teachers were happy to make comments and my essays were marked in red with useful contributions such as, 'Dear God!', 'Garbage' and on one occasion 'Pencil-brained.'

My disruptive behaviour led to my being banished to the 'Annex' of St Vesuvius, situated directly opposite the main building. This was the last refuge of the school's scoundrels; it had a population of about twelve pupils, the ones that the school had completely given up on. The surprise was that I was put there, not for hospitalising a fellow pupil, but because of my poor spelling and grammar. My teachers, rather than recognising that I was dyslexic, thought that I was brain damaged and therefore a potential psychopath. This meant that I received no specialist tutoring to help me with my English. Instead I was interned with the most violent pupils in North London. The nearest I got to specialist attention during my time at the Annex, was when the school nurse stitched up my head. This occurred when a very dangerous pupil named Ant smashed a desk over my head. He later defended himself, saying that I had been carrying a "threatening object in my hand". His hatred for books was on a par with that of Adolf Hitler.

As for girls, if you did not get emotionally scarred trying to negotiate these emotional minefields, then you were either gay or Cockey, but more of him later. Puberty is hell. Common sense goes out the window. Your hormones kick in and your emotions are pulled in every direction. My first love at school, and therefore my first romantic scar, was Maggie O'Donovan. I nicknamed her MOD, as in Ministry of Defence, because she would never let me near her breasts. As she was my first school love, she naturally hated my guts and in hindsight giving her a nickname of MOD did not help. The pain of my unrequited love resulted in a physical ache across my sternum that I experienced then and never again. I don't know if it was heartbreak, a longing for the unattainable or that I was wearing a new shirt a size too small to highlight imaginary muscles in my arms.

I was thirteen years old and on Maggie's birthday, I gelled back my hair, bought some bath salts as a present, along with a card from Boots the chemist and summoned the courage to go to her house. Her mother answered and said I was very nice to deliver presents in person, but that her daughter's birthday had been three months earlier. If that was not enough to destroy me, she then told me Maggie was out with her eighteen-year-old boyfriend in his car. My 'nice' label was quickly removed when I spluttered out, "Eighteen - Fucking sex-case!"

I have never bought bath salts since and *Brylcream* lost a client that day. I remember Maggie's last ever words to me, "Go out with you, I wouldn't cross the world to piss on you if you were on fire." The fragility of adolescence is always exposed in the cruellest way.

After that, things picked up on the girlfriend front. At school I started to realise that my sense of humour and natural aversion to joining any gang made me attractive to some girls; I was 'acceptably' different in an independent rather than in a geeky or psychotic sense. I did not discuss any details of my relationships with my schoolmates, as I felt this would be disloyal. Girls knew this and I was trusted because I respected them. I hated how some boys treated girls, using them as conquests to boast about and enhance their social status amongst their peers. Even when no intimacy had taken place, as with Maggie and me, some boys still boast about their non-existent conquests. De Bon was a classic example of such a boy. He claimed he had had sex with every girl in the school and he was despised by both sexes. It became clear that all of his stories were fabricated as no girl could physically do the things he described, unless she had a jaw like an anaconda.

De Bon was sitting on a wall watching the rest of us on the football pitch, when he shouted over to a girl he fancied. We stopped kicking lumps out of each other and turned to look who was the object of his advances this time, just as he threw what appeared to be a large stone at her. I think that was his idea of courtship. Unfortunately, his aim was good and he scored a direct hit, catching her on the side of her head, knocking her over and rendering her unconscious. Her two brothers promptly ran from the pitch to beat the crap out of De Bon. He was on another planet when it came to the female sex, even more so than the rest of us; a remarkable achievement.

A year later, De Bon was again boasting about his sexual exploits in the playground, this time while looking at a topless page three girl in *The Sun*.

"I've had her," he said.

No one responded. In what must have been a sudden moment of personal reflection, he finally said something that made everyone take notice. In an attempt to establish some integrity, he bravely tried to exhibit some vulnerability to prove that he was not really the bull-shitter we all knew him to be.

"I have done everything with girls," he proclaimed, looking around to make sure he had an audience. "But I have never sucked a woman's cock."

This time he got a reaction. Unfortunately, for him he simultaneously became the number one subject for graffiti on the toilet walls of both sexes, so I hear, for the rest of his days at St Vesuvius.

A particularly violent episode occurred when the first Chinese kid enrolled at our school. It was about the same time as Bruce Lee's film *Enter the Dragon* was released and David Carradine was appearing in the TV series *Kung Fu*. Everyone thought that, as he was Chinese, he could make a martial arts weapon such as a *Nunchaku* out of two pieces of a broom-handle held together with a toilet-chain and beat the crap out of us pasty-faced kids all at once. He quickly caught on to this and took full advantage of the kudos it gave him. He set about using his newfound infamy to bully others in the school for their pocket-money. He played up to this with various hand and leg kicking exhibitions daily in the playground, followed by the little kids queuing up to hand over their tuck shop money.

That was until he encountered Niall, or Nail, as he was known. That day the Chinese kid performed his usual constipated movements with added grunting noises to enhance the effect. Nail had been off sick with pneumonia since the new kid had first joined the school, so had not been party to the impact this twelve-year-old 'killing machine', as the Chinese kid had termed himself, had had on most of his schoolmates. Nail proceeded to batter the crap out of him and the new kid left the playground unconscious on a stretcher.

Perhaps the kid's lack of superhuman powers wouldn't have been so publicly exposed, if Nail's parents had had a television or had taken him to see one of Bruce Lee's Kung Fu films at the cinema. However, I doubt that this was much comfort to the Chinese kid, whose balls had probably only recently dropped, just to be kicked back up again by Nail. The story has an epilogue. Six months later, the Chinese kid, who had been moved to another school, suddenly entered our classroom waving a machete. Miss Martinez, the history teacher, went hysterical shouting to Nail to do something. The teachers only ever referred to him by his nickname when violence broke out, which was often. Nail duly responded, breaking a chair over the intruder. Once again, the 'killing machine', was removed from the school premises on a stretcher. Miss Martinez, still in a state of hysteria, carried on with her lecture but she didn't seem to be in the right frame of mind to be teaching us about Mahatma Gandhi's policy of 'non-violence' as an instrument of change.

Another new addition who enrolled at St Vesuvius to study advanced idiocy around this time was Ant, who I've mentioned. This six-foot-two exhibit had a face as long and miserable as horse that had jammed its head in the gates of a glue-factory. Everyone kept him at a discrete distance, about a mile if possible, as he was benchmarked as 'Best in Class' against the other dangerous lunatics at St Vesuvius. One day, this monument of pent-up violence walked into our class, or rather he entered sliding his feet along the floor as if his steel toe-caps were magnetised to it. In his gigantic hand, he held a message for the teacher. The standard ruse teachers used to avoid having him in their class was to send him around the school all day delivering messages. One day I got hold of one of these notes from Miss Martinez's handbag, hidden beneath four empty packets of an early version of *Prozac*. The message was always the same, and as he was never in class long enough to overcome his illiteracy he was unable to read the message: "Give him back this piece of paper and get the lunatic off out of the class, as quickly as possible."

That morning, he entered our class and not a word was uttered as he slid slowly up to Miss Martinez's desk. Then after Miss Martinez's returned the note, her hand trembling uncontrollably, he turned slowly around and began his journey back towards the door. Just as he turned the doorknob to exit the classroom, Eamon uttered in a low, deep voice; "You rang?"

This was the catchphrase of the gormless, elongated butler Lurch from the television series *The Adams' Family*. Lurch, as he was known from that moment, slowly turned around and collapsed onto the whole front row of pupils. Eamon took the brunt of the onslaught and his face was completely pummelled by Lurch's fists, which were a bit bigger than Eamon's head. Despite our best efforts to save him, Eamon ended up in hospital for two weeks. I went to see him and his first words were "Fucking hilarious, eh?" He had to repeat that for me, as it's hard to talk when your nose is broken and your teeth are replaced with swabs.

All kids go through a rebellious phase in their teens and I was no different. But there was no tattoo on my fist or bolt through my penis for me; I was far too much of a rebel to go down such a conventional path. Rather, I went to the extreme of wanting to join Her Majesty's Special Boat Section (SBS), the naval equivalent of the SAS. I had seen too many war films and was carried away with the excitement of it; I sent off an application to the Royal Navy to join the officer corps of the SBS. The experience did nothing for my career prospects, but it taught me a lot about snobbery and the class system.

With the help of one of the teachers, Brother Jacob, a good man who tried to help the most disruptive pupils in the school, I filled in my form and sent it off. An interview was arranged and a naïve fifteen-year-old scrapper set off with every confidence that I would be a suitable candidate.

I entered the Royal Navy recruitment offices in High Holborn wearing my school uniform, apart from a black tie that Paddy kept for the funerals of patients he owed money to; I believe he attended out of some ingrained Catholic moral guilt. Here I was met by a James Robinson-Justice type character straight out of the old *Doctor in the House* movies. I liked the actor and so thought that if he was anything like him, then I was off to a great start.

"Sit down!" he barked.

I fought back my natural instinct to retaliate when anyone in authority started to order me, but I was uneasy when I took my seat. He asked me of my ambitions to be an officer in the

SBS and then some general knowledge questions. My ability to answer most of them seemed to impress him.

"I have a glowing reference from Brother Jacob that you could be, academically, the most successful student your school has ever had."

"I am pleased Brother Jacob has such confidence in me," I replied, though I was bewildered that anyone had faith in me at all.

"St Vesuvius is an excellent educational establishment," he continued.

Now I was truly confused, as I had never thought of St Vesuvius as an educational establishment.

"Its sub aqua diving is the best in the country, how deep have you dived?" he asked.

I was beginning to believe he had someone else's application form, "Well, the deep end of the pool in Kentish Town is fourteen feet ten inches deep and I have touched the bottom," I replied.

Now it was his turn to be confused. He picked up and re-read my application.

"Good Lord, your school is" … he could barely utter the words, "a comprehensive!"

"Yes, why does that matter?" I replied, knowing that the interview had taken a turn for the worst.

"My word, your parents are Catholics! That's Northern Ireland out of the question." Now my hackles were up too, with both of us snarling at each other. It was my turn.

"I'm sure the Duke of Wellington would never have got passed you, as he was born in Ireland."

He leant his considerable bulk forward, "I am not happy with your attitude."

I leant forward, "Well, I apologise if at any time I gave you the impression that I give a flying fuck what you think."

"I have never been so insulted," he said, with a look of indignation and shock.

"Well try this. You can stick your officer post up your well-serviced naval arse."

Back at school, everyone knew what a disaster it would be, except me and perhaps Brother Jacob. I stood outside the recruitment office bewildered as to which way my life was to go now. I did learn a lot from that four-minute interview. I learnt not to plan so much that I would ever think the future was a certain one but to always have a Plan B. The back-up plan was usually the pub. More importantly, I was no longer naïve about the class system dividing the country in which I lived.

Brother Jacob called me into his office when my O-level results were out to say that the Royal Navy had been in contact. They were impressed with my results and would in fact consider me for a non-commissioned officer post. I had never sworn in front of Brother Jacob before then, but that was too much.

"Fuck em! I wouldn't join them if they wanted me to be an Admiral."

He understood how I felt, but he still thought I should go to confession to repent for my bad language.

To the school's great embarrassment, I had indeed passed my exams. I'm one of those people who luckily perform better under pressure; there were others who were far smarter than I and failed. The headmaster, Mr Connelly, called me in to say that the teachers had signed a petition, refusing to teach me, as I was a disruptive influence. I replied that I perfectly understood that none of the signatories could teach beyond the basics and I appreciated their predicament. I asked him to pass on those thoughts. There was an element of truth in my response; very few pupils had gone beyond O-level and, to my knowledge, no one had ever gone on to university from St Vesuvius. I was to be the first and only one to do so. I guess I was seen as disruptive by my teachers, as they might have had to get off their sedans in the teacher's lounge and learn how to teach subjects at a higher level. Luckily, a technical college agreed to take me on with no idea of my disruptive tendencies, only my grades.

On my last day at St. Vesuvius, I went to thank Brother Jacob for having faith in me. I also apologised for my classmates not coming to say goodbye personally, as a bunch of them had been arrested fleeing up a down escalator after nicking toys and alcohol from a shop on the concourse of Euston Station. Brother Jacob just threw his hands up with good grace, shook my hand and gave a very weary smile. He then added, "The recruitment office sent back your documents and I don't remember writing in your reference that, 'you were smarter than those teaching you.'". I gave him a wry smile and shook his hand as it really was time to leave St. Vesuvius.

My agreement to take my A-Levels elsewhere saved the school severe embarrassment and left its teachers to carry on their unwritten campaign to bring the entire education system to its knees. Eventually, the government had to close St Vesuvius down, as along with the worst academic record in the country and the highest rates of school violence, it had expanded into new areas of delinquent behaviour and within a few years recorded the highest truancy and drug-taking levels amongst its pupils in the country.

Poor educational standards were not the only issue during the seventies, it was a time of serious racial violence that ruined lives and ended others. A kid from our school, Brendan, whose parents ran the local pub, stabbed an elderly Asian to death in Manor House in North London. He had walked up to the old man and asked him for a light and the old man responded that he was sorry, but he did not smoke. Brendan, like many other impressionable kids in those times was caught up in the rhetoric of right-wing parties like the National Front (NF), preaching that all of society's problems were due to race. It offered a simple solution, rather than addressing other factors such as a crippled economy, local deprivation, poor education and lack of jobs.

Another schoolmate, Andy, joined a right-wing punk band. I knew of Andy's activities at the time and, in turn, he knew that I actively opposed them. Years later, we bumped into each other. He had married, had children and turned his back on his right-wing past. The attraction for Andy, he confessed, had been the violence. The NF offered an outlet for all that testosterone pumping through his veins and provided a vehicle to react against all the wrongs he perceived in the world. Like Brendan, he regretted his affiliation with the right. To my surprise, he told me he was a social worker in a refuge centre in one of the most diverse authorities in London. The surprises continued when he should me a picture of his Nigerian wife and her children whom he had adopted. It was hard for me to take that Andy, whom I first got to know as we shared a passion for the Napoleonic Wars at school, who became a fascist, was actively promoting integration and that his own immediate family was proof of it. A leopard never changes its spots goes the cliché, perhaps so, but maybe his fascist years were the aberration as he soon reverted back to being a good man.

I hated school. St Vesuvius offered no hope to its kids and its only expectation was for its pupils to turn out to be failures. I had expressed my views to the teachers during my time there, so the petition to discontinue my education was no real surprise. At least it showed the staff could organise something. I think the staff got a taste for petitions; when the government did decide to close down the school, the teachers organised another petition to keep it open. Unfortunately, this was easily surpassed by the local residents organising a petition to support its closure.

Ironically, the petition was additional evidence of another of the school's dubious traits; fraud. Few of the signatures had any correlation to the attendance list of past pupils; there were no records of the large Disney family attending the school, let alone its members Arsey, Crabby and Pervy. It also provided further evidence that the school's illiteracy rates were amongst the worst in the country, as one signatory spelt Snow White without the n, the h or the e. When it was announced that the school would be turned into a home for refugees, such was the reputation of St Vesuvius, that the local residents, some known for their xenophobic views, welcomed the new influx of immigrants with open arms.

The school's last futile attempt to rally support was an invitation for ex-pupils to attend a school reunion. This did not have the desired effect, for even though a number of ex-pupils were either dead or in jail, there were enough left to turn it into an almighty brawl. This resulted in a letter to the local newspaper with the following headline:

'WHY IS THIS ENTRANCE TO HADES STILL OPEN?'

I would guess that few of the teachers knew where Hades was, but some probably applied to see if there were any vacancies.

ADOLESCENT ROGUES

"There are only two rules in this prison, Rudge. Are you listening to me?

One: You do not write on the walls.

Two: You obey all the rules"

Mr Mackay, *Porridge*

Chapter 4 – *The Joker is Wild*

The technical college opened up a whole new world of opportunity, though not in an academic sense. In particular, I discovered that gambling was the order of the day. My fellow students, mainly Iranians and Indonesians, would bet on the time of day that the drunk sleeping on the bench outside the college would fall off it. Welcome to England. For an English boy with no money to his name, but with an interest in playing cards for money, this was a very rewarding environment. Two other characters became prominent in my life at this time. Loftus, who had also been kicked out of school, was attending a neighbouring technical college along with his best friend Hammond. We frequented the ramshackle Falcon pub in Maida Vale, which just about stood between our two colleges; it held nightly card schools. The Falcon specialised in cheap piss-water lager, voluptuous strippers and regular poker nights. At the poker table I formed a bond with Loftus and Hammond who, though good players, liked to enjoy all the other pleasures that London offered. We became a formidable team, playing cards to fund our way through college, partying hard and on occasion raising bail for Loftus to cover his numerous misdeeds.

Each of us had a different approach to how we played the game. Hammond's main interest was women, so his genuine bored expression at the table produced a very effective poker face, meaning it was impossible for his opponents to read his hand. Loftus would comment on everything, especially ridiculing his opponents. In the absence of anything witty to say about his opponents, he would utter general comments about anyone who happened to walk past the table. He was always animated, whether he was ordering drinks, laughing or telling stories. Kenny Rogers had a line in his song *The Gambler*; 'You never count your money when you're sittin' at the table.' Not Loftus, he counted out aloud every coin and note in front of his opponents, especially when he was winning. His method was distraction, leading to a loss of concentration by his opponents, which was soon followed by their money.

When I sat at the poker table, I focused, not on the cards, but on the other players' mannerisms. Facial and bodily movements; variations in tones of voice, enabled me to read the cards they held. I was fascinated by human nature, I still am.

Bad card players away from the table droned on about how they would spend their winnings; usually on women, cars, drugs and weekends in Barbados. The professional types never mentioned a world beyond the card-table. They remained totally focused on the game and possessed incredible mental stamina. If you possessed neither, you'd better be prepared for a long walk home after losing your cab fare. They played the long game, waiting for their opponents to tire and lose concentration, and then they pounced on the mistakes that followed, reclaiming any losses and more.

The difference between a professional card player and someone with a gambling addiction is a hard hand even for me to read. But, I knew that for those I encountered with alcohol and gambling addictions their families were secondary. The intoxication of alcohol or the playing table was an addict's reality. Nothing else mattered; only at the bar or the table were they truly alive.

I was going out with Veronica, who was training to be a teacher. One night she gave us her perspective of our different approaches to the game,

"Johnny you have what the French call *coup d'oeil.*"

"Yes, can we have a bottle of that please barman," Loftus shouted over to the bemused bar staff.

Veronica ignored him and continued. "It means the power of the glance, just like a General who can size up a battlefield and develop a plan of attack," she then gave me a proud look while I wondered who the hell she was talking about.

"Poor loved-up lunatic," whispered Loftus to Hammond, making sure Veronica and I could hear.

Veronica ignored Loftus and continued.

"Hammond your brains are in your pants. As a result, you can't read your cards, so no one can read your mind."

"Well you can't say fairer than that," said Hammond, sitting back proudly.

Loftus was eager for his turn on the couch, "What about me Sweet-cheeks?"

"You could start an argument in an empty room," she had more to say on the matter; "You're a crazed circus act, and the audience sits back in complete bewilderment while you extract money from their pockets."

"Sometimes underwear too," added Loftus.

"Only mine," interjected Loftus's girlfriend Suzy, which was true, as Loftus loved her and, to Hammond's bewilderment, often rejected the advances of other women.

Veronica was prim, proper and well educated. Whereas we were rough and ready, with edges that would gouge a passing buffalo, but she loved our company, the humour and the excitement that came with it. Loftus and Hammond loved teasing her and Sweet-cheeks was one of the many nicknames they would use to make her angry or embarrassed, either would make her blush. I would try to step in when I thought they were going too far, only for Veronica to rein me back, she enjoyed being a part of the banter. Veronica and Suzy, though from different ends of the spectrum in terms of class and education, became good friends, but they always overdid the welcome for Hammond's girlfriends, knowing he would shortly dump them and break their hearts.

During this time we survived a few skirmishes, disarming a Greek who had lost his wife's fur coat in a game and escaping through a toilet window when a losing restaurant owner set fire to his own premises in a rage. Loftus's view was never to take anything seriously and even in life-threatening situations his standard response was "Bollocks to em!" Hammond was once stabbed in the arse with a corkscrew, the weapon of choice for a cocktail waiter who's discovered someone seducing his girlfriend. Loftus offered the following words of guidance for the young nurse treating Hammond:

"You're spoilt for choice, so make sure you stitch up the right arsehole."

It was never dull. Hammond had the looks to captivate women, Loftus the mischievous nature to get us into scrapes and me the guile to get us out of them, sometimes.

Our poker adventures continued for a couple of years and I amassed enough money to buy a flat in Clapham. Loftus continued to live life to the full by wining and dining Suzy and pursuing his passion in driving the best cars, though it was not a requirement that he owned them. Hammond seduced women at an exhausting rate, yet maintained an easy going attitude to life.

However, my card-playing days ended abruptly one night when I was playing in an exclusive poker game in an illustrious apartment opposite Regent's Park. I was so bored with my wealthy public school opponents that I raised the pot by one hundred pounds to break the monotony. I won, but then it struck me that this was more than Biddy earned in a month as a cleaner. It was at that moment I realised the value of money and that there were better ways to make it and spend it. I retired from the gambling-table at the grand old age of nineteen, as my passion, or edge as it was known to players, was gone and the game no longer held any interest for me.

Loftus meanwhile, had his own problems. Unlike Hammond and me, he lived on the borders of the criminal world, particularly and unsurprisingly involving stolen cars, which resulted in numerous scrapes with the law. The police tried to muscle in on Hammond and me to get Loftus, but without success. They wanted him badly, not just because of his criminal activities, but because he would humiliate them. He frequently would seek out a police squad car and drive alongside them in a 'ringed-car', a term for when two different bodies of similar cars are welded together in such a way that you cannot trace its origins. He would then blast the car horn, wave and lead them on a merry dance across the city, but they could never catch him. As you can imagine, he was much sought after by bank robbers as a driver.

Matters came to a head when Loftus, Suzy, Veronica and I were out one Saturday night. Hammond was elsewhere in a threesome with two Dutch barmaids. We were coming out of a club in Cricklewood, when a big, burly, red-headed Detective Sergeant, known as Kerrigan, came up and punched Suzy in the face. Loftus was Kerrigan's *bête noire* and it was common knowledge that he had sworn to do whatever it took to nail him. If he wanted to goad Loftus and me into retaliating and thereby put us behind bars for the night, it worked. Loftus immediately kicked Kerrigan in the privates and followed up with one sledgehammer punch to his chin, while I just caught a now unconscious Suzy before her head hit the pavement.

That police threw Loftus and me into the cells and kicked the living shit out of both of us throughout the night. Each time the door opened I lashed out, knowing that I was going to get a kicking anyway; I thought I might as well be 'done for a sheep as a lamb' as they say in New Zealand. Loftus was doing the same a few doors down so we were both a battered mess by the morning. Funnily enough, that is what saved us.

The police station was notorious for hospitalising its 'guests' and had received so much media attention that the Commissioner of the Metropolitan Police had issued public statements refuting this accusation and others that his officers were dealing in drugs. Given the state we were in, and knowing that Veronica had called a local journalist, a deal was done to release us if we agreed to keep our mouths shut. This was not too hard since my jaw was locked swollen from the punches. As a back-up, Hammond had summoned the press and they were outside waiting for us. Actually it was his uncle with a Polaroid camera, but the police were not to know that. The station closed down a few years later and a number of officers were allowed to 'retire' after being found guilty of drug-dealing. A party was thrown when the station was finally closed, but somebody crashed it and replaced a picture of the Chief Inspector with one of Loftus posing in a stolen police car.

Shortly after our release, once our injuries had healed, the police reconsidered our deal and Loftus was charged with assaulting a policeman. I took him to see Dickens, a lawyer I knew. Loftus joined me in Dickens' chambers on the Strand. He outlined the events of that evening to the lawyer and his legal assistants. Well, his version.

"Well, Detective Sergeant Kerrigan, a most dedicated and principled member of the Metropolitan Police Service, encountered us after a night out at a local club and came up to our party for no obvious reason that I can think of." as an afterthought, he added, "Perhaps he had lost his way and needed directions." Loftus continued. "As he did so, he stumbled, perhaps he had had one too many, which is not unknown for Her Majesty's boys, as they have a very difficult job to do you know."

Dickens sat expressionless as Loftus narrated his tale, though he did occasionally raise an eyebrow.

"Naturally, I moved forward to grab him before he hit the floor. Unfortunately, just at that point, I slipped in some dog shit and my foot came into contact with his meat and two veg."

"You mean his genitals?" interjected Dickens.

"Well, if you wish to be crude about it, yes," added an affronted Loftus.

Dickens did not respond but turned to me with a look of increasing incredulity. Loftus was oblivious to the skepticism around him and continued:

"The case is a complete travesty of justice, the police investigation should be centered on putting the owner of the dog, whose crap led to the recent removal of Sergeant Kerrigan's left testicle, behind bars," continued Loftus.

Dickens had warned me earlier that Detective Sergeant Kerrigan's recent operation had not tempered his mood.

"Well it didn't do much for Hitler's sunny disposition, when he had one lopped off in the thirties," I acknowledged.

Everyone in the room, with the exception of Loftus, knew that he was going to have the book thrown at him. Knowing Loftus as I did, I was only surprised that he did not push for a commendation from the police for assisting of one of their officers in an arrest, even though it was his own.

Dickens sat back when Loftus had finished his version of the events, pondered for a while and then he expressed his view of the defence case.

"The best course of action in this case, would be to leave the country immediately or you'll be picking up soap at Her Majesty's Pleasure for the next five years."

I knew Dickens was using language that Loftus would understand to try and make him realise that he was in serious trouble and to take the trial seriously. But that night Loftus met me in the Falcon to say that he and Suzy were leaving the country.

"No jail for me," he said, "I want to be a republican, so I am not spending time at Her Majesty's Pleasure.

"I think you mean publican?" I replied.

"Well same thing, where we drink" he added.

I was depressed at the thought of both of them leaving, but so as not to end the night with everyone crying into their beer, I stood up and gave a toast to my departing friends:

"My advice to you, my good friend, is to drink and drink heavily."

We stood up and raised our pints to each other and sank them in one.

Veronica and Suzy then joined us and it was plain to see that Suzy had been crying, but not as a result of the fractured cheek-bone that Kerrigan had given her. She told us that she didn't want to leave her family, but she loved Loftus and was not going to stay in England without him. I gave her a big hug, lifting her off the floor: (she was far shorter than I). My best intentions made things worse, for my attempt to provide solace led to more tears.

"Come with us Johnny," she said.

Loftus looked up, smiled and nodded in the affirmative.

I looked at her and said, "I am sure you two will bollocks up your future together without my help." Little did I know that my words were an accurate prediction of the fate that awaited them.

I refused to let our last night together get all melancholy, so while Loftus got another round of drinks in, I decided we should all play pool and deposited a coin into the slot for the pool-table, and loaded the pool balls into the wooden triangle to start the game. At that point a gigantic specimen of the 'common red-haired mountain gorilla', a frequent native of the watering-holes of north London's Kilburn, bounded over to the pool-table. He lifted it up at one end with of his giant hairy paws. Now at a forty-five degree angle, the pool-balls rolled down the table into the lower pockets. I wanted to batter the gorilla with my cue, but realised I'd have been dancing around for ages breaking one cue after another on his ginger nut with little effect.

The gorilla then made the usual introduction in the polite conversational style that was typical of this part of London.

"Wat's your fecking name, yer little kunt?"

"Shamus O'Dumpatruck, and if you want to play, you need to put a coin on the table. Oh and you better stick it down with gum, the table has a slight slope to it" I replied.

I was trying to get into the right position to land a direct strike to the middle of his temple with the cue now gripped firmly in my hands like a Samurai sword. Loftus meanwhile, was also circling the gorilla with a cue in his hand, but it looked like we would also have to deal with the rest of the troop at the bar, which had started to take an interest in us.

"Tat's a good O'rish name!" he cheerily replied and, as if I had provided the special password to remain on this earth a little longer, he lowered the table back onto the floor and walked back to his drink on the bar.

I could have pressed him to repay us the pound for the game he had brought to an abrupt end, but Loftus and I agreed that allowing us to continue to breathe without the aid of a respirator was payment enough.

After that, Veronica, Suzy, Loftus and I worked at getting drunk out of our skulls in what was becoming a very lively pub. During the evening, two separate fights broke out in the lounge area. The first was nothing special for that part of north London, with one man receiving a hay-maker punch from the barman which broke his nose. The barman then wrestled him out of the pub and threw what was now a sleeveless shirt after him, followed by the cry "And don't fucken come back, you fucken kiddie fiddler." The second fight started off under the Marquis of Queensbury's rules; not the rules for how gentlemen should engage in the art of boxing, but those of the Marquis of Queensbury in Neasden. In fact, there was only one rule and that was that 'No fighting should take place at the bar which may hinder other customers buying drink'. Apart from that, it was a free for all. The fight descended into a big cartoon-like ball of mayhem, with the odd shoe or fist appearing briefly outside the revolving ball of bodies, which was rolling about, trying to find an Irish version of Indiana Jones to flatten.

With the exception of our party, everyone got involved, including the bar staff, apart from a shrivelled-up little man drinking on his own at the end of the bar. His nickname was Grumpy. When the fight was in full swing, so to speak, Grumpy suddenly leapt onto the bar, ran its full length and dived towards the ball of fighting bodies on the floor. As he flew through the air he cried out Tarzan-like, 'Fuckersssssss!' the fighting miraculously stopped and the bodies that comprised the cartoon-ball dispersed. Grumpy, though, was still flying through the air, until he landed on a table and bounced himself head first into the wall, before sliding unconscious to the floor.

Later that night, the gorilla came over to us and introduced his girlfriend She was equally ginger and hairy, but shorter than her partner so she had to walk around the unconscious Grumpy rather than step over him. She had a mournful face, looking like a depressed orangutan that had just been given a miniature travellers' chess-set instead of a bunch of bananas.

"Don't let the colouring fool you, you big lumbering lunatic, it's against nature to mix the species," Loftus informed the gorilla.

"I'll make a new fucken species of yer, by sticking yer smug face up yer arse and kicking you across the floor on all fours," replied the gorilla, his face only inches away from Loftus.

The little mournful woman turned to Veronica and Suzy, "How can I get a good looking man like you two?"

"Try a sh..." Loftus then butted in, but I blocked his mouth with my hand and finished his sentence for him.

"Scent, a little perfume does wonders, as you have everything else'.

At this she suddenly opened up a wonderful beaming smile exposing a missing front tooth, hopped onto the gorilla's lap and gave him big sloppy kiss. Loftus leant towards me and whispered in my ear:

"I still think you should have let me say 'shave'."

"Shut up, we will be lucky to get out of here alive as it is, without a suicide speech from you," I said, in a low voice, so as not to be murdered.

In the meantime, everyone was back at the bar carrying on with their drinking. No one appeared interested in Grumpy, who was snoring in one corner of the bar with his legs in the air. I briefly wondered what had triggered Grumpy to leap on to the bar and set off on his run, but I soon returned to the sad thought that Loftus and Suzy would be fleeing the country in the morning.

Loftus married Suzy a few years later in Switzerland, by which time they had two mischievous and incredibly cute children. I was his best man and a fantastic day was had by all at their wedding. Sadly, their story was not to end in happiness. Loftus died a few months later in a car-crash, supposedly while drunk at the wheel. I knew that to be a lie. Loftus was a social drinker and had no interest in alcohol unless he was among friends and he had made few friends in Switzerland. That turned out to be the problem.

I went to Switzerland to help Suzy and the kids return to England. Now that Loftus was dead, the British police had no further interest in them. I also wanted to find out the truth behind Loftus's 'accident'. Through one of his few friends, I discovered that Loftus was making a living stealing and 'ringing' expensive cars on the continent, which had upset some local gang leaders. But, as much as I wanted to discover more and avenge Loftus, I had enough sense to know that the priority was to get Suzy and the kids safely back to England.

Shortly after returning to England, Suzy died from ovarian cancer at the age of twenty-one. To my shock, Loftus and Suzy had thought of such an eventuality and they had arranged legally that I should be their children's guardian if anything should happen to them. I was honoured to be given such a responsibility. Fortunately, as it seemed at the time, Loftus and Suzy's family pointed out that I was an idiot who could barely look after myself, let alone be entrusted with raising two children. If I had been with Veronica, I might have had a chance. Veronica's stability might have compensated for my maverick tendencies, but after we split I think her experience with me must have turned her to God, as she married her vicar. Monica, Loftus's sister, agreed to raise them with her husband Tom. At the time, that seemed to be a perfect solution, Monica and Tom could not have children and they promised to love them as their own. I would have tried my best but I have no doubt that I would have failed miserably trying to raise them on my own. As you will read later, it was a decision that I came to regret deeply.

As for Hammond, he could not bring himself to attend either funeral. When Loftus left the country and I threw my cards in, literally and figuratively, we both lost contact with Hammond. I was to learn later that life had taken an unfortunate turn for him too.

Hammond had departed for the States and had met a woman who appeared to have mended his rakish ways. Unfortunately, he had not mended hers and she not only had numerous lovers, but blatantly flaunted them in front of him. One day, he discovered his girlfriend in bed with her latest beau; he threw him out of the window. Since New York is not known for its bungalows, the man died instantly. Hammond also broke his girlfriend's jaw. None of us would ever have laid a hand on a woman, so whatever had happened, this was a different Hammond to the man that Loftus and I had shared many an adventure with.

After failing to discover all the facts relating to Loftus death, I was determined to learn what had happened to Hammond. I visited Hammond in jail in the state penitentiary, which sadly confirmed that the Hammond on the other side of the security glass was unrecognisable as the carefree guy I had known. He was vacant, gaunt and obviously high on something. Again, this was new; none of us had ever touched drugs. All Hammond could say was that he had lost Mercedes, his girlfriend, for good. He never mentioned his family at home, Loftus or Suzy or how his beloved West Ham United was getting on. Like his team's fortunes, he was indeed lost. When I said goodbye and asked him to let me know if there was anything I could do for him, he didn't reply. I never saw him again. I did write to him, but I never received a reply.

Back in England, I traced his brother who provided me with more background to Hammond's demise. The love of Hammond's life had introduced him to drugs and he became dependent on her, as she was his supplier. She was also suing his family for her loss of earnings as a prostitute.

Alfred Lord Tennyson once famously said "It is better to have loved and lost than never to have loved at all." In Loftus and Suzy's case he was right, they loved each other more in their short time together than many other couples will experience in a lifetime. As for Hammond - sorry Alfred - but sometimes some people are better off left in blissful ignorance of love.

I had now lost both of my best friends: Hammond was as lost to my world as Loftus, along with Paddy who died that year. Thomas Hobbes wrote, "The life of mankind is nasty, brutal and short". My response was not to be bitter, nor to wallow in self-pity. I wanted some fun; in a few months' time I would reach the grand old age of twenty-one.

Only years later did I realise that I had learnt a lot from those rogues. Paddy was always looking for an angle, even with a burnt carpet under his arm; Hammond was never deterred by a challenge, in his case, having sex with beautiful women, even though he had two holes where only one should be and Loftus would readily change the rules of the game, even bringing an extra pack of cards to the table.

If there is a hereafter then rogues will be found climbing over the gates of hell, as heaven sounds, how can I put this... like hell. If heaven and hell exist, maybe God and the Devil have created images of an afterlife offering so little enjoyment so that we are in no rush to leave our world so as to upset the Karma of theirs. I like to think, as The Rolling Stones song *Sympathy for the Devil* suggests, that even Satan may not have it all his own way. Indeed, if he owned a flame-retardant car, it would need a good lock on it or it would have gone missing on the day of Loftus's arrival.

Chapter 5 – *Nurse!*

After college, I got a job in a local hospital, working in the warehouse. I had never met the boss, but had heard that she was a small, grey-haired, very pleasant Scottish woman called Mrs Burnett, though no one remembered meeting her personally. On my first week, we heard she was touring the hospital, so our foreman Mr Arbuthnot, who insisted we emphasized the Mr as you would the 'cunt' in Scunthorpe, was in a state of anxiety in case she paid us a visit. He was an obsequious individual; he even bought a large selection of cakes from the shop across the road in case Mrs Burnett had a sweet tooth.

Each tea-time, a card game was in progress. Though I did not participate, I enjoyed watching the characters play, which was now more interesting to me than the money. One day Mr Arbuthnot suggested we carry on the game in the back of Timmy's hospital delivery van, as it was best that we were out of the way. He said he would cover for us and say we were delivering stores to various wards, which was suspiciously civil of him I thought. The others sat playing cards in the van, while I went off to get the sandwiches and beer.

"That's a strange box to play on, what's in it?" I asked Tim as I entered the smoke-filled van, laden with provisions, with every player puffing away on rolled up cigarettes and the game was in full flow.

"Yeah, lung-cancer and looks like they completely gutted him during the autopsy," replied Timmy, he pushed the lid off the box, revealing a corpse that looked like it had been shredded.

As if his week had not been bad enough, the poor wretch interned in the box was subjected to streams of vomiting from some of our younger colleagues who had never seen a corpse before. Timmy was a sick bastard, who obviously did not take losing well.

At that point, the doors of the van were thrown open and standing there was Mr Arbuthnot with a little old woman, whom we assumed was Mrs Burnett. He pointed at us and turned to the little woman next to him and declared, "I'm the only one who does any work in this place," and slammed the door shut again. When we returned later, Mr Arbuthnot said we had a lot of explaining to do and told us we were all to report to Mrs Burnett's office at nine o'clock the following morning. He had set us up, it was a devious strategy, but it was flawed.

We reported to Mrs Burnett's office the next morning, an hour before she was due to meet us. This meant my colleagues had time for a few hands of gin rummy. Then at ten o'clock on the dot, just after Mr Arbuthnot sneaked in to join us, a very tall, severe looking woman entered the office,

"Who are you and what are you doing in my office?" she demanded.

"Your office?" spluttered Mr Arbuthnot.

The real Mrs Burnett demanded an explanation and when a stuttering Mr Arbuthnot explained it was a simple case of mistaken identity she responded sympathetically as only a Scottish woman of high position would:

"Yer arseholes, how could you mistake me for Howling Doris from the psychiatric unit next door? Now, get out of my office, you half-wits."

"I sensed anger," I told the others as we all headed to The Arch, the local pub down the hill, for the rest of the day.

Afterwards Mr Arbuthnot turned on us and said that as punishment for upsetting the 'official' Mrs Burnett, we would all have to work on Saturday without pay. The amazing thing is that someone actually must have gone in that Saturday, even more so as it was the afternoon of the FA Cup Final, as the evidence that someone had gone to work was there for all to see on the wall on the following Monday morning.

Mr Arbuthnot liked to sit by the door of the stores' office on his favorite cushion, so he could spy on everyone. He also liked to look up the skirts of the secretaries walking across the gangway above. As usual, on that Saturday, and on double-overtime, he had assumed his position in 'pervert's corner', which someone had scrawled on the wall above, and just as he was about to bite into his cheese sandwich, a bucket of excrement came flying in through the door, totally covering him. The stain produced a very artistic silhouette of his body on the wall. It might have been mistaken as an early work of the urban artist Banksy, except usually he works in black and white.

When Mr Arbuthnot came into work the following Monday morning, for some reason he stormed over to me and accused me of the unwarranted attack on his personage.

"Was ever a man more misunderstood?" I replied.

I refused to lower myself to his level and offered Howling Doris another cream cake, as she put down five assorted playing cards and howled "Bingooo!" to a pitch only to be found amongst Argentinean football commentators. Even though I had only worked for two weeks, I felt I had earned my pay. Though the only work I had done was emptying a bucket from the tropical diseases ward around the corner.

In celebration of my final day, we all headed to The Arch. As we entered the main bar we were greeted by a line of workers, comprised of painters, chippies and cleaners working illegally in the Home Office building across the road. All were queuing in single file to cash their unemployment benefit cheque at the 'hole in the wall' for a fee of two pounds, no questions asked. After which, they took a ninety-degree turn to the right towards the bar and pissed all their earnings up the wall. This scene was a testament to the wise words of the then Conservative Secretary of State for Employment, Norman Tebbit MP, who told the unemployed "to get on their bikes and look for work". I looked at the long queue of unemployed people in front of me, some covered in a splattering of paint, others with cement gelled to their hair. This was proof that Tebbit's message had been heard, it was just that they had forgotten to 'sign-off' the unemployment register, as they were no doubt too busy pedalling to their next job.

The queue comprised mainly of young Irishmen, Filipinos and Jamaicans, who were doing manual work that 'locals' would not demean themselves to do, or jobs that were regarded as too dangerous. Immigrants usually took on these tasks; someone had to build big infrastructure projects such as the London Underground and carry out dangerous construction work like scaffolding. People like my father. Many decent tax-payers knew of the abuse of the system, but were happy to pay cash in hand when it came to paying the bill. We were not in any way at the level of Southern Europe in abusing the tax system, but our system relied then, as it still does in Britain, Germany and the United States, on cheap disposable labour to drive our economies forward. For many of those who worked as part of the 'black economy', drink was the main release after the working week. I remember reading at the time, that young Irishmen had the highest rate of suicide of any ethnic group in Britain.

The Arch had an upstairs disco called Chicks. It was the human equivalent of the 'alien' bar in *Star Wars*, though it lacked its sophistication. Occasional violence would break out, resulting in a surprise punch to the side of your head, though perhaps the use of the adjective 'surprise' isn't quite right. If you were punched, the only thing that kept you on your feet was the Velcro-like texture of the sticky beer-stained carpet fastening your feet to the floor.

If you were not suitably dressed and could not meet the dress code, which was 'Wear Shoes' you need not worry. The burger van outside, 'The Burger Run', (some comedian had scrawled in permanent black marker 'to make you' in between 'Burger' and 'Run'), could provide you with a pair of shoes for the price of a cheeseburger along with a five pound deposit. A friend of mine, Dermot, had to pay an extra two pounds for returning his hired shoes two weeks late. He judged that his hired shoes were more appropriate than the trainers he had deposited with the burger-van for the funeral he had to attend the next day and his wedding the following weekend.

The dress code did become an issue even for The Arch, when it had a new manager. I turned up in jeans to be confronted by a sign on the door declaring,

'SMART DRESS ONLY'

While I pondered whether I should break in though the toilet window or go home, I was knocked out of the way by a burly Irishman in a donkey jacket, with a safety pin holding the arse of his trousers together, who disappeared into The Arch. I put my doubts to one side and followed him in. However, when I got inside I saw one of the regulars, Magnus, standing at the bar in a very fashionable flowery frock, in accordance with the new dress code. It's a wonder the lengths an Irishman will go to secure a pint of Guinness. It also renewed my faith in the entrepreneurial spirit of our country, for clearly The Burger Run was branching out and providing customers with their summer wardrobe.

My leaving do in The Arch carried on into the evening. I bumped into Finley, a schoolmate who was working as a trainee undertaker.

"Apprentice? Do you have to take a final exam at the end to see if you can tell the difference between an open grave and a hole dug by Thames Water?" I enquired.

"Fuck off," Finley replied; which I guessed was part of his training for greeting bereaved families.

He stood there in his long black Crombie coat, immaculately polished black Royal Brogues, a starched white shirt, black tie and waist-coat and a magnificent black top hat adorned with a black satin ribbon.

"Many funerals today?" I asked.

"No, I had the day off," Finley was a very stylish rogue.

I thought it best not to carry on that topic of conversation. I guess I was tired as I do love conversing with 'life's eccentrics', one of the many English euphemisms for 'lunatic'.

"My brother Martin has disgraced himself again," Finley began again.

"What has he done this time?"

"Well, my Uncle Tom died and we held the wake here, but Martin was barred for life a few years ago for riding his bicycle through the bar, smashed off his tits."

"A bit harsh, getting barred, old Dougal set fire to the toilets last week and all he got was banned from bingo on Thursday night, even though he can't stand the game and he'll be in hospital for at least another month."

"Anyway we begged the new manager, Grizzly Adams over there, to let Martin in," nodding in the direction of the fat, bushy-faced, very angry-looking man standing behind the bar who was staring at us. "He eventually relented, provided we all chipped in ten pounds as security in case Martin did any damage."

I surveyed the establishment and I could not envisage any disaster that could result in damage costing anywhere near that amount, including flood, fire and earthquake combined. Finley then added that it included injury cover for the barmaid, which made the amount seem very reasonable. Finley continued, "Well, we all sat here getting pissed after the funeral and started to wonder where Martin was, considering all the trouble we went through to get him in. At about ten o'clock, he came through the side door, riding his moped past Grizzly Adams who was standing behind the bar and giving him the 'fuck-you' sign. He took the other side door off its hinges on his way out and we haven't seen the mad bastard since."

I now understood why Grizzly Adams' eyes were trained on us, as I looked down at the tyre marks on the floor.

"A bit late, but a ten pound bail-bond well spent, here's a pound," I replied, handing over the money in front of the increasingly flustered Grizzly Adams.

The evening's entertainment arrived, with a special guest appearance by 'Mr Personality-Himself' – seriously that was his stage name. He had a huge black wig that looked like a chewed dog's blanket had been thrown out of a window and landed on his head. Added to that, with his the stomach bursting through his shirt he looked like he was smuggling a bag of stolen cement to be exchanged for a drink at the bar later. His act comprised of singing Elvis Presley songs, but strung out three times longer than it had taken 'the King' to sing the originals. Every syllable required a pause as he surveyed the audience, his false teeth gleaming, giving a knowing nod and a wink. He clearly thought that each lyric he wrung to death was the greatest sound anyone had ever heard. He worshipped himself and leered at the women in the audience like a greased up cuckoo entering a nest of sparrow chicks. Later, one of the women threw up over his hush puppies when he slipped his arm around her waist.

The audience paid him no attention whatsoever, except on one occasion, when he fell off the stage after a female punk violinist, (part of the Partridge Family tribute band), stabbed him in the arse with her bow, when he attempted to stroke the six inch nail pierced through her right nostril. The indifference of the audience changed when the support act arrived on stage: 'Jimmy in the Saddle'. Many a visitor was enticed into the bar by the poster in the window, showing 'Jimmy in the Saddle' looking hot and bothered and apparently mounting something. The 'something' was unclear as the bottom of the poster had been torn off. Finley had admitted to that wanton act of vandalism, when the toilet paper had run out. Thankfully, the suggestion that a seventy-year-old, grizzled Irishman would be shagging something animal, vegetable or mineral proved unfounded; his act was to provide commentary to a video of the previous year's Grand National horse race.

'Jimmy in the Saddle' frequently collapsed during his performance and had to be carried off and resuscitated with a pint of Guinness at regular intervals. His act, however, was a success and was applauded by male and female alike. His ability to remember every detail of the race, his passion, along with his determination to sink a whole pint in one go, was a sight to behold and received rapturous applause from the appreciative audience.

Later, they put on a video of *The Quiet Man*. It was projected on to a double-sized bed sheet on the side wall of the bar. To this day, the film continues to provide a stereotypical view of Ireland and is remembered for its grandstanding punch-up and mammoth drinking session. True to type and almost as a tribute to the film, a huge fight broke out in the pub. This was due to Jimmy's act; one guy had placed a twenty pound bet with a friend on the outcome of the race, not realising until the end that the race was not 'live'. This act of audience participation brought a kind of 3D dimension to the film which would have the director James Cameron, the director of *Avatar,* foaming at the mouth.

When the fighting finished, ten minutes after the film, the whole affair was rounded off by a drunken midget leaping from one rock-concert-sized amplifier to another, while the pub's bouncers tried to grab him. Amazingly, they failed to catch him even when he dropped his shorts to urinate on them. It was a scene that would have been at home in the film, as the shrivelled actor Barry Fitzgerald stole the movie, playing a similar drunken rapscallion. Perhaps, the scene will appear when the studio releases the one hundredth year Anniversary Edition, though some CGI will be required to change Fitzgerald's bowler hat into Finley's top hat, which the drunken midget wore as he swung on the pub's chandelier.

Chapter 6 – *Teenage Kicks*

After my short stint at the hospital, I landed myself a bar job in a pub in Hanover Square, called The Organ Inn. It was a perfect job for me. I was fascinated, watching the characteristics and mannerisms of the customers, as the drink slowly took possession of them, exaggerating some traits while subduing others; turning them into strange, loud, sometimes dangerous creatures.

It wasn't a gay pub, but it had a small group of homosexual customers. At that time, as a result of our upbringing, my friends and I were homophobic, which was exhibited through wariness of something we did not understand rather than in an aggressive or violent way. I had, after all, been raised as a Catholic and I remembered one priest declaring homosexuality as an abomination, even that homosexuals were a threat to children. Yet, the irony of it was that years later it was some of their brethren who were found to be guilty of such abuses.

In my first week, I worked with Sid. Sid was an altogether different rogue to those I had met before, but he stood out from the crowd. He challenged everything, was unpredictable and lived by his our moral code. He was employed in many establishments around the West End as a relief barman, and in The Organ Inn he had become a regular stand-in for the manager, known as the Phantom, as we never saw him; he was always upstairs stricken with food-poisoning after eating the pub's food. Sid would turn up for work dressed in leather trousers, biker-boots and a stretched net string vest exposing nipple-chains. Such attire would not attract a second glance nowadays, well maybe the array of sexual aids hanging from his 'utility-belt' might, but this was a time when the public images of Freddie Mercury and Elton John were that of heterosexuals. Elton even got married. I worked with Sid a few times and I learnt to admire his honesty, indeed his bravery, in coming out about his sexuality at a time when homophobic violence was prevalent in eighties England.

On the first night I worked with him, Sid jumped on the bar and proclaimed, "I'm a sadist. So all those who want a pipe under your tail get your gum-guards in."

If the sinister-looking co-worker's announcement caused a few furrowed foreheads among the heterosexuals in the bar, it was nothing compared to the gay husband contingent who were terrified of him.

After my first evening, I was in tears laughing at his antics and at the end of the night we shook hands and he said to me, "Fucking heteros, you're fucking sick the lot of you."

He took his payment of two full bottles of Smirnoff vodka off the shelf, necked half the contents of one of them and sauntered out of the bar with a pair of handcuffs securing the other bottle now hanging from his 'utility-belt'.

I felt sorry for some of the gay husbands as together in the company of fellow homosexuals this was probably the only time they could be open about their sexuality. I could hear them talking about their spouses and for many of them an 'arrangement' had been agreed with their wives. What the women got from the relationship was a mystery to me; maybe they were gay too, had no interest in sex, settled for a comfortable life and had lovers elsewhere. One of the group made everyone even more uncomfortable than Sid; he was a leading lawyer who regularly made loud and often vile remarks about his wife. One evening he came in to buy drinks for the group to celebrate the birth of his first child; the first of many he told Cat, the barmaid. An hour later, Sid caught him having full-on penetrative sex with a young man in the toilets. Sid went ballistic and beat him with a plastic truncheon from his 'utility-belt' until the police arrived and arrested him.

I was going out with Cat and she telephoned me at home to say that Sid needed help. I met Cat at the police station and we pooled what money we had, to put up the bail for his release. When Sid was let out the next morning, the three of us went for a drink in one of the twenty-four hour illicit clubs in Soho that Sid wasn't barred from. Sid loved women, obviously not in a sexual sense, and he often talked of his mother and sisters and he was very paternal to Cat when she first started. Though the lawyer was a dangerous misogynist, we wanted to know why Sid attacked him in such a violent manner. Sid was still furious when he explained that the 'bastard' was putting his family at risk of catching a sexual disease, perhaps HIV or full-blown Aids, so he decided to sort him out himself.

"Who would have thought you were such a moralist, knocking the lawyer's teeth out with a dildo," I said to Sid, while he sank a half pint of his 'usual', neat vodka.

"Bloody heteros boring as fuck, I guess you would have just punched him in the mouth. No imagination you lot," added Sid.

Cat gave me a hug and turned to Sid. "True, but you know where you are with guys like Johnny; enemies just get punched in the head, he doesn't die of mortification from his daily mistakes, he doesn't mind cuddles and no deep meaningful conversations, lovely." An unusual complement, but I take them where I can get them.

"Fucking weirdoes you two, I hope no one sees me with you, I have my image to think of you know."

We then got up to leave and Cat turned to Sid and said, "Don't worry I found your dildo and hid it under the bar-towel, when the police arrived. Night bunny," she bent forward and gave Sid a kiss on his tattooed cheek which read 'COCK HERE'.

"Disgusting," reacted Sid shuddering at the gentle touch of her lips.

I shook hands with Sid and it was the last time we ever saw him. Cat and I did get an envelope a few weeks later addressed to 'Cat & Johnny'. It contained the bail money we lost due to Sid absconding. Enclosed was a note which just read:

'FUCKING HETERO WEIRDOS X'

Apart from respect for women, Sid and I agreed on nothing during the short time we worked together; he was a staunch Conservative and a Millwall fan which I guess explains the sadism. However, I would call Sid a benevolent rogue; he hospitalized the lawyer for a month for putting his own wife and family at risk. The act of violence solved nothing and cost Sid his livelihood, but though he did not know the victims personally he did what he felt was right when he took on the abuser.

The Organ Inn was located around the corner from the renowned red-light area of Soho. We would sometimes get high-class 'ladies of the night' popping in for a quick drink, or, as one called it, 'mouthwash'. Most of these women catered for up-market business clients and were brain-dead due to drugs.

Two young women came into the bar one evening and proceeded to tease myself and Glen, an Aussie, who was also bartending that summer. Crystal was the smallest and the cheekier. Tina was the taller of the two, sterner and therefore to me unattractive. They said they 'worked tables', so Glen and I assumed they were waitresses. By the end of the night they said they had to go to work, but gave us an address and said that we should come and meet them later and go dancing. Cat and I had parted; she had returned to University so I was single again. The pub was never cleared, cleaned and locked up as fast as it was that night. Glen and I headed off to what we expected would be a very posh restaurant. Instead, we found ourselves outside a nightclub welcomed by a huge, grinning doorman who then led us two nervous teenagers to a place behind the stage.

There we found our waitresses, in lingerie, dancing on a large wooden table. The two women kept looking over at us, winking and licking their lips, like two wild cats ready to devour their prey, us. We sat uncomfortably on little wooden chairs, while the audience danced like epileptics on speed.

After the club, we joined the girls for a drink in a late night bar around the corner, but Glen and I sat down nervous, open-mouthed and still speechless. Glen stayed that way all night, but after a few belts of Bacardi rum, I started to join in the fun that the girls were having at our expense. The big bouncer joined us, bellowing with laughter, he has obviously been in on the joke from the start. Unfortunately, Glen was worried now that the bouncer had joined us; he pretended to go to the toilet and escaped out of the fire door. As for myself, I didn't care if Crystal was happy, I was going nowhere.

A few years later, I bumped into Tina one night at a club in Green Park. Tina was working for an agency escort catering for rich clientéle from Park Lane's nearby hotels. Tina told me that Crystal, if that was her real name, had met a guy and she believed she had gone back up North to raise a family with him. I was relieved to hear that Crystal had broken away from Soho. Tina was turning tricks to support her coke habit and offered me a quickie at half price for old time sake, which I politely refused. Though I had never really known her, it was sad to see how life had taken its toll on someone who was once a very beautiful, independent young woman.

As a young man, the influences on me were many. I grew up in a culture that was racist, homophobic and sexist. There was also pressure to become part of a gang on the estates where I lived. I also witnessed men and women who lived solely to feed their addictions, some alcohol, others gambling and those like Tina, drugs. These influences pull at you; some mould you, others corrupt you. I tried to avoid them all but it wasn't until I found myself in Romania that I had to work out who I really was, what I stood for and not just what I was against.

The benevolent rogues I met in Romania, Bosnia and South America, knew who they were including their failings and made no apologies for them. They had their vices, but were never consumed by them. They were my influences. You will also see that each was unique, except they shared one trait: the ability to surprise.

Chapter 7 – *A Grave Matter*

After the bright lights of the West End, I got a job as a gravedigger. I was never squeamish and, if an employment opportunity put money in my pocket at the end of the week, then I was your man.

It was one of the strangest jobs I ever had. The gravediggers, known as 'Lifers', were divorced from the real world almost as much as the dearly departed. You earned the label 'Lifer' when you had done the job for six months. You could tell who was a 'Lifer' by their reading material. When he was not six foot down digging graves, Ned 'Ambulance chaser' Murdock could be found reading books about the war from the Nazis perspective. Short-handle would spend his salary on anything linked to serial-killing psychopaths, particularly biographies of The Yorkshire Ripper or Denis Neilson. If Panini, of football trading card fame, had expanded, offering a new range entitled 'Murdering Nutjobs of the Twentieth-century'; Short-handle would have blown his bonus purchasing the full set, plus album.

They were quite civilised when conversation centred on mass murder, but tensions would arise if anyone strayed on to the sensitive subject of bonus payments for digging graves. In the three months I worked there, apart from when he leapt tombstones in pursuit of a hearse, Murdock only once exhibited emotion and that was when he threw his copy of *The Daily Mail* into the air which had the headline:

'INDIAN AIRCRAFT CRASHES OFF THE CORNISH COAST.'

"Cornwall will never cope with all that lot. We will be shovelling the corpses until Christmas," he exclaimed, gleefully thinking of the massive bonus payment he would receive.

Later that night, he was found sobbing in the local pub. The barmaid, 'Up on the slab Babs' (a name derived from an incident following the gravediggers' Christmas party), said she thought the passengers would probably be Hindu and therefore likely to be cremated.

"Why is God so cruel?" he cried.

Bonus payments were made per foot of grave dug; the deeper the grave, the greater the payment. This was open to abuse by the likes of Short-handle who always added a few virtual feet to his daily worksheet, hence his name. He secured more bonus than he was due and moved on to start the next plot as soon as possible, but his system did lead to an inherent problem; as I found out. One day, I was working on my own, when, digging three feet into a family plot; I plunged straight through a coffin. According to the plan of the plot, it should have been a further four feet deep. With a scream, I shot straight out of the grave without touching the sides. This was a regular occurrence, with Short-handle knowing his deception was unlikely to be discovered until the next family members dropped some years later. He was unpopular with his colleagues not just for this, but also for altering the work roster and securing all the lucrative family-plot jobs, which paid additional bonuses.

I later heard he died of suffocation when the walls of a grave fell in on him while opening a deep family plot. In his usual haste, he had not shored up the walls properly with wooden supports. Even in death, he buried himself, thus, and as he had in life tried to deny his colleagues a bonus.

There were two other diggers in the team, who were not 'Lifers' but just passing through like myself. There was Gene, who was a devotee of Gene Simmons of the glam rock band Kiss. I was more into the Clash and The Jam, but I respected his devotion to his passion; he wore the complete Kiss outfit, with his face covered in their trademark black and white make-up. Once, in a rush to see a Kiss retro band playing, he accidentally dropped his wallet in the last grave he had dug. He ran back to retrieve his wallet and emerged from the grave laughing waving his ticket in his hand. This came as a considerable shook to mourners at a nearby grave and led to the collapse of an elderly woman when they saw Gene looking up out of a hole in complete Kiss attire.

The other 'Non-Lifer' was Stevie, who, like Gene, became a good friend. We started on the same week and at the end of it we went for a drink. Stevie threw down the gauntlet and challenged Gene and me to a vodka session. That was the last thing Stevie remembered, before waking up the next day in the doorway of a Dixon's electrical shop, wrapped in the arms of a tramp.

Stevie had two claims to fame, firstly he was a local legend in his home town of Glasgow, for getting drunk and stealing a disabled three-wheeler from a car pound to return to a handicapped friend who had had it impounded for non-payment of tax. His notoriety was further enhanced, as it took two hours and three police cars to catch him, which was impressive if you realised that the car he'd stolen could only reach a maximum speed of fifteen miles per hour. That can only be exceeded if you tip one off a cliff. For this unselfish act of drunken idiocy, I firmly class him as a benevolent rogue. But, it was another claim for which he was really renowned: that he was completely in love with a petite blonde German girl called Mika, one of the most beautiful women ever to grace our shores.

In Stevie's eyes Mika was perfect, but it was this perfection that unsettled the lovestruck Scotsman; in all other respects he was fearless. Stevie believed that Mika was too good for him; he was a big lumbering, balding, whisky-swigging, drug-taking, chain-smoking man with no prospects. He was also practically illiterate. I often had to find the gravestone or plot for him and lead him to where he was to work. This insecurity led him to believe that he was not good enough for Mika. It came to a head one evening after work when we were having a drink in the Elgin Arms in Ladbroke Grove. He turned to me and said he needed my advice about relationships.

"Christ, you're asking me, it's not that bad, surely?" I queried.

"Look, Mika's parents are arriving from Germany tomorrow. I need your help," he stuttered, as he took another big gulp from his pint.

"Happy to help. If you're too busy to meet them, I can put on a good Scottish accent, secure a kilt, dye my hair red and punch the old man in the face at 'Arrivals'."

"English bastard, I don't need you to take my place, I need advice," Stevie replied.

"Well my standard advice, my good friend, on these occasions, indeed any occasion, is to drink and drink heavily."

"That's the problem. I'm terrified and when I'm like this I need a drink. That's when it all goes wrong."

"Then, I never thought I would say this and I would deny this if it ever got to court, but stay sober, smile a lot and grin like an idiot," I continued, "It's standard procedure for boyfriends when they meet their girlfriend's parents; mind you her dad will always think you're a sex-case anyway for sleeping with his daughter."

"I will try," he said and I wished him luck but I feared what would happen next. My fears were justified.

Two days later, I got a call from a very distraught Stevie asking me to meet him in a pub. This did not sound good. When I entered, there was Stevie in a terrible state leaning against the bar. He told me his sorry tale. Stevie met Mika at the airport as planned, but that was the only part that actually went to plan. The flight was delayed and Mika persuaded him to have a drink to steady his obvious nerves. Unfortunately, his nerves continued to get the best of him and he was paralytic by the time the plane landed five hours later. On the journey back to London, the conversation was conducted in German and Stevie became increasingly isolated from the clearly delighted Mika reunited with her family. At dinner, the wine flowed and the conversation continued in German and nobody noticed the drunken Scotsman who, in addition to the table wine, was helping himself to cans of Tennent's Extra Strong Lager hidden under his chair.

"I'm not sure I want to hear the rest," I said, "but I take it the beer kicked in and the confidence grew, which meant you opened your big Scottish gob. Christ, didn't Culloden teach you lot anything?"

"As much as I would love to punch your lights out, you English bastard, you're right," he said head bent low towards the bar. He continued, "I suddenly thought I should say something to Mika's parents, as by now I had been in their company for eight hours and I hadn't uttered a word. The problem was not that I spoke no German; Mika said that her parents did speak a little English, but for the life of me I could not think of anything to say. Then I suddenly remembered something about their homeland, which I thought would show that I at least had a thought in my head."

"Ah well, I warned you not to start thinking," I added.

He could barely look at me, let alone speak.

"Come on, it can't have been that bad," I lied, knowing it was probably far worse.

Just then, I was distracted by Gene walking into the pub in his standard Kiss regalia, thinking to myself 'I hope Stevie didn't ask Gene what to wear at the airport'. I ordered a round of three pints and three whisky chasers.

"What did you say, sorry I missed it?" I asked, turning back to Stevie.

Stevie took a long pause, followed by more muttering, which Gene and I had trouble comprehending through his broad Scottish accent.

"You can't get a lot of fuel in Germany," I said, trying to translate his mutterings.

"No," he shouted at me in anger and frustration.

"You have to pay your dues in Germany," added Gene.

Then Stevie shouted, "No! I said ...You don't get a lot of Jews in Germany!"

It was easy to gage the effect this had on a lovely young German girl and her family, particularly as we looked around at the shocked non-Germanic faces in the bar.

I just sighed, "God help you, Stevie, as I always say about your country, evolution before devolution."

"Fucken tell me about it," he said, as Gene ordered another round. Whisky was a drink I hated, but on this occasion our poor friend needed support and this was the only way we knew to provide it.

The three of us were in a drunken state by the end of the evening and Stevie and I loaded a profusely drunken and sweaty Gene into a cab. He looked like a big Daliésque painting of a melted domino; his black and white face paint had run. I slipped the driver five pounds to drop Gene off at his house nearby which was next door to a Blues Bar. Then I had to get the guilt-ridden, inebriated, boyfriend back to, I hoped, his understanding girlfriend. When I carried Stevie back home, Mika had indeed been worried sick about Stevie and was glad to see 'her man' again.

I had met Mika before and I made our apologies, "He's an idiot of the highest order, but he means no harm and he loves you, so please forgive him."

"I do and I always will," she said.

Her father joined us and, while he helped me steady Stevie, I shook his hand and said, "Achtung". That was the only German word I knew. It was the most common word uttered by German soldiers in war films; I thought it was a greeting along the lines of 'Hello'. He just shook his head from side to side and gave me a resigned smile as he took Stevie by the other arm and helped me carry his inebriated and now incontinent future son-in-law up the stairs to leave him to sleep in the bath.

Mika kissed me on the cheek and then went back to cleaning up her man. I left, closing the door gently behind me. Now that is true love, I thought, and I wondered if one day I would find someone like Mika who would love me despite my faults. After a moment's contemplation, I concluded, 'No chance', and I set off to the Blues Bar for a late drink. Despite our drinking efforts that evening, I doubted that anyone could reach Stevie's level of intoxication. I was wrong; as I entered the Blues Bar, I saw Gene on stage, still in full costume, singing the Kiss rock hit *Rock And Roll All Nite* to a bewildered audience.

Three months later I decided it was time to move on, so I left the job choosing life to that of becoming a 'Lifer.'. My only memory of my last day was of an elderly man who came up to me as I was mowing the grass on the area designated for Polish burials.

"Here's five pounds for looking after that patch," said the old Polish gentleman as he slipped me a note.

"Thanks," I replied, thinking it would get me into Chicks later, plus a few lagers that night, "But why?" I asked.

"You see, just next to where you are is my wife's grave and one day soon I will be buried where you are standing now," he replied solemnly.

They say one person's grief is another person's gain. I carried on that evening, dancing in Chicks with a drunken blonde just arrived from the Emerald Isle. The grief-stricken man probably sat at home looking at photos of his wife, just waiting to join her. I woke up the next morning and watched the blonde fall over laughing as she tried to put two legs in the one leg of her knickers. As a young man you give little thought to death. You believe you are indestructible. It's better that way, there is plenty of time to contemplate one's mortality.

My next engagement was one of the most mind-numbing jobs I have ever had. I got a job as a postman to cover the Christmas rush. I quickly understood why the people who worked there had one shared goal, namely to finish work and get to the pub as quickly as possible. No one appeared to be interested in anything else, even in having a conversation. One postman did have a nickname: The Frenchman. Perhaps the man of mystery could be an ex-Legionnaire, a fluent French speaker, who knows he might even be French. So I approached him.

"Why do they call you the Frenchman?" I asked.

"I went to Calais one afternoon on a school trip. It was shit," he replied, and carried on sorting the mail at a rate of one letter every five minutes.
After that, I told the post office to put my cheque in the post. I never received it.

After a few driving and labouring jobs, I managed to get a temporary stint at the Alexandra Beer Festival which, for that year, had to be held under a huge marquee, as someone had burnt down the Alexandra Palace. My first customers were a group of young men trying to impress the young women with them.

"What's the strongest beer you have?" asked the bravest.

"Depth-charge," I replied.

After an hour of drinking, his quest to become the dominant male of the pack had reduced him to being down on all fours barking at everyone. Thirty minutes later, he had passed out and defecated in his pants, though I could not swear for sure which had happened first.

Later, as most of the customers descended into a drunken stupor, I heard a huge crash. It sounded like a herd of elephants falling head first into a mud bath. I was not far wrong. The portable toilets had flooded early in the evening, one exploding due to a combination of a lighted cigarette, a leaking gas cylinder, and no doubt, an abundance of methane. As bladders filled, the patience of the drunken revellers was exhausted. The fence at the bottom of the hill, leading to residential gardens, became the alternative toilet facility and a common resting point as fat-bellied, full-bladdered ale specialists positioned themselves with right-arms stretched out against the fence, like a column of leaning Nazis, to urinate perfectly over their shoes. Under the weight of now fifty drunks, the fence collapsed into a moat of stagnant urine. It was one of the most disgusting sights I have ever seen. A mass of beer-bellied drunks tried to get to their feet in this man-made cesspit only to fall back on each other. The nearest comparison I could think of was a TV wildlife programme filming wildebeest, surrounded by crocodiles and leaping over each other in panic trying to get on to the bank of a river. If David Attenborough, that great narrator of animal behaviour, were commentating, he would have had little choice but to finish his voiceover with; "While tragically drowning in their own piss."

I left them to it as my shift had ended. On my way out, I passed two drunks trying to drive home on an old Triumph Bonneville motorbike. They would have missed the concrete boulder at the exit road on the way out, if they had remembered that they had a sidecar. The sidecar was now wrapped around the bollard, causing the bike, its attachment and both passengers to revolve around the concrete post as if continually negotiating the smallest roundabout in Britain. Reminiscent of Max Sennett's Keystone Cops, the police ran around the rotating bike trying to turn off its engine, while it spun them off in various directions. I would have loved to stay and watch them lift the doughnut-shaped motorbike and sidecar off the boulder, but it was nearly nine o'clock in the evening and I needed my first drink of the day.

My next job was as a tree surgeon in a London zoo. On my first day, I was told my predecessor had cut down a branch. Nothing extraordinary about that, you might think, apart from the fact that he was sitting on it. The zoo's safety inspector suspended all 'tree monkey activities', while they launched an investigation into how such a thick bastard had been allowed up a tree in the first place.

I was pointed in the direction of the communal hut, until they could work out what to do with me. As I entered, I was promptly greeted with "You Bollocks!" All those sitting inside looked a little sheepish. I did think it was a little pre-emptive of someone who had never met me to greet me in such an intimate manner, but it did make my new place of work feel kind of homely.

For my first task, I was sent off to hoe weeds around the chimpanzee cage. Unbeknown to me, this was the zoo's initiation ceremony for new recruits. Chimpanzees are angry bastards first thing in the morning, and who can blame them? They sat, surrounded by bars, with a spare tyre hanging from a rope for comfort. They vented their rage by defecating into their hands and lobbing it at the first human they saw each morning. That morning it was my turn. I had been welcomed to my new work place with "You Bollocks!" and was covered in the shit of a species that was officially below me in the food chain and it was not even nine o'clock in the morning. I have had better days.

Jed was the zoo bully, as well as the resident religious fruitcake. He was under qualified for the first position I thought, as he was a midget. I don't know if the others were genuinely afraid of him, or if no one had low enough scruples to hit a midget. Fortunately, I had signed up to a policy of equal opportunities for all. When he threatened to hospitalise me with a shovel for not joining in during grace over our breakfast of bacon sandwiches, I refused to patronise my little friend. Instead, I pinned him halfway up the side of our hut with his shovel. I extracted some information and discovered that the originator of the verbal abuse I received that morning was the hut's mynah bird.

Everyone was embarrassed by the abusive bird, to the point that nobody wanted to be associated with it, so the poor little feathered abuser didn't even have a name. Call me an old sentimentalist, but I christened the bird Loftus as his expressive manner was uncannily like that of my friend. I broadened the bird's vocabulary. To be greeted with "Fuck em!" each morning as you opened the door to the hut was a perfect way to start the day. As the bird had a name and a friend in me, the rest of the team started to find our dark-feathered friend endearing. That is with the exception of Jed, as I had enlarged the bird's repertoire, not only by added "Bollocks to em!", but I taught it to end its obscene sentences with the words "Yer, god-bothering Midget."

If, by the end of my first week, I thought I was back in control of the situation. I was wrong. Nothing could have prepared me for week two.

"Today, Mr Righten, we have work for you at the incinerator," said my supervisor, as if he were announcing the third of the 'Trials of Hercules'. The first trial had been to avoid being killed by simian shit throwers, the second to deal with a god-fearing midget armed with a shovel.

"Who do I ask for?" I asked.

"John Doe is the name on his wage packet," he said, pointing me to the south side of the zoo.

I headed in the direction of a small, non-descript windowless building with a chimney emitting plumes of dense smoke. I opened the door and entered my new place of work. The heat was blistering and it was so dark that I immediately fell over something that turned out to be a dead impala.

"Shut de fuck kin door!" I heard as I lay face down on the floor. Despite the open furnace, it was so dark that I could not see anything; the voice might have belonged to a huge mynah bird with a larger vocabulary than I expected, but I could tell by the accent and the warm greeting that whatever it was, it was Scottish. That was the most sense I got out of my colleague in the short time we worked together. When I asked around to find out more about my new colleague, everyone said they kept their distance from him and just simply referred to him as the Jailer. His name derived from the Monty Python character from the film *The Life of Brian*; all he did was grin and grunt a kind of laugh similar to Terry Gillian's creation. During my week there, I never saw the Jailer emerge into the sunlight. The only time he stopped shovelling objects into the furnace was when he ate his sandwiches. He did so like a macabre Doctor Doolittle, surrounded by a spellbound audience standing stiffly on their hind legs due to rigor mortis.

Come to think of it, he was there when I opened the door in the morning and still there when I slammed it shut at night. He must have burrowed out under the fence at night to scavenge at a local twenty-four hour store, as he always had a supply of Mother's Pride bread. There were never any empty cans around or wrappers, so where he got his sandwich contents I guess was no big secret. We agreed an unwritten pact, that I would leave by the fire exit at lunchtime for the pub and he could carry on unimpeded as he put his box of chopping tools to work.

The Jailer loved his job and approached it from a different angle each time, to ensure his task never became monotonous. As he tore open each body bag delivered from the zoo's mortuary, his eyes would light up under his long, grey, unwashed hair and he would let out a strained giggle. Then he would open the furnace door and swing in the corpse, or on occasion lob it over arm. That week we received about fifty dead penguins, the victims of an outbreak of a rare strain of botulism. After a week of autopsies, they were completely solid, so he dropkicked the frozen colony into the furnace one by one. This was no mean feat, as the aperture was only slightly bigger than each bird. Most reached the furnace at the first attempt, so he had obviously been practicing over the years. When one bounced off the metal surround, he would occasionally get it on the rebound. Ah, if only he had played for the Scottish football team. Mind you, they would have to tie his sleeves at the back in a makeshift straitjacket to stop him terminating any player he found diving to get a penalty.

A few weeks later, we received an orangutan, frozen solid after being a month in the mortuary freezer, awaiting autopsy. The 'ginger stiff' was ten times the size of the opening to the furnace and I observed The Jailer rooting around in his toolbox. To transport Benjamin, as that was the name on his toe-tag, to the next world was clearly not going to be a dignified operation. Even worse, it looked like Benjamin must have died during an epileptic seizure while on a disco floor; he was contorted in the most unusual position with limbs rigidly pointing in all directions. At that point, the Jailer produced a sledgehammer and hatchet. I decided this was the right moment to say my goodbyes while I still had some remnants of sanity left. I leapt the fence, headed straight to the pub and only returned at the end of the week to collect my possessions and my wage-packet.

On that last day, I packed my bag, collected my pay cheque and fed Jed's sandwiches to our feathered Loftus for the last time. As I walked towards the main gate, I witnessed an example of Anglo-Chinese relations. The Chinese had requested that Her Majesty's Government send one of our pandas back to China to mate. The Government must have thought this was an excellent opportunity to get some good press for Britain's foreign relations and agreed for the zoo to send Randy, the panda, back home.

I watched as they loaded the hairy Casanova into his transportation crate noting with surprise that he was not sedated. His agitated state indicated that this might be a mistake. There was a further little problem with this Anglo-Chinese exchange; it was Friday and the zoo payday was Thursday. This meant that the zoo carpenter had pissed up his wages the night before and the badly hung-over craftsman had only that very morning nailed the panda's crate together. He had hammered the nails upwards, rather than sideways, into the sides of the crate to secure the base. This structural fault was not evident to the eight equally hung-over 'panda-bearers' as they carried the horny panda to the waiting photographers at the main gate.

At this point, the slats on the bottom of Randy's love carriage cracked and his backside appeared unceremoniously through the base of the crate. As the flash bulbs exploded, I walked off leaving the panda unconscious with its erection wedged into his belly and his huge hairy arse dropped further through the wooden slats. Thinking of that moment, I must ensure that Bear's coffin is reinforced with ship rivets. Again, I ask for your patience, as you will hear more of this similarly frustrated lothario later.

During all this time and after meeting so many characters, I had only met one person whom I would call a benevolent rogue. That was Stevie. Rogues and eccentrics are as common as birds, but benevolent ones are as rare as those chirping profanities. Why is this? Well Stevie epitomises the core skills required to stand alongside Loftus and Sid. These include a mischievous nature, a strong belief in what you think is right, fearlessness in the face of authority, and using their strength to help and support the vulnerable. And finally, that trait which Stevie had in abundance, the ability to commit the stupidest act or say the most inappropriate thing at exactly the worst moment. If you meet a 'benevolent rogue', savour it, for due to this last trait, they have a high mortality rate.

Chapter 8 – *Please, There's No Need for Tanks*

I continued drifting from one job to another, one of which was as a cocktail barman in Camden Lock. The head barman was a Frenchman, called Jean. He made it clear to me right from the start that he would serve at the top end of the bar where the wealthy customers sat, as the tips were bigger. He intimidated the other bar staff so that they remained at the lower end of the bar. I was having none of that and, after the experience of the recent Anglo-Chinese relations debacle in the zoo; it was time to challenge our Anglo-French *entente cordiale*.

On our next shift I asked him, "Why are there trees along the Champs Elysees?"

"I do not know and I do not give a *mung-keys* for your English humour."

After a few hours of ignoring each other, Jean's curiosity got the better of him and he demanded an answer.

"Because the Germans like their shade," I replied.

As expected, Jean launched a torrent of abuse at me. My strategy of diversion worked; I was able to slip past him and take over his end of the bar. He realised that, like the Germans had done to his country in World War One, I had circumvented his French *Maginot line* and he became apoplectic with rage. After that night, the manageress of the bar barred us from working together again, which suited me fine. I suggested to the other bar staff when they worked with me, that if they wanted they could work the top end of the bar, as I didn't like the people drinking at that end anyway. I was never interested in the tips; I just don't like people who throw their weight around.

I knew Jean was out to get me sacked, but luckily, he was sacked first by the manageress. She caught him wrestling with a delighted drunk who had discovered the bottle of Veuve Clicquot champagne that Jean smuggled out in the rubbish after each shift, and until that night, retrieved when everyone had gone home.

I then joined a London hospital as a theatre porter. Mary, a nurse I knew, told me before I took the job that a typical candidate should possess a large beer gut, chain-smoke, gamble away their wages and be sexually active with the strippers from The Plump Bird public house across the road. I lacked the necessary qualifications in all these categories, but Mary put a word in the right ear so I got the job. Amongst my new colleagues, one porter called Nigel stood out; he was forty years old, a declared virgin with no obvious vices and was therefore judged by staff as completely untrustworthy. This was confirmed when it came to light that he was the subject of several restraining orders, involving nurses and female patients at the hospital.

A further restraining order looked likely when I heard screams coming from the maternity waiting room. I ran in and found a very large Latino woman on the verge of labour, holding Nigel by his throat and reshaping a bedpan through a basic process of beating him about his head with it. I allowed her to carry on battering him, until I thought I should stop her for the sake of the baby who was making his own way out. Nigel broke free, but she ran out of the waiting room like a bandy-legged sumo wrestler, in pursuit of the battered orderly. 'God,' I thought, 'I love this woman'; she was not to be distracted from her mission by the mere act of childbirth. I watched her catch him square on the back of his little bald head with the battered bedpan, before I led her back to the maternity room to give birth.

If an unborn child is influenced by the actions of its mother, then little Manu should, by the age of ten, regularly have put his neighbourhood police station under siege. Afterwards, when she had calmed down and given birth to a particularly ugly armed robber to be, I asked what had caused her to launch an attack on such a defenceless pervert.

"Well, I was in the anti-room on my own, just carrying out my breathing exercises, as I had been taught to do. Then all of a sudden this little specky geek's head appeared from under the end of the table between my legs, with a horrible grin on his face." She continued, "The little bastard said, "Don't worry, I will be the one putting your feet in the stirrups", and stroked my thigh, so I whacked him across the head with a bedpan."

"Quite right," I replied, "When you're ready to be discharged, I'll show you where he sleeps during shifts, so you can carry on where you left off."

She thanked me; I wished her all the best and presented her with a top-of-the-range stainless steel bedpan from the private ward, which reduced her to hysterical laughter. When I next saw Nigel, he had a tightly banged head and his horrible grin exposed a few missing teeth.

Despite the official complaint registered by Manu's mother, Nigel remained in his job. His role as the regional shop steward may have given him some immunity, but there had to be more to it, for whatever reason the management regularly turned a blind eye to his abuses.

The night before I left my employment there, someone had thrown Nigel down the laundry shaft. The guy in the loading bay was expecting the usual consignment of laundry to fall into the waiting container; but he didn't expect to hear Nigel's screams as his freckly skin was burnt off his little bony knees due to the speed he was travelling down the metal chute. That was followed by a loud thud when his cranium met the metal container, when he was launched forth from the metal duct.

It was an interesting few months and, again, the job gave me insight into the lives of others. On a monthly basis, one man would consistently have more of what remained of his limbs amputated. The reason was not due to the spread of some kind of cancer, but his continuous smoking affected the healing of his skin graft following each amputation. I expected to one-day find only his head on his wheelchair, still asking all who passed to "Spare a fag mate."

However, the incident that I remember most and the reason I had to leave involved Alice, a girl I was asked to watch over one day. She had just had a mastectomy and was recovering from having both of her breasts removed. I thought this was as a result of breast cancer. That was until later, in the local pub I found Mary hitting the drink hard; I understood why when she told me that the girl had slashed at her own breasts with a knife. Mary had grown close to Alice; there were only a couple of years difference in their age. One day, Alice had confided to Mary that her father sexually abused her and that she tried to free herself from his repeated abuse by cutting off her own breasts in an attempt to remove her sexuality.

"Fucking men are pigs," said Mary angrily, as she wiped away a tear.

I said nothing; I had to agree.

When I heard screaming a few days later, I entered Alice's hospital room and found her father struggling with her. I smashed him into the wall and held him there. I promised to find him, if his daughter was re-admitted to the hospital for even a cold. Unsurprisingly, it cost me my job, but I made the hospital agree to report the case to the appropriate child protection authorities. Even then, the authorities, the hospital, as well as the police, did not want to press charges against the father. This was a time when the authorities did not want to get involved in domestic disputes involving sexual or physical abuse, as the children were rarely believed. Mary kept in touch with Alice for a while, but she eventually lost contact; my wish that she remained free from her father, amounts to just that.

I am not a violent man and you can count the fights I have had on one hand; well, in my early years, perhaps one holding a calculator. Moments where I have used force, such as with Alice's abuser, are rare. When I have, I have the scars to prove it, such as when I disarmed a mugger, but was too late to stop him leaving me a fourteen inch wound when he sliced through my thigh with a cut-throat razor.

Life is filled with moments where our response truly defines us or exposes us for who we really are, not who we are expected to be or who we want to be. These moments are rare and when something unexpected happens there is usually an acceptable response. We react in the manner that we are trained to do or that is laid down by law. But what happens when the situation is so unique to our experience that no precedent exists, and there is no one to uphold the law? Do you do nothing, or act? In Alice's case and later when I was doing my relief work, I acted. But, the real test of one's character is to think of the consequences of any action you take and not simply act because it makes you feel good.

ADVENTUROUS ROGUES

"I naturally gravitate to London, that great cesspool into which all the loungers and idlers of the empire are irresistibly driven."

Sherlock Holmes

A Study in Scarlet.

Chapter 9 – *"There are No Homosexuals in the Soviet Union"*

I didn't just meet rogues through work; I also met many on my travels. Indeed rogues by their nature are nomads. My first independent foray was to volunteer to restore a castle in the south of France. I did not have much money, but as long as I toiled hard, the charity organising the project would cover the flight costs and provide a daily allowance for food. So off I went. It was a fun time, as the volunteers were a mixture of Spanish, Swiss French and a beautiful Dutch girl called Elaina. Each morning we took turns to cook a traditional breakfast. When I produced a big English fried breakfast, the recipients were amazed to discover any Brit lived beyond their thirties.

After our first week restoring the French fortress, we went down to the local river. It had a number of small locks to protect the crops from flooding. Elaina stripped off down to her underwear to unveil an incredibly beautiful body. Mind you, we were young men so any woman's body that was symmetrical and not over twenty stone would have been attractive. The other males strutted around her like peacocks or fawned at her feet. I decided to play the aloof cool Englishman; I continued to act nonchalantly, climbed onto one of the locks and dived into the stream. As I did so, all the sleepers that were used to dam the waters collapsed under me and the river waters flooded the fields. I had come to the south of France to restore part of their historic culture, but I had done more damage to that part of the landscape in one sunny morning than German bombers had managed during the war. Throughout this book, romantic liaisons are more tales of my ineptitude and idiocy than boasts of sexual conquest. Woody Allen once said that if he was reincarnated he would like to come back as Warren Beatty's fingertips; the best I could hope for is to come back as Woody Allen's ears. As the rest of her suitors ran off, I grabbed Elaine's hand to make sure she was not caught by the screaming farmers now in pursuit.

My guilt subsided a little when Elaina suggested that we form a pact and together work even harder to help to restore the castle. This newfound passion for our task completely evaporated when we were all invited to the Town's Mayoral Office for a presentation to thank us for our help in restoring the local castle. As we entered the Mayor's main reception, we were confronted by a huge portrait of the French leader of the National Front, Le Pen, which held pride of place above his mantelpiece. The Mayor then took great pleasure in informing me, the only English person there, that the castle I had been working was a notorious prison used for torturing captured English troops during the Napoleonic Wars. I had no problem with the town's history, but I did have a problem with its politics; the Mayor then made a speech declaring that he was elected on a manifesto of deportation of all immigrants, blaming them for the country's woes. I have to concede that in my case he was right, as one English immigrant had probably wiped out their surplus crops, but I cheered up a bit thinking that it might reduce donations to the Mayor's political coffers that year.

Elaina and I remained together after that in England and later on in Holland, so it was more than just a holiday fling. But, we were too young to make any long-term commitment and we both happily settled for the fun we had.

As a boy, I was intrigued by the Cold War and read every thriller from my local library, so now, as a young man, it made sense that my next adventure had to be crossing through the Iron Curtain and into the Soviet Union. Well it made sense to me. It was virtually impossible to get there; all tourists were closely vetted and all visits rigorously controlled by their 'Intourist' tourist agency. 'Intourist' was more a security escort operation than tour-guide service. If you wanted to know what the main cathedral in Red Square was called, there was not an Intourist guide to be found, but if you went for a piss in a field, you would, more than likely, see one of their faces looking up at you as spray bounced off their forehead.

I managed to pass through the Iron Curtain by finding a coach company called Cosmos Tours, which provided trips for elderly tourists who were not seen as a threat by Soviet authorities. As I bordered the coach, I realised that the only threat was that their colostomy bags might explode en-route, black out the coach windows and result in a head-on collision with the Kremlin. If Uncle Paddy had dropped me off that morning and seen the poor health of some of my travelling companions, he would have quickly devised a scheme to sell their soon-to-be-vacant seats for the return journey.

Apart from myself, there were two other 'youngsters' amongst the grey-haired clientéle; two Australians in their late twenties who did not appear to be travelling together. For whatever reason, I guess Australians were classed alongside coffin-dodgers as a low threat by the KGB, the Politburo's Secret Police. I hit it off straight away with one of them, Ryan, who 'ripped the piss' out of everyone, including this solitary 'pom'. The other Aussie, never spoke a word to anyone all the way through our journey and never expressed any interest in any of the countries we travelled through, until we got to Russia and when he did, he did so with gusto. By the way he walked and his extravagant gestures, he was most definitely gay. He also had a ritual of dashing into the toilet at every coach stop, firmly pressing his horn-rimmed glasses to the top of his head with one hand and clutching his shoulder bag tightly close to his chest with the other as he ran off ahead of everyone. He was always first into the lavatories and the last to emerge, always with a huge silly grin under his moustache. His toilet visits were not, as we were to discover, anything to do with the George Michael's practice of 'cottaging' where one could meet strangers, shake hands and then shake something else, before the police arrived. Our mysterious fellow passenger had a different craving, which we were soon to discover along with the Moscow police. Even then he remained an enigma.

My first experience of the communist bloc was East Berlin, a unique city. Following partition after World War II Berlin was split between the allies: the US, France and Britain had the Western half of the City and the Soviets had the East. Both the Americans and the Russians used it as a showcase to exhibit their wares and the benefits of their system over the other. It was a one-sided campaign. The capitalist West's shops were packed with goodies and the streets full of young liberated Berliners in full voice questioning authority and wanting even greater liberty.

Trying not to be outdone, East Berlin too had its youth, deliriously happy, striding forth under the Soviet banner. The difference was that these happy citizens were visualised on huge posters; while below them walked its inhabitants, miserable and subdued. This dramatic split screen contrast was also evident in its shops, where posters offered a large variety of produce for sale, while the shelves below stocked only stale bread and tins of food, stacked next to tractor components. How their Soviet masters thought that this sales pitch would win the hearts and minds of East Berliners was a puzzle, as all it did was expose the poverty of the communist system.

After a late night in a West Berlin club with Ryan and two female students from Sweden, I made my way into East Berlin the next morning. Ryan had decided he wanted to marry one of the Swedish women, Anna, from the previous evening, and set off to find a registry office to secure a marriage licence. This was a surprise to Anna, who was travelling across Europe having a good time with her friend Anis, especially as Anis was the one with whom Ryan had spent the night.

To enter East Berlin I made my way through 'Checkpoint Charlie'; the only above ground entrance into East Berlin. To enter that part of the city, you needed your passport and had to change about thirteen pounds sterling into East German marks. Having done so, I found myself in a dirty mirror reflection of West Berlin, one without the colour, vitality or smiling faces.

I walked around East Berlin for a few hours, but each time I discovered anything of interest I would be immediately moved on by the Stasi, East Germany's plain-clothed secret police. It was a frustrating day; I did the only sensible thing and hit the nearest café, which only sold inedible food, but I was happy to settle for a beer. My most vivid memory of that divided city was the appearance of the waitress, the most beautiful woman I had ever seen. She stood out from the drab non-descript canvas, as if Marilyn Monroe had been superimposed on a 1930's black and white cartoon world.

I never found out her name, as her English was on a par with my non-existent German. However, our lack of language skills did not appear to be a complete barrier; she laughed often as she spoke German to me. It appeared that having a young western man in the establishment was a novelty; she continued laughing whenever she looked over at me. Perhaps she enjoyed visual humour. The cafe's clientele watched me constantly and the children whispered amongst themselves and occasionally looked over wide-eyed in my direction. I opened Fleming's, *On Her Majesty's Secret Service,* which was banned by the communists and drank the high octane diesel masquerading as coffee. Though we could not communicate, the smile of my waitress and her laugh as she picked up my book was a greater act of rebellion than my banned novel. Hell, her laugh was gloriously illegal. I knew that if her fellow Berliners emitted that same fearlessness in their laughter, then one day this would find the resonance point in the wall that divided their country and bring it down.

I could not buy anything else with my remaining East German marks or change my money back into any western currency, so I left her the twelve pounds I had left as a tip. To me, she represented the hope that the Berliners would one day take back their city and live as every other teenager did in a free world, discovering their own pleasures and making their own mistakes. Years later, when the Berlin Wall was torn down in 1989 by the Berliners on both sides, I watched *BBC's Newsnight* and hoped that the beautiful waitress was scaling the rubble to spend her accumulated tips on a night of uninhibited celebration on the streets of her newly united city.

After Berlin, my journey towards Moscow continued. Ryan and I disappeared each night to live it up in the nearest city and took turns each morning to purchase a case of beer for the day's journey ahead. If you are wondering what happened to Ryan's fiancée, don't; that was an almost daily event. Each girl he met resulted in him taking a trip to the local registry office, followed by a trip to the flower shop and then later on joining me at the bar to drown his sorrows; his prospective fiancées always absconded, usually with his wallet.

"Ryan," I advised, "I don't think you can blame them if they don't accept your proposal after one night together; no woman in her right mind would expect you to return with a marriage licence, flowers and a wedding-ring."

"But if it was to be a surprise, anyway where will I find a woman in her right mind?"

"Well, Ryan, you've got problems, mate. By the way, next time you get down on one knee and propose marriage, can you make sure that I still don't have my arm around her, as Anna and I were saying goodbye at the time?"

A few elderly coach passengers must have had offspring that were in some way off the pace with the rest of their class, as they were particularly protective of my good self, the youngest of the ensemble. A strange thing I observed about elderly people was that any idiotic comment or sexual innuendo uttered by us younger passengers was simply ignored. I remember when we were in Warsaw and over lunch I heard an elderly woman from our coach say to Ryan that her meat was a little bit chewy.

"It's as tough as an elephant's foreskin," replied Ryan.

The woman just smiled and nodded, although five minutes later, her dentures fell out of her mouth and landed on to her plate, producing a rather surreal grin on her main course, which we were later informed was pickled pig's snout.

When we crossed into Soviet territory, I noticed that the other Australian was starting to get fidgety at the back of the coach. As the Intourist guide took over from his Polish counterpart at the Russian border, I saw his Polish colleague give him a one finger gesture. Our guide, like our silent Aussie, I guessed was gay, due to the way he caressed the driver's cheek when he got on the bus.

As the next stage of the journey began, our Intourist guide began to describe his country in glowing terms. Following an hour of this propaganda, he declared that he would be happy to answer any questions we had about his glorious country. He then added, "Which always welcomed free engagement with all visitors to the Soviet Union," without any hint of irony detectable in his voice.

After a further hour of banal questions from our coach party ranging from "What is the currency you use in Russia?" to "Do you show *Star Trek* here?" a hand was finally raised excitedly by our mysterious Australian friend in the back seat.

"Yes, what is your question," asked the guide.

"Do you have homosexuals in the Soviet Union?" demanded the Australian.

"No, there are no homosexuals in the Soviet Union," replied our guide, who in doing so denied his very existence.

"Well there fucking are now!" said the disgruntled Australian, who turned back to look out of the window, clutching his bag even tighter to his chest.

The banal questions continued, but I had no doubt that things were going to get a bit livelier on the next leg of our journey as we headed into Moscow.

As soon as we got to Moscow, our mysterious Australian, who we now learnt was booked under the name of Gilbert Penisbury, darted off as usual to the toilet; he never returned, as he had escaped through a door out the back. Later we heard he was arrested in 'Lenin's Tomb', completely out of his head on drugs. These he had concealed in his handbag and smuggled through eight separate countries and seven border checks. It made my contraband in the form of a James Bond novel a somewhat tame act of subterfuge by comparison.

I escaped from the group, along with Ryan, when we reached Moscow and went in search of any lively illicit bars. Later we heard that the Intourist agency were in complete panic in case KGB headquarters heard about the disappearance of two Aussies and an Englishman. Ryan and I were successful in our mission to find an illegal bar and set about socialising. I met one girl who was certainly not Russian, but I was not sure I believed her claim to be Salvador Dali's great-granddaughter. I didn't care if she said she was Arthur Daley's, the crafty TV character from the series that was popular at the time, *Minder*; it was good to have female company of my own age. Everyone in the bar was paralytic and I was reminded of a joke;

'What stage comes between socialism and communism?

'Alcoholism.'

I surveyed the other customers in the bar. The locals were going for whatever form of alcohol was available. Everyone in the bar was swaying three hundred and sixty-degrees like a child's wobbly clown rolling on its ball shaped bottom, strewn drunkenly across tables or face down on the floor. I noticed some unusual brands on the spirits shelf, such as Eau de Cologne. The *Soviet Politburo* was desperately trying to sober up its population by increasing the cost of alcohol, only for Muscovites to purchase cheaper, but often lethal forms of perfume as a kind of alcohol substitute. This was like substituting Mogadon with one hundred percent pure uncut strychnine to wean addicts off their coke habit.

This was the era prior to Mikhail Gorbachev, *perestroika* and *glasnost*. Russia may have been top dog in the Communist Eastern European regime, but it had little to offer that I could see apart from the extraordinary Leningrad now the St. Petersburg Museum.

If the Soviet Union showcased anything, it was what was wrong with communism when you actually put it into practice. It was also in a very surreal time warp. When I turned on the TV in my minimalist hotel room, all I could watch were black and white documentaries on the Soviet armies fighting the Nazis, apparently on their own, with no mention of the Allies. This was interrupted every now and then with clips of President Reagan who, through some dodgy editing, seemed to be declaring war on the 'Evil Empire', this being how he referred to Russia, apparently on an hourly basis. In fact, Reagan only did that in his annual address to the Senate. Mind you even if it's only once a year, when the world's leading superpower threatens to eradicate your country it might explain why everyone I met on my visit was so bloody miserable. Those regular appearances of President Reagan on Russian television screens appeared to replace adverts, which was the only benefit I could see from my experience of communism in practice.

Nevertheless, the Soviet state system trying to subjugate its neighbours fascinated me, because its grip was so tenuous. Everyone I met could not wait for the overthrow of their Soviet masters. People covertly made transactions in American dollars, traded in western goods and made no secret of the fact that they wished they were elsewhere. It made no difference if they were East German or Czech; what gave me a real feeling that change would come, was that this is what most of the young Russians I met over those days wanted too. Though I did not realise it at the time, I also learnt many lessons from this and further visits to the eastern bloc, such as how to bypass the restrictions imposed by the authorities, which helped me years later in getting convoys of medical aid across hostile borders.

When I left Russia, I jumped ship, or rather leaped out of the coach and decided to make my own way back to England. Describing my fellow passengers as coffin-dodgers is ungracious, as all the people on the coach were very kind to me. When I said my goodbyes my American and Canadian fellow travellers organised a whip-round for me and gave me a case of beer. Elspeth, the eldest of the Canadian contingent made a farewell speech and thanked me for "For giving them all a laugh." Elspeth's speech was very funny and centred on ribbing me and my exploits over the trip. Embarrassingly, it seemed they did understand some of the risqué comments after all. Ryan was not there for my departure, as he was off to get another marriage certificate. The other Aussie, who, according to his deportation documents was called Ophelia Bumgardener, had been put on a plane at Moscow. We heard from a Polish guide that he was still clutching his handbag as the KGB man-handled him onto his seat. What had set Gilbert on a mission to get high, not just behind the 'Iron Curtain' but in Lenin's Tomb, rather than just going to Highgate Cemetery, and taking copious amounts of drugs while sitting on Karl Marx's grave remains a mystery.

I made my way north and got a ferry to Finland. If I thought the bar in Moscow was an alcoholic's paradise, it did not prepare me for the twelve-hour cruise across the Baltic Straits. Years later, I have yet to witness such copious amounts of alcohol being consumed as I did that night sharing a bar floor with the Finns. English, Germans, even Muscovites, should all hang our heads in shame and tip our glasses to the Finns, but with care, as the Finns will swipe the drinks out of our hands.

The passengers, who were all Finnish, would launch into a fight every thirty minutes or so. As they did so, the Russian bar-staff would close the bar and pull down a large metal security grill, until everyone had quietened down or most of the combatants were knocked unconscious. Then the metal grill was raised and drinks were served. Though the system operated on the ship was supposedly a communist one, the Russians knew how to turn a rouble or should I say a US dollar or two. They were also quite matter of fact about the frequent brawls; when the metal grate was pulled down you could see them carrying on talking while on the other side glasses, chairs and tables were been thrown across the bar. Unfortunately, I was on the wrong side of the metal grate. In between the fights, a member of the bar-staff wanted to practice her English with me, as she had never met an Englishman before. Give or take the odd interruption when the Finns set off on another fight, she explained to me through the metal grill that taxes on alcohol in Finland were the highest in the world and their only chance to get drunk was on the tax-free ships like this one. Fortunately, I appeared to be invisible to the Finns, so I escaped the night's voyage unscathed.

When I returned home, I was desperate to experience more of the world, but no package holidays for me, I wanted the freedom to discover the world myself. I got work on a construction site and as I was working double shifts I was able to earn enough money to go travelling again; not knowing it at the time, I ended up making my first of many visits to the Balkans. Things were starting to warm up there following General Tito's death; all the various countries that formed Yugoslavia were struggling to break free of their Serbian rulers. My visit to the region was by accident, as I was heading through Austria and I met a girl called Karina, who was a little older than I being in her late twenties. She was clearly in a distressed state and not only due to me going over to her in a bar and offering to buy her a drink. Karina told me she had been working in Vienna as a secretary, but she had to get back to her family in Zagreb. Her family were Croatian and she feared for their safety, as violent skirmishes were breaking out across the Balkans. I phoned up my job, as I was working on a building site in west London, and said I had a cold and would be in bed for a few days. I told Karina I would take her to Zagreb.

I was a young man desperate for adventure. Karina was beautiful, so despite her concern for my safety, it made sense to me to take her home even if it meant crossing into a potential war zone. To be honest, if there was a chance of making love to her, I would have happily let her jump up on my back and carried her all the way to Zagreb. Fortunately for us both, the trains were still running. After three days moving from one train to another we made it to her home city; she called her parents who were delighted and would collect her later that night. In Zagreb, it was I rather than Karina, who was drawing the curiosity of the locals, as foreigners were held with deep suspicion. I realised that I had a responsibility to her and I should forgo any chance of a kiss, let alone sex and not bring any further undue attention to her in those last hours before her family arrived. She rejected my offer to leave her and said we should stay together. Not for the first time, my heart overruled by head and I agreed that we should remain together until nightfall. We kept to the side streets to keep out of sight in the meantime.

We came across a dingy little cinema showing an early Rambo movie. We bought the only two tickets sold for the performance and entered the deserted little cinema. Having shared the last few days together, I put my arm around Karina and she cuddled up to me for what we thought might be our last time together. Stallone was dubbed in a camp German accent, that I almost expected him to straighten his opponents tie before he garrotted him with it. This along with the growing fear on the streets due to the impending war explained the poor box office takings that day, but for the first time in three days we forgot the dangers that threatened us and both exhausted we fell into a deep sleep.

When the film finished, we were rudely awakened, and I mean rudely as the usher pushed his torch directly into my face. Thinking it best not to draw attention to ourselves, I restrained from turning his torch into a suppository for him. It was now dark, and we mingled with the people rushing home before nightfall and made our way to the main square. Karina saw her parents' car and her father standing anxiously next to it. She then turned and pressed her mouth to mine. We quickly said our goodbyes and promised each other that we would keep in touch. I watched as Karina ran off to meet her family and then I turned away to make my way back to the train station. My letters to Karina remained unanswered. I just hope that Karina and her family made it through the war.

My adventure was not over and a young man about the same age as me, wearing some kind of home guard uniform, demanded to see my papers at Zagreb train station. I had no time for that. I answered his question with the appropriate amount of physical force, broke free and jumped on a train as it was pulling out. Unfortunately for me, the train was going deeper into Yugoslavia and I had to change trains again. I had to jump on many more trains over the next few days, as I desperately tried to circumvent Zagreb after my altercation with the toy soldier. It added an extra four days to my journey and it was impossible to sleep in the knowledge that I had no papers allowing me into the region and no one knew I was there. As with my visit to the Soviet Union, what I learnt in circumventing the official authorities would help me many years later in similar situations. The skills gained during these early adventures would help me later to carry a far greater responsibility than just my own life. Before I escaped Yugoslavia, someone blew up a water tower and it hit the train I was on; I had to get another train and found out too late that it was heading to Zagreb. I learnt to hate that town even more than my childhood day trips to Southend.

Chapter 10 – *Foreign Affairs*

My passion for adventure took me to India. I started on my journey by purchasing an old British Enfield Bullet motorbike. I soon realised that travelling by truck, which I would do later, was far safer, plus I should have waited until they had built roads. To travel the roads of India at night was an accident in progress, but to find that sections of tarmac had been removed for resurfacing made it a foregone conclusion. To place a light or a fluorescent sign on the road to warn traffic of the hazard had not occurred to anyone. Maybe they thought someone might crash into it. That night I ended up lying face down in a cow-shit-filled ditch in the middle of no man's land, my motorbike resting on my back. Seeking adventure, I thought, was not always what it's cracked up to be.

I then joined a group travelling on a converted US army truck. It was like being in a stage version of E.M. Forster's, *A Passage to India,* though played for laughs, with an international cast of passengers. There was the obligatory Aussie, Gary, who on my first night with them was punched out cold by a woman. Gary was inebriated, and falling on to the women at the bar, as he was for most of the trip when he wasn't driving. Thinking it was best he directed his attentions elsewhere; I told him that there was a woman across the dance floor winking at him. He asked me which one as he could hardly focus. The one with 'NYMPHOMANIAC' tattooed on her bicep, I informed him. As he staggered over to her, I realised that he was not only blind-drunk but short-sighted; he failed to notice that the club's female bouncer's tattoo read 'LIVE FOR PUSSY'. In my defence, I didn't know that his first drunken action when he reached her would be to try and stick his tongue down her throat.

Then there was Barnaby, who had lost a leg through having cancer as a child, but who kept losing his prosthetic one through forgetfulness. He had lost his spare leg by the time we left Mumbai, on the first leg of our journey so to speak. Barnaby did not realise he had lost his spare prosthetic, until three days later when he got off the bus for a piss and fell in a ditch; his primary wooden leg floated off down the river. Fortunately, I was able to dive in and rescue it.

Next on the passenger list was Alex, who was as deaf as a post when asked to help with the washing up or other communal tasks; his hearing was fine otherwise. He also had problems with his eyesight, and was perpetually squinting his eyes as he leant towards women's breasts, until they responded with the standard, "Bugger off, you lecherous old git."

Each day on our journey, we would try to visit one of India's incredible temples. For one of these treks, Barnaby and I went high up in the mountains near Jaipur. Barnaby read in his guide book that no man-made objects of the human-form were allowed into this particular sacred temple, so he removed his prosthetic leg to enter. I tried to reason with him that it meant idols of other gods or other religions icons, but Barnaby would not take the chance of causing offence to the temple hosts. Despite his best intentions, other worshippers were nevertheless shocked to find a prosthetic leg in a pirate Adidas trainer, next to their flip-flops. This was matched by the look of astonishment on the face of the temple guardian, as the one-legged Englishman hopped around the sacred shrine, snapping away happily with his disposable camera.

When we left the temple, he could not find his prosthetic limb. We formed a search party with the temple priests, and later wrestled it free from the jaws of very possessive mongrel. It occurred to me that, as my new friend could not hang on to anything, my last image of him would be of him standing on his one good leg in just his underpants waving me off at Kathmandu airport.

Then there was Belinda, an American woman who was travelling around the world. She was an independent woman, never let it be said that Americans only travel in bunches, protected by tanks and view foreign lands through the cross-wires of a rifle lens. Belinda had fantastic spirit, particularly evident when Alex would make a move on her. Alex too was an American citizen, though originally born in Ireland and Belinda knew how to keep her amorous countryman at bay simply with a disdainful look.

On our journey across northern India, our truck was involved in many an incident. In the second week, our vehicle got speared by a spike protruding from a roadside fence. Whenever we encountered such a problem, the truck's passengers would immediately spring into action. Alex would seek out a goat to stroke, perhaps to subdue it in case it attacked us as we worked, though the cynics thought it was to avoid helping. Barnaby would place his one good foot in cow shit, which he would, as if by instinct, use to mark out the area around the bus. Perhaps, it was a makeshift cordon warning passers-by of the danger, 'clowns at work'. Gary, if he was still conscious, would pick up the sledgehammer and swing it at whatever had damaged the truck. On this occasion, it was the spike from a metal railing, which he swung at and missed. As the head of the sledgehammer hit the bus, the spike shot straight into my leg. At the same time, the metal end of the sledgehammer came off and hit Gary on the head which meant, not for the last time, we had to load him back into the truck in a state of unconsciousness. To the onlookers, this exhibition showed that all westerners were either brain-damaged, or, through our performance, desperately wanted to be.

I remember, some patronizing member of the god-squad telling me once, and I made sure it was only once, that people with physical disabilities were somehow superior to us fully-limbed beings, as God had given them a greater goodness to compensate for their physical loss. I am not convinced that there are many bonuses to losing a limb or having any other form of disability. As with any section of society, there is a mixture of good people and some right horrible bastards. A disabled man with no legs lives near me; he is infamous for driving his motorized wheel-chair directly at buses to cause the maximum amount of disruption to the greatest number of people. He then proceeds to abuse them mercilessly as his wheelchair pens them in. I like him. If I lost an arm or an eye tomorrow, I have no doubt that the whole world would hear about it. If someone implied that I was a better person for it, I would grab a knife and give them the opportunity to try that theory out for themselves.

Barnaby was a good man and, if he had been given any enhancement by some supreme power, it was his exceptional and forgiving nature. Not an ounce of bitterness resided in his body and his compassion for the Indian people was beyond that of any of his able-bodied travelling companions, including me.

Whereas Barnaby's disability was something that occurred later in life, I have a friend, Sullivan, who was born with only a third of his arms and the upper parts of both legs. To make it worse, as no 'official' medical link could be established to thalidomide as the cause of his disability, though that or a similar drug was administered during his mother's pregnancy, he did not receive the compensation paid to those who were officially diagnosed. Sullivan was similar to Barnaby in the sense that he was not bitter at the cards that life dealt him. Rather, like Barnaby, he had a positive attitude to life; he wanted to change society for the better. Barnaby's approach was to try and make his travelling companions aware of the positive aspects of Asia, while engaging wholeheartedly with its people, while Sullivan was a leading member of a socialist group wanting an egalitarian society and a strong welfare state. Both believed in the greater good of mankind and both were better men that I, not because of their disability, but because of their humanity.

On one of my evening walks with Barnaby, we met an old Indian man sitting by a campfire. The old man, whose name was Arman, had a vast knowledge of world affairs and therefore a jaundiced view of the future. We got talking about religion, always a good place to start, well at least if you want to start a war. Whereas Barnaby was the more spiritual one and I the pragmatist, Barnaby judged that the old man was more of my persuasion and so he sat down with a beer to view the 'sparring' as he later called it on our walk back to camp.

I gave the old man a can of Tiger beer and as we took our first sips of warm alcohol the old man opened the conversation.

"What do you think of Hinduism?" he asked.

I would have preferred an easier question to start with along the lines of, "Are Jordan's breasts real?" But, I was a young man in the presence of an older, clearly wiser, more experienced one and I was always up for a challenge, even if my knowledge of the battlefield was shaky.

"In Hinduism, if I have it right, the more someone suffers in this world the better it will be for that person in the next." I said. If my understanding was correct, then that was good news for Tottenham fans, I thought.

"Basically yes, nirvana is a way for our leaders to subjugate the poor and maintain the caste system. If you rebel or show ambition, particularly if you are at the lower end of the caste system, this is somehow blasphemous and you are accused of not believing in the afterlife."

"It's the ones who do believe in a better world in the hereafter, that worry me," I said.

"Why?" he asked.

"Well they never give way on the road and if they crash into you, why should they care? A better world awaits them. Meanwhile my motor bike and I, are arse over tit under an elephant." I had come off more than once in my first week.

"Ah, the English, when it comes to faith, as with your talk of liberty and equality, in the end it's all down to how it affects you."

"Are you any different?" I asked.

"Only in the sense, that I…," he fumbled for the next word.

"Believe there is more to life?" I added helpfully.

"No, I am not stupid enough to get on a motorbike," he said. I laughed out loud for the first of many times that night.

I liked my new friend. He was considerate to Barnaby in giving him his cushion and he made sure that my friend was comfortable as the night got colder and, at one stage, though Barnaby protested, he wrapped him in his blanket. These small acts of kindness elevated him to 'benevolent rogue' status. I had met many rogues, such as Ryan and even Gilbert, and although they were funny and certainly characters, there was nothing more to their lives than their own passions and pleasures.

Arman had a kind, bronzed face and a wide open smile which exposed large tombstone teeth when he laughed. With the traditional English acknowledgment of good company, I offered him another beer.

"You English, you are hypocrites." he said.

"It would be hypocritical of me to deny it."

"In the days before the fall of your empire, you pretended to encourage us to work and learn from you, so there would be a smooth transition towards Independence. But really you just did not want to do the work yourselves, as your chief administrators just wanted to sit on their…" he fumbled for the word.

"Laurels," I interjected.

"No, arses," he replied. "So when Gandhi stated it was time to take control of our country, you tried to subjugate us once more."

"It could have been much worse. You could have had the French!" I argued. "They desperately wanted an empire and envied us Brits, so they hung on to what colonies they had to the bitter end, shedding even more blood than us Brits."

"I like you Englishman, have you ever played poker?" I ignored his question; a poker player never shows his hand. "Here I am attacking your country's foreign policy and you turn it on your neighbour. So English," he took a sip of beer and continued. "Yet what you say is true. Look at Algeria, the Ivory Coast and Vietnam. In Vietnam it was only when they were on the verge of complete capitulation, they…" again he fumbled for the word.

"Soiled their undergarments," I added.

"No, they called into play America's Achilles heel."

"What's the septics' Achilles heel?" I said using the slang septic tank, meaning Yank.

Barnaby laughed, obviously enjoying the banter, as were I and the old man. The old man even knew what 'septics' meant.

"Communism."

"Yes, that usually gets them to throw reason out of the window," I agreed.

"The French dragged the Americans in by telling them that the war was about stopping the spread of communism, rather than it was about hanging on to the remnants of their empire, which was..." he fumbled again.

"Nonsense," I added.

"Bollocks is a better word," he corrected me.

What a superb master he was of our quaint old English folk expressions. The old man was well into his stride now.

"The Americans got caught in that fiasco for years, while the French withdrew their forces and their people organized anti-American demonstrations on the streets of Paris."

"Ah, you have to admire the French for that," and I meant it, what superb political engineering by the old world taking advantage of the naivety of the new.

Up to now you may have thought I disliked the French. Far from it, I love their blatant self-interest and mischievous nature. They are always at the forefront of the European Union, unless of course it makes a decision that adversely affects them. At that point you get their standard "Furk you." Even their *Non* is not simply a No. There pronunciation seems to say, "I can't believe you are so stupid as to ask that question in the first place." Once I saw a mass demonstration of dock-workers on the streets of Paris. I noticed that, during the march, two demonstrators were cutting off the heads of parking meters with a gigantic bolt-cutter and putting them and the money they contained into a sack to finance a decent lunch before they stormed the gates of the Palace of Justice. Even their riots have a certain cool aloofness to them.

We carried on our tour of global politics late into the night, laughing and drinking like old friends around Arman's fire. Then I realised that Barnaby was asleep, so we said our goodbyes and, after I found his prosthetic which another stray dog had carried away and partially buried, we all stood up and the three of us shook hands. We never saw the old man again, but I am sure that many other travellers would be as enthralled by the wise old rogue as Barnaby and I had been.

In a club, on the last night of our Indian journey, Alex was sandwiched between two large Australian women, called Yenta and Anna. No doubt as a result of his hearing difficulties, he referred to them as 'Yeti' and 'Anaconda'. These women cannot have had sex for years; they were practically de-bagging him in full view of the entire club.

"Alex, have you got enough medical insurance?" I shouted across the dance-floor.

"I will be a pressed flower in the morning," he retorted.

Even though they could hear his derogatory comments, it made no difference to them, as they continued their joint efforts to undo his trousers.

"Any port in a storm," I yelled to my semi-naked friend.

"Sometimes it is better to sink," he shouted in response.

Later than night, Alex managed to pull himself from between the female bookends and joined me as I headed back to camp. We found that the main gate was shut. Alex, who was twenty-five years older than I, then attempted to climb over it, so I gave him a leg up onto the gate and then I pushed it open, for which he has never forgiven me.

I am still in touch with Alex. Last Christmas he sent me photos of his 'Henna Tattoo Emporium' in Long Beach, California, and, if the photos were anything to go by, his specialism was applying his art to the breasts and arses of young blonde female clients. I wonder what he has written for 'Profession' on his passport? Alex is now in his seventies but the rogue's passion for life, his wit, his adventures and accompanying tales continue to make me smile. He still manages to get to some of the most isolated parts of the world, nearly two hundred countries to date, and at little cost as he usually travels as 'cargo'. He really did have a hearing problem and parts of his body really did shut down at various times. All of this was because he was in a car crash years before, which killed his fiancée and put him in a coma for three months. Even rogues have a scar on their heart somewhere.

Chapter 11 – *The Special Relationship*

I then worked my way across America and I fell in love with the country, though as in any love affair each has to accept the others' faults. After working as a barman in New York, I started to take cheap flights across the country and one of my first flights landed me late at night in Orlando. Unfortunately, the youth hostel was closed; a female tourist had been stabbed the week before by a local boy for refusing his advances. As in all towns, even quiet towns as Orlando was then, there is always a darker side. When a schoolyard gets wiped out in some little American town, it is usually the result of some misfit expressing himself through his Playstation game: *Die Head Principal, Die. III.* This does not develop into a noticeable problem, until the batteries run out on his console. This escalates his psychotic tendencies up beyond level seven, 'strangling an armoured mammoth', to reach level eight, which is 'killing his teachers and school friends' with a bazooka that he kept under his bed for self-defence. Maybe longer life batteries distributed free with Mogadon at the school gates would reduce the number of violent incidents.

After walking the streets of Orlando, I discovered one place that might offer me a room for the night. You need to remember that this was the seventies and tourism had not taken off; Sea World was only just being built and I think Disneyworld was just an itch in Mickey's shorts. My search led me to the front door of a residential nursing home. I had nowhere else to go and having read that some of the locals had beaten up a tourist recently, I took a room for the night. Later, I found out that my room had only become available as its resident had died in it earlier that day. The TV in the communal lounge showed early episodes of *Star Trek* on a continuous loop, so after a dinner of pureed chicken legs, I retired to bed. I did not sleep a wink; all night I could hear the residents either moaning or gasping for breath. Christ, if I ever end up in one of those places, shoot me or give me a copy of Piers Morgan's biography and I will do the job myself.

Breakfast was a strange affair. As I tucked into my pureed meatloaf, everything appeared to be pureed (I think there was only one set of dentures passed between the residents), another corpse was carried through the breakfast room. All the residents looked up and groaned. The guy at my table looked up.

"Lucky son of a bitch," he mumbled.

The general consensus was that death was an acceptable method of escape, rather than an unfortunate event. My fellow diners then returned to their breakfast, dropping their heads towards their lumpy porridge, but a gentleman on my left went head first into it. The momentum, helped by his weakened turkey-neck muscles, had got the better of him. I think the lumps in his cereal cushioned the impact and saved him from serious injury. I lifted his head out of his congealed soggy oats, settled him back into his chair and wiped his face with his bib. He thanked me for saving him from drowning with the word, "Bastard."

"Time to leave," I thought, but my one and only day in Orlando had only just begun.

I headed through the deserted town centre on my way to the airport and came across an old gentleman of about seventy trying to change a flat-tyre on his dilapidated pickup truck. I offered to help him and shared my tale of the night before as I put the spare wheel on. He laughed, as he told me no way was he ever going to enter that hellhole.

"I fought in World War Two and Korea. If the authorities try to put me in there I am going to start blasting away with my shot-gun."

"Do you have a games console?" I asked.

He just looked at me puzzled, but he seemed harmless, so when he offered me a lift to the airport in return for changing his tyre, I thought why not. A big mistake, as I was soon to discover.

Due to sleep deprivation, I fell asleep as soon as I got into the truck. I woke up an hour later to find myself being driven through the desert.

"Are we near the airport?" I asked.

"Soon be there," he replied.

I thought I better had make conversation after rudely falling asleep; the poor old guy had seemed pleased to see me and chatted incessantly when I helped him earlier.

"Did you ever serve in England during World War Two," I began.

"Oh yes, it was a night with two guys from the RAF that made me realise I was gay," he said. I must have misheard him, as I was still half-asleep. I ignored what I thought he said and changed the conversation.

"What is there to do in this town," I continued.

"Nothing, I just sit at home and watch gay videos," he said, he looked at me and smiled.

Well, at that point I looked out the window and muttered under my breath, "Bollocks!" as we passed a sign for the airport, pointing in the opposite direction.

Now a number of thoughts were going through my mind, one of them being that he was taking me to meet his mountain men mates and I would never be seen again. I was not sure if there were any mountains, let alone mountain men nearby, but I had seen what nearly happened to Jon Voight in the film *Deliverance* and no way was any hairy lunatic going to put a piece under my tail. I noticed that there was a huge screwdriver in the door pocket on my side. If he had a gun, I'd better move fast. I grabbed the screwdriver and pressed its broad flat end into the turkey skin folds of his neck.

"If you don't turn this thing around and take me to the airport, I will stick this through your head and nail you to the fucking wall," I shouted at him.

He threw his hands up and the truck spun off the dirt road and came to an abrupt halt.

"Please don't hurt me. I thought you were gay. I didn't mean any harm," he said, now completely terrified. He started crying.

"Why the fuck did you think I was gay?" I demanded.

"Well, I only ever met two Englishmen and they were, so I thought all Englishmen were," he said, pleading.

I removed the screwdriver from his windpipe, put him in the passenger's seat, took the wheel and drove to the airport. When we arrived, he seemed calmer with all the people around, but he continued to apologise. I left him there.

"Just don't go giving lifts to young men again, or I will come back and stick that screwdriver through your head, so you look like a fucking Dalek."

As I sat on the plane, I went cold thinking of the headlines if the old man had attacked me and I had skewered him with the screwdriver. No one would have believed it was self-defence. I would have been on Death Row by Christmas, with sparks shooting out of every orifice before New Year. I had spent less than twenty-four hours in Orlando and was never going back, even if Mickey offered me unlimited free rides on Space Mountain, in the front seat with a full mini-bar.

I headed south and secured a job delivering newly purchased cars to their buyers, not in a Loftus sense as this was legitimate, or so I thought. It was a great way to travel around the States, provided with free transport in the form of luxury cars. When I got to Miami, I found that my next assignment was to courier a brand new Audi Quattro to the town of Key West, which was the southernmost point of the United States, right on the tip of the Everglades, fifty miles from Fidel Castro's Cuba.

The drive through the Florida Everglades is one of the most beautiful journeys I have ever experienced. Having arrived at Key West, I looked up the house of that notorious adventurer, author and alcoholic, Ernest Hemingway. Benevolent rogues are not new, for despite, or maybe because of his roguish nature, he was one of the few war correspondents filing reports from the front line during the Spanish Civil war.

On my first night in Key West I parked myself in the tourist hot spot, Sloppy Joe's Bar, where Hemingway supposedly spent many a day and night knocking back mojitos with the locals. I could understand why Hemingway got into so many fights, if he met anyone like the guy I got into an argument with who described my language as un-American. This connoisseur of the American language had turned to me and in a very threatening manner corrected me for using words like motorway and lift. I apologised and promised him I would say "balls" rather than "bollocks" next time he said anything. At that point a young woman came between the two of us and her forward approach seemed to instantly placate my tutor. She was much shorter than us and wore cut-down jean shorts and a T-shirt with YES! YOU HAVE SEEN ME IN THE TABLOIDS, blazoned across it, which was so tight over her curves that it looked like it was sprayed on.

I was entranced not just by her natural beauty, in a land where you have to excavate an inch of cosmetics to find out who is underneath, but her confidence in standing between me and the other guy and saving me getting a good hiding. We stayed together at the bar as my English tutor made his apologies and left. She was well known and liked by the locals and I was clearly marked as her guy for the rest of the evening. This courageous woman was called Casey and the bar-owner told me, that her dad was a famous republican senator and she was often in press as she was known as a wild child, hence her T-shirt. This might have contributed to her influence in the bar, but I believed the real reason was because she could handle herself.

I asked Casey where all the wealth in Key West came from, as all I could see was the occasional tourist and no industry to speak of. She enlightened me,

"Two income streams make this town wealthy, the transportation of drugs or any association with the pink dollar, as it has a very large gay community. What brings you here?" she then asked.

"Neither. I just delivered a car down here today," as soon as I said it I went cold, as it suddenly dawned on me that something might have been smuggled in the car.

I had checked the boot and all the obvious places before I signed for the car, but I had recently seen a car stripped apart in the Gene Hackman film *The French Connection* and cocaine had been welded into the body of the car, so anything was possible. Casey then did some checking for me on whom I delivered the car to and found out it was to a policeman.

"So it was drugs," I declared.

She thought the opposite as she had every confidence in the police,

"Well I know a few coppers back in Stoke Newington who would change your view," I added.

We had no proof that anything illegal happened, but I thought it was time I left Key West. Casey joined me and we both hitched our way back to Florida the next day, as she refused an allowance from her father because she rejected his politics. We spent some time together in Miami and though I loved being with Casey, I didn't like the city. Casey got a job waiting tables in a bar, but I decided it was time to move on again. Once again with such temporary encounters there were no tears, no promises to meet, as she was a young free spirit like me, who just wanted some fun and no ties. God knows there would be plenty of time for all that later.

I headed west to California and, after seeing the sights in San Francisco, I took a bus to Las Vegas. As a young man with little money, I took full advantage of any freebies on offer and tore out the betting coupons from the free papers given out on the Vegas sunset strip. The idea of these promotions was to provide you with a few free bets and get you hooked at the gambling tables. The promoter's strategy had a flaw; the success of the promotion relied on the user having the normal emotions of shame and embarrassment and not the nerve to lose at one table then use a fresh set of vouchers at the next table to start all over again. I had no notion of shame or embarrassment, let alone that they had acceptable levels, so I moved from one table to the next, landing vouchers on gambling tables with unabashed brazenness. Though I had retired from the poker-table, I didn't regard this as gambling as I was playing with the Las Vegas version of monopoly money. I only bet real money only when the game was at a stage where I had weighed up that the odds were in my favour; only then I raised my bet. As a result,

I was able to earn enough money to extend my stay by three weeks.

Unlike their po-faced male counterparts, my cheeky approach made one or two female croupiers laugh, due to the fact I didn't give a damn and always greeted them with a big unashamed smile. However, I always gave the croupier, male or female, a healthy cash tip irrespective of whether I had won or lost, as wages for those in the service industry were nominal. I had no interest in adding to the profits of the Casino owners, but having relied on tips myself as I worked my way across the States I respected the fact that my poverty was no excuse for their staff being deprived of a living from my time at the tables.

Las Vegas is the shallowest place on earth outside of the Houses of Parliament. I settled in perfectly. I was staying in a hotel where you received a free night's board if you produced an international plane ticket. This was another ruse to get you in the hotel's casino to spend your dollars at the tables. I was more interested in the hotel pool that had a bar in the middle of it; I swam out to it the next morning and had a great time drinking with a party of casino workers. The party was a mixture of English, Germans, Scots, Australians and one Finn, all of whom were in a drunken state by the end of the afternoon. As dusk descended, we were just about able to swim back from the bar. The Finn refused to move and the next morning I saw him still drinking at the bar, even though the pool had been drained for cleaning and the bar was now a six foot drop on to the pool's tiled floor.

I headed off to a local diner for breakfast, where you will find Americans at their most relaxed. To discover what makes England tick, go to a bar, it's where the English unwind as all pretence is cast aside. In the States, bars are very different; they are more formal and many customers strut and preen, as these establishments are good places to 'network' in.

On that particular morning, a tall blonde waitress wearing a bikini top and a tropical wrap, poured coffee into my vase-sized mug. That is another great contribution from America to the civilised world, bottomless coffee cups. My waitress started chatting to me. "You look like you had a great night," she said, no doubt noticing that my hair looked like the result of an explosion in a mattress factory.

"Well I thought I had until I realised you weren't there," I replied, with an extra-large cheesy Las Vegas welcoming smile.

My poor attempt at humour was lost on her.

"Ah, that's a nice thing to say, are you Canadian?"

"No, do I sound Canadian?" I replied.

"No, it's just that Canadians say nice things."

"I am English," I said.

"Wow that is great. Were you always English?"

Oh dear, this might bring out the worst in me, I thought to myself.

"Err. Yes, except for when I was Irish. When I get old, bitter and really hate the world I become Scottish, unless my brain goes first and I turn Welsh."

"I meant silly, is your heritage English?" she corrected me and then added, "Hey, don't you also have rabies in England? I read a story on that last week."

"Yes, if we get bitten by wild poodles we go completely mental, or French as we call it; if we go into paralysis and become void of all emotion, then that's what is medically termed Germanic. Well I think that concludes our tour of Europe."

"You are funny, I love the nonsense you Brits come out with," she continued. "What are you doing here in our quaint little town?" she added.

Quaint! Not the word I would use to describe the gambling capital of the world.

"I was working back in England, serving my apprenticeship shovelling shit for midgets. Now I am doing my PhD in Dungology, so for research I'm following the elephants around the 'Circus Circus' casino with a wheel-barrow and a shovel."

"That's great, people should have ambitions," again ignoring my puerile attempts at humour she added, "One day I hope to become a Mom."

In this artificial of all towns, I found her honest response quite touching. I suddenly felt quite ashamed for having mocked the one person I had met since I arrived, who was a genuinely nice person.

I got to know this woman; her name was Josey. I remembered the waitress I met in Berlin, a natural beauty, a slight frame and a smile that would make even the most cynical of men grin like an idiot. Josey was the opposite in appearance; her long Farah Fawcett blonde hair took hours to train and her supine perfectly formed statuesque body was the result of a two-hour workout every day. But, like the waitress in Berlin, she projected an uninhibited optimism and she too had a smile, exposing chemically whitened teeth, which would still melt the hardest heart.

To my shame, I judged her purely by her looks, not realising that this was standard 'war paint' in the battle for a young woman to secure an honest day's pay in Las Vegas. Her blind faith in a better tomorrow, coupled with her 'pneumatic' body originally brought me to a view that she was a few chips short of a full stack. I was to find out that I was wrong and, through her, learn so much more about women trying to survive in the most cut-throat of male worlds.

I moved in with Josey a few days later. She frowned on me obtaining free accommodation using my flight-ticket at different hotels and she said she could 'take care of me'. No one had ever wanted to take care of me before and I certainly didn't need it, but I was not going to say no to a Californian beauty. During those few short weeks, my favourite moments were when we caught a rare chance to have breakfast together in her one room apartment that she shared with Andrea; they had a rota for the use of its one small bed. She would sit at the table reading her paper without her war paint. I guess I was one of the few people that she felt comfortable enough with that she didn't need it; at least I'd like to think so.

The more I got to know Josey, the more I realised that I was the shallow one for mocking her that morning in the diner. If I now met that same smartarse cocky runt I was then, I would slap the shit out of myself. Through Josey, over those next weeks, I was given a glimpse of the American work ethic; she and her friends sometimes did two or three shifts a day to earn enough money to survive. There were many, like Josey, working to pay their way through college. Andrea who was studying law, was making pizzas fourteen hours a day and Jimmy, a policeman, worked with Josie in the bar at nights, even when earlier one day he had been 'sliced' by a blade while apprehending a pickpocket.

Though she appeared to know little beyond her immediate world and she never got my jokes, she was nobody's fool. She knew when someone was trying to take advantage of her; despite her appearance, she had a steely edge. Each night, usually about three in the morning, I would take her for dinner and that was the only time we ever clashed. I would not let her split the bill, as I was staying at hers free. She would baulk at this, but we were both very independent and making up after our clashes was always a fantastic way to welcome daybreak. Time is of no consequence in Las Vegas; despite the boasts of other cities, to me, it is still the only twenty-four-hour town in the world.

One of Josey's many jobs was as a cocktail waitress on the floor of the Sahara casino. Often I would play the poker tables nearby, look over, smile at her and receive a big heart-warming smile in return. It was a simple relationship, but we had grown quite attached to each other in a very short time.

I learnt so much about America from Josey and her friends; family is the cornerstone of America, greater than allegiance to its flag. Everyday Josey would telephone 'mom and dad'. I would disappear around the world for months, but the nearest I came to contacting anyone would be an illegible postcard to Biddy or H saying 'it was a beautiful country,' 'great bars' or 'I hear Chelsea got beat by Reading.'

When it was time for me to move on again, Josey cried like a big beautiful, baby. Though she was a few years older than me, in her early thirties with a fiancé back in Los Angeles, amazingly we had grown close, which was certainly rare for me. Josey and I had some great times and plenty of laughs, but, more than that, I had learnt not ridicule those I saw as innocent or naive. I realised that until then, when I encountered innocence I saw it as stupidity and mocked it. My view changed then and, though I still kept the view that life will fuck us all in the end, I would not tease or mock the Joseys of this world; an optimistic nature was something to applaud, not ridicule. Once optimism is crushed it can never come back, and it wasn't my job to crush it.

So I grew up a little. I was still impossible to embarrass in polite society, but I was a little ashamed of my behaviour when I first met Josey. I also opened up slightly, as the object of my affection, whom I had teased on our first meeting, was a woman who over a very short time I grew to love and respect.

CAREFREE ROGUES

"God! What on earth was I drinking last night?

My head feels as if there is a Frenchman living in it."

Edmund Blackadder. *Blackadder II*

Chapter 12 – *The Greenhouse Effect*

After America, I came home and secured a summer job in a large London park. That summer job stretched to five years and became a big part of my life and the people I met then still are. I had family working there, so on my first day at work, I popped in to see my relatives. Just as I walked in to the kitchen, my Uncle Pat was saying to my cousins that if they followed a life of godlessness and debauchery they would end up just like, well he did not say anymore he just pointed directly at me.

I met the governor of the park, Blakey. He was a big man, with a red face due to high blood-pressure, which I helped to raise further during my time there. My first task was to mark out the cricket pitches for the blind. Well, to hell with that. I had fallen for these initiation tests before, just ask a certain chimpanzee troop. I was not to be fooled into marking out a pitch for people who could not see, so I disappeared off to the pub. The slight flaw in my plan was that there were blind cricketers who did indeed play with a sighted umpire. They were able to pinpoint the cricket ball, as the bell inside it would ring when it was in motion. I had this image of a lost Morris Dancer getting battered to death having accidentally strayed across the pitch. The next morning, I received the first of many summonses to Blakey's office,

"Do you realise the pandemonium you caused last night? Without a pitch the blind cricketers were bashing balls all over the place, we even found one fielding in the car park. The blind cricketers are livid," he screamed at me.

"I think you're picking on the wrong man here. You should be having a word with whoever told them there were no lines, as they seem to be having a great time up to then."

That was it. Blakey exploded and ordered me out of his office. I stood outside and thought; "I didn't get sacked, I think I am going to like it here."

I termed the park the Arkham Asylum in honour of the establishment where Batman's enemies were interned, with Blakey as head warden. At the time, the park operated on what was called a 'green card' system, which ensured that one person in every hundred employed by the government, our employer, had a physical or mental disability of some sort. I believed this was turned on its head by Blakey in that only one in a hundred in his employment was normal, though in the five years I was in Arkham, I never encountered this exceptional employee.

By the end of my second day, I viewed the other inmates and reached the conclusion that this was legalised mayhem, paid for by the state and all that the institution needed was a little bit of fundraising to put a roof on it. I stayed there for five years and I never had one normal day in that green turfed looney-bin. For the first time, I was not one of the main players, but just one in a vast collection of colourful characters. I had finally joined the highest echelons of Britain's most dysfunctional lunatics; my chest swelled with pride.

Blakey didn't appear to like me, though I always held him in the highest esteem as no one else had brought together such a unique collection of allsorts since the inventors of liquorice. By the end of the first week, Blakey had already designated me for punishment duties and had me sweeping out the tractor pit. That morning, I looked up to see Blakey and two very distinct silhouettes with the sun rising behind them. One was over five-foot-ten, but was wide as he was long, constantly shuffling around expectantly like a bulimic waiting for the dinner bell. The other stood menacingly, he was well over six foot tall, lean, standing motionless with his fists clenched; I guessed that a welcoming handshake was some way off. They reminded me of a couple of Bond villains, even more so when Blakey introduced them as Mr White and Mr Brown. I soon came to know them as Bear and H and, if I can stretch the meaning of the term to its very boundaries, as friends.

When I first got to know H he would only acknowledge me with a nod of the head or a muffled "Alright" and only then because I knew his accomplice in crime, the Bear. H was not one for idle chitchat and only spoke when he had something to say, which would normally be something of an extremely perverted nature. I once asked him about an ex-girlfriend.

"How's it going with Lynn?" I asked when I caught up with him for a pint.

H took a sip from his pint his and replied, "Well it's OK, but we are not at the cling-film stage yet."

Best not to go down that path I thought, and moved on to talk about football or Bear's latest misadventure. H became one of my best friends; a more honest, genuine individual I have yet to meet and he has helped me many times over the years with humanitarian convoys. He is a true 'benevolent rogue'.

Then there was Bear and, I say this without any fear of contradiction, a complete fruit-loop of the highest order, who even dressed the part. While H seemed perfectly normal, that is until you got to know him, with Bear you knew you were in the presence of insanity right from the offset. As for the relationship between H and Bear, each thought the other was a lunatic, and in that, each was entirely correct.

In the time I knew him, his unofficial remit seemed to be to annihilate all flora and fauna within the bounds of the park. Bear's scorched earth policy was on a par with the deforestation of Vietnam by the US Air Force spraying crops with napalm. His chosen attire for this work was a bumblebee jumper, which looked like it had been knitted by a blind auntie with arthritic fingers as she rode the Helter Skelter ride in Margate.

I remember in those early days, when we first worked together, Bear used to come into our hut and sit down to read *Le Monde*. I asked him if he could speak French, he replied. "Not a word mate."

Some of the workers, if I can again abuse that term, were affected by the seasons. Bear became exceptionally manic when the sun came out. Blakey decided that the best way to deal with Bear's frenzied state was to put him in the driving seat of a very powerful turbo lawn mower. I watched Bear doing hand-brake turns down gravel paths; the rotating blades sending stone pellets in all directions. Members of the public, along with us, ran for cover, with many diving over hedges and into bushes. Bear was stopped in his tracks when he rammed a park bench from behind with his motorised mower, sending three nuns across the tarmac and ripping the knees out of their habits.

As punishment, Bear found himself once again banished to the greenhouses. This was known as the 'observation block', next to Blakey's office so he could keep you under surveillance. Bear had been banished there the previous week. His crime that time was that he had been sent to deadhead some roses, but he cut off all the 'brightly infected parts', as he termed it, of the surrounding flowers, which stood out because they were red or yellow; as Bear explained later, everything else was green. Plus, he said later to Blakey in his defence, "And they were on the top too."

At the end of the week Blakey came up to Bear declaring, "I have some good news for you lad."

Even Bear knew the signs were not good, Blakey was laughing so it was bound to be at Bear's expense.

"Once you finish that little task laddie, you can go home early," he said chuckling away.

The reason he was laughing was that the little task in front of Bear was to water the half mile long greenhouse of geraniums.

"Early": this was excellent news, thought Bear, but as he told me the following day.

"If I had watered the plants the traditional way, I wouldn't have been home until Christmas."

For Bear at least, the solution to such a delicate task was obvious. He linked up the main fire hose and turned on the water at full pressure. It immediately blew the plants into the air, straight through the glass windows. Occasionally, with the force of the water pressure, Bear was lifted off the ground himself, as if grappling with an anaconda who was attempting to eat Bear's last chocolate finger.

Once Bear had completed his task, to his satisfaction rather than anyone else's, he set off home for lunch to watch his favourite TV show *Rainbow,* a TV show aimed at the under-fives. Bear was a big fan of the lead character, a children's presenter dressed as a bear called Bungle. Bear was hairier, though a little less acceptable at the dinner table than Bungle. When Blakey returned, the greenhouse looked like the Pamplona bull-run had tried to flee a pursuing tidal wave.

Bear, H and I now formed a new anarchic triumvirate. This new model was different to the original with Loftus and Hammond, as there were no cards: Bear would have eaten them. Each, however, shared common goals. To bring chaos where there was order, a very common expression of being British and have fun without hurting anyone. In the pub one day, Bear got excited about our new friendship and attempted to follow Alexander Dumas' *The Three Musketeers* rallying cry,

"All for one and one for me mum."

"This isn't Italy, this is England," I said and stood up and repeated an old toast, "My advice to you my friends is to drink and drink heavily."

We all raised our glasses, but Bear missed toasting his glass with ours and as a result threw his beer all over H. It all kicked off. I sat there watching the beer and chairs flying as they tore lumps off each other. I said to myself, 'I was right, I am going to enjoy it here.'

Chapter 13 – *A Flock of Cuckoos Flying Over the Nest*

There was one other character, who although not an active part of this three-ringed circus, certainly played his part. If we were the Marx Brothers, he was our Zippo, looking on from the side lines playing it cool, while my Groucho, H's Chico and Bear's Harpo ripped the set apart.

"Bear and H, are insane you know," Cockey informed me at our first encounter.

H who was sitting next to Cockey at the time then turned to me.

"OK Bear and I are off the wall, but Cockey is a genuine all round psychopath," he said. Cockey said nothing and just wore his standard fixed smile. Amongst the other inmates of Arkham he was known as the Terminator.

Perversely, Bear and Cockey were paired together when working. Bear revelled in idiocy, while Cockey was a deadly serious character, who never laughed and who was thoroughly diligent in all that he did. Pairing two such opposite characters could only result in disaster, once again pure genius on the part of Blakey.

One day both of them were sent to mark out the football pitches, which involved no more than pushing a paint-roller up and down the green marking out a few white lines. I was putting a chainsaw through a hedge nearby, so I was able to watch how the dynamic duo set about their simple task. Cockey marked out the pitch meticulously with a measuring tape, a compass, even a magnifying-glass as he double-checked his steps to make sure everything was exactly right. Cockey then banged in the marker peg with a mallet to designate where the lines had to be drawn. Only then did Cockey release Bear from the caged trailer moored to the back of the little tractor to perform the simple task of pushing the paint roller from one peg to the next. Cockey then moved on to mark out the next pitch.

It was at this point that Cockey's strategy began to unravel. Bear set off with his paint roller, while waving and making obscene gestures in my direction. Only when he stopped gesturing, did Bear discover that he had missed his intended mark by some twenty feet.

Quickly he checked to see if Cockey was looking and when he saw he was on his knees with his magnifying-glass, he bounded over to the peg he had missed, pulled it out and replanted it at the end of the white line. He then made another obscene gesture in my direction, with his tongue in his cheek, indicating movement with his right hand in Cockey's direction and set off to the next peg. This time he missed the designated peg by an even greater distance.

At the end of the day, Blakey arrived to survey the pitches from the top of the hill and went ballistic as he surveyed what looked like long sloping ladders rather than football pitches.

He screamed at Cockey, "You idiot I wanted you to mark out four full-sized football pitches, not create a modern art exhibit."

Only at that point, did Cockey turn around and survey how Bear had adapted his work. I watched as Blakey screamed at Cockey, who was then sent home for being drunk. Meantime, I could see Bear in the distance waving happily at Blakey, adopting his standard childlike innocence when everything around him went tits up. As soon as Blakey and Cockey's backs were turned, Bear began making obscene gestures again, this time looking like he was moving an invisible tree trunk back and forward between his thighs with his hands.

Blakey kept the two on the same job all week, as I say he was a genius. The end result was a huge mural of what appeared to be randomly interconnecting ladders. I was still tasked with chain sawing hedges, and I watched Cockey finally snap and chase Bear with a peg-hammer. After the incident involving the three nuns and the park bench, Bear was banned from driving the little tractor, but as he was now being chased by a mallet-waving Cockey, he jumped into the little tractor and sped off. Unfortunately, attached to it was a trailer with a dustbin full of white paint which tipped over as Bear drove off. I sat on my lorry eating my sandwich and watched Bear zigzagging across all the pitches creating long curved white streaks across the 'mural', with Cockey in pursuit. The paint was mixed with paraquat, a deadly weed killer, so evidence of their debacle was seen for years from the air, a kind of English version of Peru's Nasca Lines. My final thought, as Bear drove out on to the main road with Cockey still in pursuit, was that Bear was one of our greatest natural artists, very much in the style of Jackson Pollock, as he had now added the final slithering white reptiles to his snakes and ladders landscape.

Cockey's animosity towards Bear was not limited to their working environment. Any social interaction between the two also resulted in some display of psychotic violence. One Saturday night, we all went for a drink before heading to the Lyceum in Covent Garden. Cockey proudly showed off his new 'Indestructible' shoes, which had cost him a week's wages. 'Indestructible?' noted Bear. You do not have to be a genius to realise the challenge that this presented to my hairy blond friend. No sooner had Cockey confirmed that his new shoes were indeed 'Indestructible', than there was a loud crack, as the heel of Bear's boot came down hard on one of Cockey's shoes. The steel toecap detached itself from the body and shot across the pub floor.

"You have a hole in your sock," Bear said to Cockey, who was now blood red with suppressed rage, his acne bubbling like volcanoes on the verge of eruption.

Cockey took every opportunity to avenge himself on Bear. There was a firing-range on the outskirts of the park, where we used to have our lunch so as to watch Cockey shoot to pieces his personally made cardboard cutouts of Bear. When one day Bear joined us carrying his Mr. Men lunchbox packed with custard doughnuts, Cockey asked him to change the target. For the twenty minutes he shot at Bear with his air-rifle, while Bear dived behind anything providing cover. H and I ate sandwiches as we watched Cockey's continued attempts to murder Bear.

H said, "See I told you Cockey was psychotic."

"I never doubted you," I said, helping myself to another of Bear's doughnuts.

Later that week, Cockey waited in the yard behind the wheel of his immaculately maintained Mark 3 Cortina wearing his white Michael Jackson gloves. Cockey told me that on average it took him fifty-six minute to get home, including eight minutes to put on his gloves, finger by finger; psychos are always sticklers for detail. As I went into the main yard with Bear, I could see Cockey revving the engine more than usual, the back wheels went into a spin and the vehicle took off at top speed in Bear's direction. I really thought Cockey was going to kill Bear, but at the last minute, he pulled up within an inch of where Bear was standing. Cockey's objective was to scare the life out of Bear. As he told me later, it might induce a coronary. "Too many witnesses to kill him outright," he added.

Most people in the line of a speeding car would jump out of the way or freeze in a state of paralysis, but not Bear. Only Bear would think the best thing to do was jump into the air in the direction of the car, causing maximum impact and certain death. The result was that when Cockey slammed on the brakes, Bear's considerable bulk landed on the new bonnet of Cockey's car, resulting in three hundred pounds worth of damage. Bear's face was pressed against the windscreen, six inches from Cockey, whose spots were once again pulsating, but otherwise he showed no obvious signs of emotion. Bear lay for a while on the redesigned bonnet, seemingly winded as a result of his landing.

His only comment was, "I think I've shat me pants."

Bourgeois worked with H and he was quite a nervous kid, but seemed to grow in confidence when he was with H, Bear and me, though it may have been the drink. He joined us on many of our adventures and he was a good friend of H's, but I found out years later that he had been the only one in a family of twelve that had been put up for adoption. Can you imagine what that would do to you? Only to find out when you turned eighteen that your birth parents and siblings lived on the same street and you had passed them every day of your life without knowing it. No wonder he drank more than any of us. When I later learnt this, I thought how lucky I have been in life. To cap it all, he absolutely worshipped one of us and it wasn't H or me, it was Bear. If the idolisation of Bear becomes a religious sect and they erect a monument of their god, I'll get my hammer.

Brian worked in the same gang as H and Bear, and he made no secret of his homosexuality. It is hard to believe that it was not too long ago that attacks on gays were prevalent in London. Brian couldn't fight his way out of a paper-bag, but like Sid, he was a brave man in coming out at a time when homophobia was rife. I wondered how such an intelligent man found himself in Arkham. When I asked him why, he told me that he was once a barrister and often represented defendants at the Old Bailey. However when he got up to speak, he sounded just like the comedian Kenneth Williams, who was gay, and the court's public gallery would collapse into fits of laughter. This ended Brian's chosen profession as the Partners in a number of law firms refused to engage him. Now an inmate of Arkham, Brian publicly declared his sexuality. No one was ever going to push him back into the closet. I admired him for it.

Bear and Brian were always acting up and Bear frequently dropping his trousers and pretending to offer himself to Brian. One day Bear was stung by a wasp on his backside, he dropped his trousers and bent over offering his bare backside to Brian.

"Any chance of sucking the sting out of this?" he laughed.

"Only if it stung you on the cock," replied Brian.

Bear stayed there, pants down, but for once had little to say, while Brian walked nonchalantly past, pushing his personalised pink barrow which he had blazoned along both sides of it, 'The Pink Oboe.'

Chapter 14 – *I Think I'm Turning Japanese*

Larry, an Australian, only worked in the park for one summer, but this rogue became one of my best friends. The first time I met him, I was asked to make up the numbers up on the park's cricket team, a game I have no interest in, but I was always up for challenge. A bowler from the opposite team had played a dangerous ball at our new Australian colleague. It was now the Australian's turn to bowl at the same guy and I knew that the guy was going to regret his action when Larry passed me muttering, "This is going right up his fucken block hole." It did and the guy was carried off, holding his pelvis. Though I could not appreciate the game, I acknowledged that he was a man of his word. We shook hands and we have been good friends ever since.

Years later, I went to Australia and met up with Larry and his new wife Laura. Larry was still as hard as nails and not taking any crap from anyone. That explained why it was so hard to get residence in Australia; Larry was an immigration officer at Sydney airport. If you did get past Larry and then misbehaved, you would meet Laura who was a prison warden.

Larry is one of the toughest rogues I have ever met, but Laura was his match. We went into town that night for dinner, it being Valentine's Day. The women of Sydney were dressed to kill, wearing tight little black dresses which made me feel I was on the set of a Robert Palmer video. Larry could not keep his eyes off the women, but as he was married to Laura, this was a life-threatening indulgence. Larry was staring at one statuesque blonde sitting at the table opposite ours.

Laura calmly said; "Why don't you go over to her and stick your head in her tits, so she can tell you to go and fuck yourself, you fat bastard."

Larry disappeared up his arse for the remainder of the evening, while Laura carried on with dinner, now she had her man back in his rightful place.

Larry lives on the opposite side of the world, and lives by his own moral code; not rules laid down by government, his peers or even his family. As one of the select benevolent rogues, he stands up for what he believes in, even if it means he stands alone. A few years ago, a US battleship docked in Sydney harbour and the troops on board disembarked and ran amok, mainly due to being off their heads on drugs. A number of women had been sexually assaulted as a result. While both governments tried to suppress the scandal, Larry as the senior immigrations officer at the scene, quarantined the ship, along with its crew and the troops. Following international pressure from the US State Department, Larry received a call from his superiors saying that his job was on the line unless he lifted the embargo immediately. Larry was confronted by the combined force of two governments and a battleship with eight hundred marines on board; it made no difference. His job was to protect the people of his country, so his response was "Fuck em!" He kept the battleship embargoed until it sailed off five days later.

Clean-cut heroes always take one step forward when their superiors ask for volunteers to 'do their duty'. Many of those in my gallery of benevolent rogues have faults. They are not politically correct, they often clash with authority and have an array of vices. They are not heroes in the traditional sense and they will never volunteer to take that one step forward. But, they will be the only ones left standing after everyone else has gone.

Even in an asylum, some are madder than others. In Arkham, Mad Gordon was head and shoulders above all the other Napoleons; he was the Joker, though without the laughs. He was part of our team, but was hardly a team player, as he refused to sit in the main hut with the rest of us. Each morning on my way to see H and Bear in the hut, I would pass him seated on a little stool in the tool shed. As anyone passed he just stared menacing at them, but as he took a real dislike to me, when I walked past he made that extra effort to shout out "Bastard." I would in turn sit by the door of the main hut, smile, wink, blow kisses and occasionally wave while I drank my tea.

Early one morning, I made an extra effort to get into work before him, which meant getting into work around dawn, as he would start work two hours before the park opened.

Immediately, I set about sawing an inch off each leg of his chair. I did this regularly and on occasion Larry would take a turn with the saw, but we only reduced the legs gradually to ensure that my aggrieved colleague's descent to ground level was not too abrupt. At the end of one week, Larry and I sat in the hut drinking our tea and surveyed the extent of our labours. We were surprised that Mad Gordon seemed oblivious that his knees were up about his ears.

He even welcomed me with the usual, "Bastard."

One day Blakey called H and me into his office.

"Mad Gordon has taken a funny turn again," he said, which meant he had refused to take his medication, "so you two better to do something."

"Do what Blakey?" I asked.

"Just do something," yelled Blakey and stormed out of his office, leaving us to fathom out what that something was.

Scratching our heads, we set off in the van to Gordon's hut. There he was still on his adapted Japanese prayer stool. Even the sight of me, did not stir any response.

"He hasn't told you to get fucked," observed H when I tapped Mad Gordon on the head to see if anyone was in.

"Do you think he's dead?" I said.

We went to lift him up, but he refused to release his grip on the stool. As his knees were bent and he was still attached to his chair, we slid him backwards at a ninety-degree angle into the back of the van. We then drove back to the yard to check with Blakey what he wanted us to do with him.

"Well here he is what do you want us to do with him?" I asked.

"I don't know, just get him out of the way," he shouted, not wishing to take responsibility for our cargo.

"We could kill him," suggested H.

I replied, "No, lost property would be best, put him next to the umbrellas in the paralysed psycho section."

H and I were making the most of winding up Blakey.

"Just get him out of here," Blakey bellowed as he walked off, waving us away with his hand.

H and I looked at each other, not quite sure what to do next, "He likes football," said H, "Well, he's never actually mentioned it, but he has that look."

"Ah, I see what you mean," I said, "the starring eyes, clenched fists and keeping a piece of furniture handy to use as a weapon. Yes, he's Millwall alright."

Well it was decided. The three of us set off for the football pitches, after a vote of two to nothing. There was no game, but we thought we should leave him propped up on his Japanese stool by the side of the pitch. We congratulated ourselves on engaging him in a sporting activity and thus contributing to care in the community. After doing our good deed for the year, we parked the van back in the yard and headed off to the pub. Unfortunately, the following morning, the redness of Blakey's face indicated that he was not in agreement with our strategy.

"You idiots, the police found Mad Gordon, at ten o'clock last night by the side of a football pitch with what was left of his chair sunk into the mud."

"Not true, as it had no legs in the first place," I said.

Blakey continued screaming, "There was a college final played there last night and their manager called off the game after half an hour, when the ball hit Gordon on the head and he didn't move even though it knocked him on to his back." Blakey continued, clearly vexed about something, "And he hates football, you pair of bastards."

The next day, the hospital had pumped Mad Gordon up to his eyeballs with horse tranquilisers and he was back to 'normal'. I use the word normal in its most liberal sense, in that he was back on his Japanese stool the following morning, staring at me with his usual intense hatred and welcoming me in the usual fashion.

"Bastard." he growled.

"Up the Chels," I added and opened my arms and offered him a hug, but my gesture of reconciliation was cruelly rebuffed as his stool flew by my head.

The Penguin managed a team based on the other side of the park. His official title was Foreman, but he was the Kim Philby of the park being a staunch communist who looked for every opportunity to bring down the capitalist system. A good friend who worked in his team was Ben, whom I knew from our lunchtime kick-abouts or 'murderball' as we called it. Ben was black and though we were both young men of the same age, education and background, his colour made him a marked man with the police. Though I had my run-ins with the law, that was nothing compared to Ben's experience. He would be picked up by the police on suspicion about once a week, simply for being a young and black. A police van drove into the park one morning and an officer tried to arrest Ben on the pretence that his description matched someone robbing a house ten minutes earlier. I went up to the police sergeant and informed him that three other men and I had been working with Ben all morning. It was easy to see that the police sergeant was showing off to the new recruits sitting in the police van; all blacks "were fair game" I heard him say as he turned to the new recruits in the van.

He turned back to me and said, "Take my advice, don't get too close to these people," and then facing Ben he just laughed and said, "Next time we will get you for something, boy."

Meanwhile, one of the others placed a broken bottle under the back tyre of the stationary police van. Later, on the other side of the park we discovered the police sergeant trying to get through to headquarters to get a mechanic to change the back wheel which was flat. He was in a hurry, trying to get back for lunch. As I drove past the police van, Ben and the others in the back of my truck noticed one or two of its uniformed passengers giving us a wry smile. I would like to believe they were letting us know that they were not all like the idiot screaming obscenities into his walkie-talkie.

Dodo, another inmate in the Penguin's team, was commonly believed to be an alien.

He was a small kid, with massive staring eyes that would fix on you. It turned out that he had a kind of Mona Lisa effect, as everyone was of the opinion that he was focused on them which unnerved them. Even when their small hut had its full complement of thirty inmates crammed into it, they would be crammed into one half, while Dodo sat alone on the far side. They only ever mixed together outside the hut, when Patrick, Penguin's chef-in-residence had set it alight. Patrick was short-sighted and as the hut had poor light in winter, he would sometimes light a blow-torch outside in daylight, so any coats hanging on the wall by the door would go up in flames whenever he came in to light the cooker. One week its inmates had to eat their sandwiches sitting under nearby trees after Patrick had razed it to the ground. It was the torrential rain that made all of us grumpy, as no one seemed to mind all the other times.

Dodo lacked Bear's hatred for flora and fauna, but he had the same impact on them as the Bear on the other side of the park. It was as though the park had been hit by two asteroids. One day I watched Dodo planting some bulbs. As part of his preparations, he would break off the sprouting shoot of each crocus the way that one would flip the pin out of a hand grenade, so that in the spring the flowerbed which should have sprouted beautiful purple and white flowers would be as barren as Bear's salad bowl. Dodo's ignorance of horticulture was also on a par with his ignorance of the park's wildlife. A member of the public ran up to him one afternoon and shouted that a duck was drowning. Dodo dived in to retrieve the bird and then, he promptly threw it back in.

"You bastard that was no duck, that was a mallard," Dodo said with genuine anger to the now dumbstruck gentleman.

I waded into the water and saved the poor bird. All I could offer the open-mouthed gentleman a shrug of the shoulders and an apologetic smile.

Following the incident with the duck, the Penguin was convinced that Dodo was the calibre of man to apply for the post of Protector of Culture of a very famous medieval castle in London, which was under the park's management. No doubt, this was an attempt to bring down the capitalist government from the inside. Guy Fawkes had gunpowder to blow up the Houses of Parliament, but we had a potentially far more lethal device in Dodo. A mock panel of eminent experts was convened to develop Dodo's interview technique. It was comprised of the Penguin and two members of his team, Paranoid (who regularly ate deadly nightshade berries to build up an immunity, as he believed that one of us would slip some in to his dandelion tea), and Patrick. I say eminent experts, but they were the only members of the team who turned up for work that day as it was a Monday. Dodo answered their one and only question, convincing them that he could plant a Beefeater if one strayed past him. As Patrick later informed me when I asked him if Dodo was a suitable candidate, "a perfect match, just like your arse and my face." Dodo approached me to help fill in his application form. Within minutes, I had completed the form, signed it for him and told him off to post it on the back of the nearest duck. To the bewilderment of the Penguin and me, his application somehow made it to the selection panel and Dodo was granted an interview.

This was the moment when things became serious. The park's board of governors were increasingly concerned that the park was getting a little out of control. They were wrong. It was completely out of control. So an ex-army officer called Coldman was despatched to bring order where anarchy thrived. At the end of his first week we received a taste of what was to come when a memo was stapled to everyone's clock card. It stated that he was to be addressed by his full military title when we met him or put his name to paper.

Coldman's first task was to go through the applications for the castle post. I am sure that Dodo's stood out from the other applicants, because his stated qualifications ranged from 'possessing a Kevin Keegan World Cup football ESSO badge', to 'a certificate for swimming a hundred yards butterfly' and ' being able to wiping his arse with either hand'. I learnt later from Maureen, that Coldman told her in a moment of anger that he thought he would be accused of discrimination if he rejected Dodo's application, thinking that Dodo was a special needs candidate. He told Maureen this, thinking that Blakey and the Welfare Officer, the two other members of the selection panel, would do his dirty work for him and reject Dodo's application. It was a flawed strategy; Blakey was far too wily to do Coldman's dirty work for him and the Welfare Officer surprisingly, did value diversity, so both sanctioned Dodo's application. His selection for interview was therefore a unanimous one.

On the day of his interview, Dodo had only one rival candidate for the position; Alf, a man whose hatred of any type of authority was well known. Alf went into the interview room and extinguished his cigarette on the carpet with his muddied boot before taking his seat. He was disqualified and thrown out. Dodo went next and sat silently for all thirty-two questions, looking blankly at his inquisitors. Question thirty-three was the final question and the one that I had prepared Dodo for; Maureen had given me the questions. The chair of the interview, Coldman, informed Dodo that he only had one question, number thirty-three, as this was on the subject of his own personal project for the park, a model of which was positioned on the table during the interview.

"In the summer I will spend a quarter of a million pounds on a rockery in the Queen Bernice gardens. I would challenge anyone to improve on the genius of my innovative design, however can you think of maybe one little thing to improve on it?" asked Coldman who now sat back believing that with Dodo's track record he would again offer no response, which I was sure Coldman would interpret as confirmation of his genius. Dodo leant forward to survey the scale model on the desk in front of him and then to the panel's surprise he provided a clear, unambiguous answer, "Bulldozer it!"

As expected, Coldman exploded and he ended the interview right there and then, so the opportunity was lost for Dodo to ask his end of interview question; "Would it be appropriate to design a garden display based on the allotment of blackheads on my nose?"

This was just another laugh to us as we carried on creating harmless mayhem. We didn't realise that Coldman was plotting to bring sanity to Arkham. In doing so he would release its biggest lunatics out into an unprepared world.

Chapter 15 – *The End of an Error*

As with all good stories, you need a villain. For our happy-go-lucky bunch of under-achievers working in the park, our villain was never Blakey; our nemesis was Coldman. Blakey was as mad as us; the only difference was that he was in a position of authority.

Coldman was one of a new breed of men who were out to make a name for themselves by 'out-sourcing', 'streamlining', and 'off-shoring', or as it is more commonly known, sacking people. In their ranks were efficiency experts, carrying out benchmarking exercises and offering a utopian dream of more efficient services by making massive reductions in staff and budgets. This was before privatisation really took off and trains came to a standstill or even worse carried on through the buffers.

On the other side were the unions who had forgotten why they were there. Their role was to help ensure decent pay and conditions for the workforce, rather than fighting political battles with governments. When a union official justified industrial action with phrases such as "This is the thin end of the wedge," and "We are standing up for all workers" or "This is an attack on every working man," then the workers they represented were fodder in a much bigger ideological struggle.

But as with all things, nothing is ever simple and before you think that we were victims crushed under the foot of capitalism and knocked aside by unions flexing their muscles, I have to acknowledge one slight weakness in my diatribe. You see, if the country was based on our model of the park, it would be anarchy, a wasteland with hardly a tree in sight, where delicately flowering plants were rare and the sounds of birds were replaced with manic laughter.

Many respected politically analysts look at the eighties as the origins of privatisation. These experts see its roots in Thatcherism and specifically in the architect of monetarism, Sir Keith Joseph. This is a common misconception, for privatisation actually originated with the Bear. As a public employee, he made as much of a contribution to public services as the

German doodle-bug did to the architectural heritage of Coventry. Bear wrought devastation throughout the public sector. After initiating the contracting out of one of the country's most famous parks, Bear's rampage continued into the Royal Mail, British Gas and British Rail, as he found temporary employment with all of them. Is it a coincidence that all then became prime targets for privatisation? I think not.

Bear's first success, as an unguided weapon of monetarism, was the park. A few years earlier he had completed his City and Guilds Certificate in Horticulture. I qualify this sentence, in that he completed three years of attending classes, rather than actually passing any exams. In fact, after his years of study, his final exam mark was two per cent, a ratio of less than one per cent per year; it was a commonly held belief that he had cheated. In his first month, Coldman brought the park's fifty-year old apprenticeship scheme to an abrupt end; the first government training scheme to close in the eighties.

Bear's exam result was not a great surprise. H who was in the same class as Bear, related how in 1977, their tutor had set them a question in honour of the Queen's Silver Jubilee. His students were asked to give an example of a miniature shrub that could be planted in a window box by an old woman living at the top of a block of flats who wished to celebrate the royal event. Everyone in the class opted for some species of miniature Japanese bonsai tree, bar one student in a large bumblebee jumper.

"What plant would you suggest and can you provide us with its Latin name please?" asked the tutor.

Bear struggled to come up with the Latin name for any plant and he was not that good on their English names either. Finally, his eureka moment arrived and he emitted in a high-pitched squeal, "Sequoiadendron giganteum".

H appeared to take issue with Bear's selection, when he shouted across to his good friend, "You fucking moron." H knew that this species was more commonly known as the Giant Canadian Redwood. This revelation was followed by a minute's pause then the class, with the exception of the teacher and Bear, collapsed into hysterical laughter. Bear had chosen the largest tree in the world, one famous for having a tunnel cut out so that cars could drive through it. If an old woman grew one of those in her window box, her building would have toppled over quicker than the Tower of Pisa if Bear paid it a visit. The dejected lecturer ended the lesson at that point and sat down in front of his class looking at his shoes for the remainder of the day.

The mantra that it is the right of everyone to do a day's work is all very laudable, but in Bear's case, it's in society's best interests that he be provided with regular benefits for life, and be relocated to some forgotten wilderness where he can cause no further damage. Perhaps Grimsby.

H's academic progress fared little better, even though he was the most knowledgeable apprentice in the horticultural college. It was just that H had little patience with his fellow human beings. This tension was self-evident when H was doing the practical part of his final examination and was demonstrating how to lay ten square yards of turf. As H was laying down the last piece, the examiner innocently asked him if he was happy with the level of the laid lawn.

H replied, "If you can do it any better, you fucking do it," and promptly threw the spade at him and walked off.

That was the end of H's three-year apprenticeship. He never completed his examinations, so in theory, Bear had secured a better mark and to this day, he has never stopped boasting to H about his superior academic qualifications.

One Friday after our shift, we headed into the yard to clock off and head to the pub for the weekend. Bear ran up to me with a note he received in his clock card.

"Coldman wants to see me, what do you think he wants?" he asked.

"Well, it's about bloody time, with your well recognised talents he wants to promote you," I replied.

"Really, do you think so?"

"Without a doubt, how could he miss your exceptional horticultural achievements, all encompassed by a simple moronic view of life," I replied.

"You're right, he might be a bit of an arsehole and not realise what a total fuckwit I am," said Bear, as he wobbled off in the direction of Coldman's office.

As it was nearly clocking off time, most of Arkham's inmates were gathered in the yard waiting for Bear to emerge from Coldman's office. Fifteen minutes later, Bear came out and to our surprise was very upbeat.

"Well, did he offer you his job?" I asked.

"He is a very nice man, he offered me a cup of tea, even some hobnobs," began Bear.

"Sounds cosy. Did he then ask you to drop your pants and bend over his desk?" H added.

"Not that I would have said no, but no, he asked me about my career prospects and ambitions," said Bear.

"He's a bit of a piss-taker then. How long did you sit in silence?" I added.

"About ten minutes," replied Bear.

"What then?" I asked.

"Bastard sacked me."

H was next to be called in and then appeared within a few minutes from Coldman's office.

"Bear had eaten all the Hobnobs," said H, but he too had been given two weeks' notice.

"What will you do?" I asked.

"Cut his nuts off and give them to Brian as earrings," replied H.

Everyone was depressed; no one to our knowledge had ever been sacked before. H and Bear were, despite their lack of social graces, the two biggest characters in the place. To lose them both was like chopping down the biggest oaks in the park, oaks planted in the wrong place and blocking out sunlight but oaks nonetheless.

Everyone expected I would be next, as my disciplinary file was always on the left-hand corner of Coldman's desk. But, I did not receive my summons to Coldman's office until a few weeks later; H and Bear's last day at the park. That morning I strolled in to the main office, to find Blakey and Coldman in their usual positions. Coldman was sitting in his chair with his neat short-cropped blond hair, which looked like it had just been cut, washed and combed by his mum, while Blakey stood next to him like a protective sumo wrestler.

Coldman began, "Mr White and Mr Brown have had their service curtailed today, but your fate will be different." Some swing from squire of the manor to B-movie villain, I thought. He continued, "They will be the first of many here to have had their employment opportunity withdrawn."

"In the real world, we say sacked," I interjected.

"When they leave today they can claim unemployment benefit from next Monday. However, if I make your life a misery so that you resign, you will be making yourself voluntarily redundant and you will not be able to claim benefit for seventeen weeks."

I could tell by Blakey's body language that he wanted no part of this pure and simple, malicious abuse of power. Blakey looked uncomfortable, but Coldman reclined into his chair clearly believing that he had finally beaten me.

"Thank you very much and I look forward to working with you, for many years to come." I said.

They both looked astonished. I think Coldman thought I would lose my temper and resign on the spot. I then saw something I had never seen before, Blakey smiled and gave me a quick nod of his head. The old bugger, good on him I thought. I then realised that Blakey was not just mad like us, he also shared our one common trait; he was not a vindictive person. Blakey quickly understood my response, for rather than Coldman intimidating me into resign, he had shown his cards and I was free to do whatever I wanted. There was a fundamental weakness in Coldman's strategy, as it was based on not sacking me. I then walked back out into the yards and said to the gathered Arkham inmates that I would join them down the pub later for H and Bear's leaving drinks.

Nearly all of Arkham's inmates turned up that evening for Bear and H's farewell drinks, now that both had been 'declared sane' and could leave Arkham to corrupt respectable society. But it was a momentous day in another respect, as it was the day that Cockey informed us that, at the age of twenty-seven, he was no longer a virgin.

H was in the pub a few hours before his final shift finished. Bear finished work even earlier and had been going around the park all morning taking pictures of himself with inmates from Arkham and, for some reason, one of him hugging a startled rabbi. Bear was so pleased with the success of his farewell tour that by lunchtime he had used all the film in his camera. I asked if I could have a look at his camera and as he handed it to me I dropped kicked it into the lake.

"It's for your own good Bear, it's a tough world out there, so take a valuable lesson from this and never let your guard down," I informed my now very angry mop-haired friend.

Bear set off on one of his runs in another fruitless attempt try to kick several shapes of shit out of me.

As that evening wore on, copious amounts of beer were consumed in the bar, with Larry, Penguin, Budgie, Dodo, Paranoid and many more in attendance, only Coldman, Blakey and Cockey were lacking. Cockey then arrived two hours later with a woman. We had never seen Cockey with a woman before, so we all sat open mouthed until H broke the silence and greeted his friend.

"Where have you been you boring twat?" demanded H.

We then recognised the woman as Sally, a big stocky summer student from New Zealand who, like Larry, was taken on as a temporary worker over the summer.

Cockey expressionless as usual replied, "Sorry I was having sex in a bush."

"Well that is the usual place," I said.

"You mean having a wank, but that doesn't take two hours," said Bear who then sat back and pondered. He then added, "Well it never takes me more than two minutes, if I'm honest."

"No, with Sally," replied Cockey which resulted in the most dropped jaws and pint glasses in history, since the Titanic hit an iceberg.

At this point, Sally raised herself above the immature banter and declared, "I'm off for a shit."

Bear was still in a state of shock and pointing in her direction, "That Sally, she's already worked her way thought most of the men in the park, with the exception of me obviously."

H added, "And me, plus I hear from Paranoid that she is a bit of a sadist. She put him in a set of stocks, made him wear a horse harness and whipped him from arsehole to breakfast time," he then added, "Mind you; I'm not saying there is anything wrong with that." Paranoid, nodded in agreement.

H then drifted off into some fantasy world somewhere, before returning to the task in hand. "You're not a virgin anymore, how the fuck did that happen?" enquired H.

Bear was still in dire straits, sitting with his head in his hands muttering, "How the fuck did that twat meet a girl, I am going to kill myself."

Cockey explained, "Last week I found a note that Sally had wrapped around my clock card, stating that she liked the look of my calves and she wanted to take me out."

"His calves! Fuck me, wait till she looks at his face," grumbled Bear still with his head in his hands.

Cockey continued to describe the experience in his usual robotic and unemotional manner, "We went for a drink, and then Sally suggested that we go back to my place for sex. I told her that would not be possible, as I live with my parents."

Bear lifted his head from his hands and looked directly at Cockey in disbelief, "I would have taken her back home and fucked the tits off her on the living room carpet while Gorgeous George and Gladys were watching *Eastenders*."

My blond friend was clearly totally distraught at the knowledge of Cockey's sexual initiation and plunged his head back into his hands.

H continued his interrogation, "What did you do then?"

"Well fortunately she regularly uses a hotel in Bayswater for various sexual encounters; we then proceeded to engage in the sexual act," he told us.

"It must have been like a lorry load of yogurt hitting a wall," declared Larry, who was as shocked as the rest of us, but was braver in visualising Cockey in the sexual act.

This must have stimulated a similar thought in Mad Lynn who worked with and despised Cockey, who immediately threw up on the middle of the pub's carpet.

H ploughed on with his interrogation, not just out of curiosity, but also from a professional perspective as he was the parks 'unofficial' sexual technical advisor, "What happened then, you spotty twat?"

"We made love for five hours," continued Cockey, which led to the sweet little innocent eighteen-year-old barmaid from Belgium on her first day in the bar, joining us all in a chorus of "Bollocks!"

At the same time, we all knew that Cockey did not have an ounce of imagination to construct a simple lie, let alone exaggerate one. H resolutely tried to focus on this unimaginable event, and believe me H has conjured up some surreal images in his time, while trying to retain his composure.

"Five hours, how in the hell did you keep going for five hours?" he asked.

Cockey then said that he realised his previous statement was factually incorrect.

"Thank fuck," commented the still distraught Bear.

"I did stop to eat an apple," qualified Cockey.

"Bollocks," muttered Bear.

H was still trying to keep some hold on reality, "How many times did you shoot?" "Shoot?" continued Cockey, "If you are referring to the act of ejaculation. I did not."

I have never seen H, as shocked as he was at that moment.

"So you fucked for five hours only to take a break to eat an apple and never came after nearly twenty-seven years a virgin. What the fuck is wrong with you?"

Everyone leant forward to hear Cockey's reply, including the demure Belgian barmaid. Cockey replied in a way that only Cockey would.

"Well, she never said I could."

Bear with his fingers now firmly clamped to his head for the rest of the evening muttered, "When I die, I'm taking that fucking cock-less cyborg with me."

Cockey was next to leave the park, with my help. Cockey had been trying to join the police force for years, but he failed as each time his aptitude tests indicated that he was psychotic. I must have been crazier than him, when for a laugh and god knows how much trouble that has caused over the years, I agreed to fill in his next application and do his written tests for him. His reapplication was approved, and he was selected for training.

Cockey's training did not go all that smoothly, particularly when he had to go through a number of scenarios that recruits may one day encounter, in a specially built pre-fab village. No problem for Cockey on day one, as all trainees had to do was shoot everything in sight, which delighted his trainers. Day two did not go as smoothly. All recruits were given a new task, where their gym teacher played the part of a drug-crazed knife-wielding maniac. The objective was for each candidate to calm down the assailant and disarm him through gentle persuasion. All went well with each recruit, until it was Cockey's turn. He promptly threw the gym instructor out the first floor window and hospitalised him for weeks, nursing a concussion and a broken collarbone. Unsurprisingly, Cockey joined the ranks of Her Majesty's Police and has made rapid progress through its ranks ever since.

As for me, I stayed at the park a few months longer and then decided to go to University. I still had some remnants of a brain and I decided to use it. I had realised that there had to be more to life than just having fun and fighting authority. As you will learn later, University turned out to be not so much of a cultural change.

Chapter 16 – *The Simple Bear Necessity*

My friendship with H and Bear continued after they left the park. I was fortunate to be welcomed into their family lives, though for obvious reasons their worlds were on different plains.

H's mum Fran always treated me as her third son. She was a fantastically warm woman and the heart of her family. Her husband Bill was a good man who shared our passion for Chelsea along with his own love of bingo. It was the first time I had come across such a strong supportive family.

I have had numerous nicknames over the years and the one that friends in south London used for me was Wombat. Fran and Bill adopted this nickname. I was always invited to join them for their family Christmas dinner. In the festive tradition, we ate until we could barely move and then drank until we all collapsed on various items of furniture. All went well until Fran suddenly asked:

"Why are you called Wombat anyway?"

There was an uncomfortable silence, only broken by H who for the first time in my experience told a lie. "It's because he is like a cuddly Australian marsupial."

Bill looked up to see if his beloved wife had fallen for it and thankfully, Fran seemed pleased with the explanation. How different the evening might have been if H had answered, "It's because he eats, roots, shoots and leaves," based on numerous broken relationships and my failure to commit.

Biddy often had H and Bear turning up at her house. The welcome though, was not the same for Bear as it was for H. H and Biddy got on very well and Biddy accepted H as one of the family, as Fran had accepted me. H would be invited to a special pre-Christmas dinner at Biddy's a few days before Christmas. The celebration of our Lord's birthday was distinguished by Biddy keeping H's glass of vodka topped up until he could barely find the door on his way out. H would always extend Biddy's invitation to me, (you could never accuse Biddy of favouritism and God love her, she does keep me on my toes), so I could also indulge in Biddy's excellent turkey and vodka Christmas lunch.

H always looked forward to his Christmas invitation to Biddy's; in particular he liked to see what latest adornments she had purchased or received as freebies with her shopping. These ranged from Power Rangers stickers, which had fallen out of a cornflake packet and ended up stuck to the fridge; to a laptop keyboard, without a computer, which Biddy attached across the handlebars of her exercise bike. H particularly liked the butterfly clock Biddy had won at bingo. The butterfly was too heavy for the second arm and it could never get past nine o' clock; H was captivated by the insect's futile struggle. My favourite item was her Pope rosette which commemorated the pontiff's visit to Wembley. The Pope had completed three laps of the track in his Pope-mobile for seventy thousand ecstatic catholic old biddies and then shot off down the tunnel back to Italy.

Bear's visits to Biddy's were more anarchic. One day I walked into Biddy's council flat, to find Bear and Biddy on their knees picking up some cards that had fallen out of Bear's pockets. These were not your standard PG-Tip tea cards or Panini football stars collector's cards, but the type you see in telephone boxes advertising 'Kinky Kirsty, KISS MY WHIP', blazoned above a 0898 telephone number.

In their own way, Biddy and Bear did get on well and would have run of the mill conversations.

"Hi Biddy how are you?"

"Very well Bear, would you like a cup of tea?"

This would carry on for a good five to ten minutes. The only strange thing about these encounters was that both were bent down on either side of Biddy's front door, talking through the letterbox. Biddy would be on one side of the door in her Woolworth's dressing gown and slippers, Bear would be on the other side dressed in a nappy and wearing a cape of some description. Biddy wouldn't let him in the house when he wore this regalia; she didn't want the neighbours to see him. I don't think Biddy really thought her argument through, as this left the Bear in full view of every passer-by for the duration of their conversation. It was also puzzling to see Biddy trying to feed a full cup of tea with a saucer through the letterbox, as if it was a test in *The Generation Game*. When Biddy moved to a new council flat, she gave me her instructions concerning the move.

"Turn off all the electrics, check I haven't left the gas on and don't tell that blond lunatic where I live."

We live in a society where nothing surprises us any more, that is until Bear walks in. Before I go further, you have to understand the impact Bear had when you first met him. Bear was then was about eighteen stone, with blond hair all over his body and a haircut modelled on Purdey, the character played by Joanna Lumley in *The New Avengers*. Occasionally, he would convince himself that he was from another country and over the years, he has dressed as a Welshman, a Turk and a Russian. During the eighties, he usually wore a nappy, no trousers and twenty-hole boots, whilst adorned in various national flags.

It was not unusual for people to run up to Bear and say 'fantastic outfit mate, what are you advertising?' Once an Italian TV producer looked up as Bear entered the pub; he promptly dropped his decaffeinated coffee all over his tangerine coloured trousers. He then leapt up to ask Bear, in Pidgin English, if he could do a TV special on him. Bear walked off, politely turning down the offer and saying he had to be home, as his mum needed help with her 'spotted dick'. It would have been interesting to see what the producer made of Bear's response when he took out his English-Italian dictionary later on.

Bear's parents were Gorgeous George and Gladys. They had conceived the Bear, "after overindulgence in Ramsbottom's Barley Wine" according to the story related to me by their blond offspring. The house where they lived looked like it had barely survived an earthquake, due to Bear moving around inside it. Whenever we went on boat trips abroad, Bear would rip our cabin apart, no doubt in an attempt to make himself feel at home. On a visit to the Bear's house, Goldilocks would have found all the pictures on the walls were at fifty-degree angles; the banisters were smashed to splinters; all the bed legs were broken and the porridge was all over the wallpaper, well, I think it was porridge. She would have said, "To hell with this for a game of soldiers" and decided she would be better off staying that night in the Wolf's house just down the road from Little Red Riding Hood.

Gladys, a lovely woman, was the maddest of the three. Bear and Gorgeous George were bonkers and their options in life were therefore limited, whereas Gladys was normal, except that the poor woman agreed to stay with them of her own free will. As proof of this, one day H and I were shown into the house by Gorgeous George, who told us that Bear would be down in a minute as Gladys was giving him a bath.

I liked Gorgeous George, as he was a lovely harmless old man from the East End of London. When H and I would pop around to meet Bear, he would be sitting in front of the television in his underpants, wearing the archetypal 'Englishman on the beach' string vest. All that was missing from this iconic image was a knotted handkerchief on his head and Bear making sandcastles on the living room carpet. None of which would have surprised me. One day I found Gorgeous George surveying the wreckage that was his front room, as he recalled a recent visit from the council.

"A guy from the council turned up and offered us the opportunity to buy this place." He then looked around again at the torn wallpaper and broken furniture and paused, shook his head and with a sigh and added, "but we didn't want it."

In the corner of the floor sat a pile of broken picture frames and ornaments and all that remained on the sideboard was a two-foot-high toy robot. Gorgeous George informed me, that Bear had run in that morning with Robbie the Robot which he had just bought from the sci-fi shop Forbidden Plant. With one sweep of his right arm, Bear had cleared all the photos documenting his family's history off the sideboard and onto the floor, mounting Robbie in their place. Gorgeous George sighed again. "Two weeks wages that cost him," and then he returned to his daytime television programme on dysfunctional families.

I was reminded of the old joke about coming from a broken home; in Bear's case he broke it.

Gorgeous George had only one tooth in his head.

"Was it as a result of the war?" I asked.

"Yes, shore leave," he had answered.

During the war, Gorgeous George had been in the merchant navy. While he was walking down the gangway to go ashore, waving to his friends on ship, he fell, hit his jaw on the side of the quay, which knocked out all his teeth but one and then plunged thirty feet into the harbour waters.

"The war, its true cost will never be known," said Gorgeous George shaking his almost toothless head.

I had seen Gorgeous George many years earlier, when he would drop his 'little boy' off for work at the park, but I had never had a chance to talk to him. Gorgeous George and Bear would drive up the main road to the park's yard, the two of them squeezed into Gorgeous George's little Robin Reliant, the type seen on TV's *Only Fools and Horses*. The cars' makers could have used the scene in an advert to show how a three-wheeled car, built for a midget, could accommodate over forty-stone of the 'Best of Britain'. They would have had to use an airbrush to cover the fact that all three wheels were at a forty-five degree angle to the road.

One time, I telephoned the Bear household.

"Yes," answered Gorgeous George.

"Is Bear there?" I asked.

"Yes, he is." He then put the phone down. I called back.

"Yes," said Gorgeous George.

"Can I speak to Bear?" I asked.

"Oh, you want to speak to him." I thought I could answer, "No, I just want to know where he is," but I could never be rude to Gorgeous George. He was definitely a one off and if Bear, H and I have one thing in common, it is that we do miss that mad, but very gentle old man.

I once received a phone call from Bear who was terribly excited as Gorgeous George and Gladys had bought him a snooker table for his birthday. H, Bourgeois and I entered Bear's bedroom, to find a full-length snooker table. It took up the entire bedroom area except for a ten-inch gap from the wall on one side of his room. Bear had had to put his bed out on the landing and his clothes in the cat's basket in the kitchen.

"Come on, I'm all excited let's have a game," said Bear gleefully.

"Good idea, give me a hand to move the table up onto the roof," I replied.

"No, it's not possible, I tried that and I got covered in pigeon shit," said Bear, pointing at the green baize, liberally sprinkled with white mounds, which didn't help smooth the path of the snooker balls.

Bear kept on at H to perform the 'break' for our first game. Eventually, after Bear offered to buy the first round, H agreed, picked up the cue and stabbed it straight through the green velvet, an action on a par with that of Peter Sellers' character Inspector Cousteau, in the first of the Pink Panther series. H then lobbed the cue straight out of Bear's sixth-floor bedroom window, bringing our first and only game on his table to a close. Bear was speechless, as his present was now useless and all he received for Christmas was scent in the form of cat piss applied to his somewhat eclectic wardrobe.

H's only comment was, "No point hanging around this shithole, come on Bear you promised to get the first round in."

That was not the only present bought for Bear that did not turn out as he wished. He once asked his sister for the Brazilian national football kit, for his birthday. Excitedly, he ripped open his present to discover a Norwich football kit instead.

"What's this shit?" Bear asked his sister.

"Well they didn't have the Brazilian kit, but Norwich have similar colours, so I didn't think it would make much of a difference," she replied.

"Well I'll be fucked," said a despondent Bear

Once, about twenty of us were sitting around a table having seen in the New Year. Bear was in a contemplative mood and was clearly a little tipsy.

"It's wonderful to have moments like this, surrounded by my best mates," he said to us all, looking around the table. He then paused a little and continued, "The funny thing is," he paused and then lifted his head up to look at all of us and added "I don't know where any of you live."

H, Bourgeois and I and other friends who could still stand looked at each other and after a pregnant pause we moved on to another subject.

One day H phoned me in a bit of an anxious state.

"What is it?" I asked.

"I went to the doctor today because I was in fucking agony and he says I have a double hiatus hernia. Do you know what that is?"

"No, but I remember Cockey talking about it once." I said, "Why don't you give him a call?"

I did not hear anything for a few days, until H called again in an even more agitated state, having discussed his condition with Cockey. H related to me what Cockey had said.

"A double hiatus hernia is a tearing of the stomach wall that requires an operation," said Cockey.

H asked if there were any side effects from the condition. H and I knew that Cockey had no sense of humour whatsoever or indeed any human emotion including sympathy or even empathy, so his response would be honest, perhaps even brutal. It was.

"Well in my experience," Cockey informed H, "it resulted in sexual desires towards one's own gender and testicles the size of coconuts."

I could understand the effect this advice had on my good friend who, as it happened, always wore Levi's, and though not homophobic, hardly showed any inclination in this area. I told H I would call Cockey and clarify that his information was right, but knowing Cockey I knew he couldn't lie. However, I made a mental note in the meantime, always to meet my soon-to-be-baggie-trouser-wearing friend away from any public toilets.

I called Cockey, "What's all this about H turning into a big-bollocked extra from the Village People?"

"Those are not my exact words, but yes the effects are a major growth in the testicular area and a desire to perform a sexual act with your own gender."

"Cockey, I have never known you to lie and I know you wouldn't take advantage of H in vulnerable state like the rest of us would, but is it really true what you told him?"

"Yes," replied Cockey.

Well H is fucked, I thought. "This happened to you?"

"No, to my horse," he replied.

H went on a massive bender, no pun intended, until I broke the good news to him a few weeks later.

H and Bear continue to be my friends and our adventures continue to this day, which to a large part explains why I am still single.

INQUISITIVE ROGUES

'Connecting to a complaint from a Mr Purdey about a large gas bill, a spokesman for North-West Gas said, "We agree it was rather high for the time of year. It's possible Mr Purdey had been charged for the gas used up during the explosion that destroyed his house.'

Daily Telegraph article

Chapter 17 – *"When I Grow Up, I Want to Join a Boy Band"*

Around this time, I attended night school where I met a lively character called Chelsea Mark. Chelsea Mark and I always argued and we still do to this day, particularly as he was then an active member of the left-wing group Equality Means Unity (EMU).

His passion though, was the pursuit of women rather than a socialist utopia and he would never miss an opportunity, as he would say, to offer 'his gift to a lady'. After one of my birthday parties in Clapham, I went to the Windmill pub the following Sunday morning to continue my birthday celebrations with Chelsea Mark, H, Bourgeois and others that had survived the night.

Chelsea Mark told us of his ambition to open up a diving school. To prove to us that he was fully qualified for the task, he jumped into the duck pond outside the pub. It contained only two inches of water and a broken bottle that Chelsea Mark managed to land on. So, taking some time out from our drinking session, off we went to St Thomas Hospital. As we entered reception, Chelsea Mark lifted his grazed foot right up on to the nurse's desk and pointed to his injury, which was not even a cut. Never one to miss an opportunity to impress, when the nurse asked him for his occupation he replied, "Pearl diver."

On other occasions, he used his height and athletic build to impress women by pretending to be the goalkeeper for Charlton Football Club. It worked for a while, until he met a girl who was the cousin of the goalie and he had to exit out of a pub's toilet window. On one of his birthdays, when he was very drunk, a young woman asked him what his plans were for the future.

"When I grow up I want to join a boy band," he replied to the bemused woman as she looked at the thirtieth birthday badge on his pink jacket.

Chelsea Mark usually had a woman somewhere, but one summer he went through a dry spell and, for some strange reason, asked me if he was doing something wrong. I told my friend that he was too self-obsessed and that he should try and engage with women more by showing a genuine interest in their lives and ambitions. He had a date that night and got back to me the next day.

"Load of old bollocks, I tried your approach and got nothing," he said.

"OK, what happened?" I asked.

"Well, I took her to a nice restaurant."

"What is her name by the way?" I interrupted.

"Fuck knows," he continued, "well, I talked about my life, just as an introduction to get things going."

"How long for?"

"Until dessert and then just as they were closing, I took your advice and said to her, 'What about you, what do you think about me?' and she said fuck all and just sat there with her mouth open." He then angrily declared that I must still be a virgin and slammed the telephone down.

Sullivan's best friend was Chelsea Mark, but they had more in common than both being members of the EMU, they both shared a love of drink. Sullivan had a disability and social services had secured him a very small three-wheeled car, slightly bigger than a tricycle and barely big enough to aid his mobility. Imagine the astonishment on the faces of the police when they stopped his car to breathalyse him and Sullivan fell out, followed by a six-foot-two Chelsea Mark and a case of lager.

With Sullivan I shared a sense of humour that upset some people. On meeting Sullivan, most people would never acknowledge his disability, while others were blatantly uncomfortable at not knowing what to say, so their silence or the occasional embarrassing look highlighted that his disability was clearly an issue for them. We had no such problems in our relationship, as Sullivan was a friend and a man and as such we shared a lively banter as mates do.

Chelsea Mark and I met him in his local pub to watch his team Spurs, as if life were not hard enough for him already, play their hated rivals Arsenal.

As I came over with the first of many rounds he asked, "What are you so happy about?"

"I got a new job as the England football manager's speechwriter," I replied.

The reference was to Glenn Hoddle who, earlier in the week, had said that disabled people were being punished for their sins in a previous life. Sullivan saw the joke, laughed and knocked the filling out of one of my back teeth with the stump of his right arm. The North London pub was full of Arsenal fans. An hour into the game one of them, a lanky individual who looked like *Where's Wally* in his scarf and bobble-hat, came over with some of his friends to take issue with Sullivan and his regular references to the marital status of their parents. A big mistake. As my dentist can confirm, Sullivan could take care of himself, and with Chelsea Mark and me in attendance, they quickly marched back to Wallyland.

You could not have found two more different characters than Bear and Chelsea Mark. However, they had two things in common; both were obsessed by their image and fashion in particular. Chelsea Mark's dress sense was based on the latest Parisian styles and Bear's on that of a circus clown with an ample wardrobe, who dressed in the dark. Chelsea Mark would consult the latest men's fashion magazines to ensure he knew what was 'this year's' new black. A waste of time, as this year's black was always black; in my case, it was blue, though I had one white shirt for funerals and weddings. Chelsea Mark wore a pink jacket, as he thought it made him look 'cool and yet reserved' – his words not mine. In fact, it often made him look the opposite: on many occasions when he was chatting up barmaids they would say, "I like your pink jacket."

"It's not fucking pink, it's salmon!" he would explode.

I never thought that a conversation on fashion could be life-threatening, unless of course it involves John Galliano, until the day Chelsea Mark burst into the Jolly Brewers' pub in Parson Green, which was full of Chelsea fans drinking before the game.

He was breathless, "Have you heard the news?" he finally managed to say.

We were all were on tenterhooks, was Captain Birdseye our beloved chairman killed in a tragic fish finger accident? Or perhaps we had a new exciting goal scorer just signed from AC Milan? We all learnt forward waiting for Chelsea Mark to get his breath back and then he broke the news.

"We have a new kit designer who is renowned for his high fashion collars."

In the eighties, many journalists regarded Chelsea fans' outbreaks of violence in the stands as mindless violence. H, who has little time for Chelsea Mark, suggested that the main cause of crowd trouble on Chelsea's terraces was simply due to fans trying to get their hands around Chelsea Mark's throat.

Chelsea Mark is not a violent man, indeed none of us are. I have only ever known him use physical force once in thirty years and even then, he did not realise it. One night he was drunkenly headed towards my house; he was too drunk to find his way home and intended to sleep in my wardrobe, which was all I was prepared to offer. On the way, he dropped his kebab, recently purchased from the soon-to-be-shut-down The Crazy World of Kebabs. As he struggled to bend down and pick it up, he felt a weight on his back. He thought his back had gone, only to find as he straightened up that a small mugger was on his shoulders with his arms wrapped around his neck. I can only think that the attacker believed that Chelsea Mark's silhouette as he was bending down was that of a much smaller man. Imagine his surprise when his victim straightened up to double his height. Chelsea Mark was confused about what was going on; he grabbed the weight behind him, swung it over his shoulder and smashed it on the pavement in front of him. Still he carried on eating his mixed lamb shish, stepped over his dazed assailant and zigzagged his way to my house.

When Bear and Chelsea Mark did occasionally meet, they tried to communicate with each other, but it was always an embarrassing disaster, as they had absolutely no shared interest in anything. Part of the problem was that Chelsea Mark, coming from a wealthy middle-class family in North London's Islington, naturally saw himself as a champion of the common man. This is better than the many others who see those who are worse off as beneath them and there to abuse: but the real issue was that he saw Bear as a representative of the common man and there isn't anything common about Bear. Chelsea Mark, H and I were drinking in a pub a few streets from Bear's home, the comically named Bunhill Row – everything about Bear is surreal, even the name of his street sounded like it came from a cartoon – when Bear trundled in. Everyone acknowledged him with a hello or a punch to his head which is H's way of saying hello.

"Alright mate." Bear greeted Chelsea Mark.

"My man!" said Chelsea Mark. It was a common ritual for each to pretend to be pleased to see the other.

Then there was a poignant moment, as Chelsea Mark offered his condolences on the death of Gorgeous George who had died earlier in the year.

"Yeah, it was a shock, even more so when we found out he hadn't bought the house that Gladys and I live in," said Bear with his head held low.

"But the council can't kick you out," said Chelsea Mark appearing very concerned.

"Don't know, we have no rights say the council, as the house is in his name," said Bear.

"But on his death it reverts to your mum," replied a now increasingly indignant Chelsea Mark, with his jaw jutting forward sensing a new crusade for EMU.

"Nar, it's not as simple as that," said Bear.

For the next fifteen minutes, Chelsea Mark perched on top of his soapbox denouncing the injustices of the capitalist system that would trample over, as he put it, "Simple folk like Bear."

At that point Bear turned to me, "He's right, you can't get simpler than me."

Chelsea Mark was on a roll and he made a declaration to the simple folk standing in front of him.

"Bear, we will all march on the town hall and rip it down brick by brick. Stand up man, we will fight alongside you and Gladys."

"Like fuck," said H, who was more concerned about getting served at the bar.

Chelsea Mark was euphoric now that in Bear he finally had a cause to champion - to him the epitome of the working class, to everyone else in the room a total nut-job.

"Nar, it's not that simple," muttered Bear once again shaking his head.

"The workers united can do it, why be afraid, my man?" declared Chelsea Mark.

"Well, my mum's still drawing his pension," said Bear, bringing the conversation to an embarrassing end.

Bear and Gladys relocated to the coast after the council seized the flat, having discovered that Gorgeous George had been buried for nearly a year.

"How is the move to Folkestone going?" I asked Bear a few months later.

"Well is a bit windy on the coast, sometimes I have to sit on Gladys to stop her blowing away."

A funny remark I thought, until I realised it was probably true.

"Do you give Gladys any warning, before your fat hairy arse lands on her?" I asked. "He who hesitates is lost," replied Bear.

Chelsea Mark preferred to work alone when it came to chatting up women, not only because he was good at it, but also because he thought the rest of us were a liability. On our way to a big cup game in Manchester, we stopped over for the night in Blackpool, where he met and spent the night with a lady from Cleethorpes. The next morning, I bumped into the woman sneaking out of our hotel with the sheets from his bed. Without my help with the door she would never have managed to get his case out. Despite being a complete rake in the style of the late great English actor Terry Thomas, Chelsea Mark would fall in love at the drop of a G-String. He had fallen in love with the bedspread-snatcher and later that night he drank himself into oblivion when she failed to meet him as arranged.

"Look she's probably just late working on her market stall and in this cold weather bed linen will be in demand." I said.

Chelsea Mark still looked depressed as he continued to drink from his pint, so I carried on with my attempts to try and cheer up my good friend.

"Look you can't just meet a girl on a Friday night and expect her to be free for the weekend."

"Oh yeah, where do you think she is?" he asked warily.

"Probably with her one true love." I took another swig of my pint and after a suitable pause added, *"The Finkley Male Voice Choir."*

"Thanks for that, now fuck off," responded my love-sick, pink-jacketed friend.

Chelsea Mark is a complete hypochondriac, always talking to anyone who will listen to the details of his latest aliment. Recently, he stormed into his doctor's surgery demanding immediate attention for what he determined were newly forming blood clots on the backs of his legs. His doctor had not the slightest interest, having seen him already three times that week. Chelsea Mark indulged in a stressed out rant on his doctor's lack of interest in, what he believed, were the last few remaining hours of his life. He believed that the clots had worked their way passed his arse and were inches from his heart.

That weekend, I met Chelsea Mark at a Chelsea home game and his tirade continued about the outrageous treatment he had received; well actually, the lack of treatment he had received. Meanwhile it was a great game with end-to-end football. The crowd jumped to their feet in response to each shot on goal. As this happened, I noticed that each time Chelsea Mark jumped up, his seat swung up and hit him on the back of his legs. The previous Saturday, Chelsea had won by four goals. I sat there, as he jumped up and his seat whacked him once more. I thought one day he might injure himself, or at the very least, it might leave a nasty bruise, or two.

Chapter 18 – *This is How Wars Begin*

Weekends away were a big part of our early twenties, and our trips were just an extension of our anarchic behaviour in London. Bear in particular was always very excited about a day at the seaside and I loved to cross the Channel. We would compromise and head to the seaside for the morning and then take a ferry to Calais or Boulogne in the afternoon. Bear loved the traditional English beach scene: the candyfloss, the sticky rock and above all, the donkey rides and how bandy their legs went when he got up on one. I loved the chaos that would erupt when Bear tried to enter a foreign country. We never once got to Calais or Boulogne.

One trip in particular stands out in my memory, although nothing unusual happened on our way to the coast, apart from Bear jumping off at the station before our final destination, Folkestone. He had seen a sign on the window of the station's café, which read 'AS A RESULT OF A BLOCKED TOILET, CAKES ARE NOW HALF-PRICE DUE TO FLOOD DAMAGE!' We arrived at the coastal town only two hours late, which is on schedule when one travels with Bear, to find that the town was as empty as Bear's address book. Bear was wearing his yellow and black striped bumblebee woolly jumper and his dad's extra-large nylon trousers he got when he was demobbed after the war. The final subtle touch, to complete the look he was after, was the Japanese national flag which he wore as a cape with a crescent of the 'Rising sun' section discreetly tucked up his backside.

A few days later when I related the details of Bear's outfit that weekend to H, he replied: "The best place for it." H's love for the Japanese was on a par with his love for the Germans.

"Well, I've had three candy flosses, four tubs of jellied-eels, two sticks of Hong Kong rock and four soggy cakes from that station café and a strawberry trifle. This day could not be any better and it's only eleven o'clock in the morning," said a very happy Bear. At that point I spotted a large neon sign which read 'DO YOU DARE TAKE A RIDE ON THE BUCKING

BRONCO?' It was perched at the entrance to a little fair that looked like it was modelled on the Magic Roundabout. I had no idea how dangerous it was, but I have always been an inquisitive person and I was happy to volunteer Bear to find out. I had a quiet word with the owner.

"How's business?" I asked.

"It died and then it crawled up my arse." I thought a simple response of 'slow', from the man, who was wearing an eye-patch, would have done.

He was right, there was not a soul around but there was another problem affecting business. For those not familiar with the mechanical horse advertised, the device first came to prominence in the John Travolta film *Urban Cowboy*, which had been a big hit five years earlier. To the owner's great delight I negotiated to pay him five pounds if he would put the next ride on maximum speed and keep it going for a full fifteen minutes.

As to what followed next, I cannot do comic justice to on the written page. Every time Bear mounted the machine, it rotated at something like twenty-miles per hour and shot him into the air, landing him with a loud thud just beyond the perimeter of the floor mat. By the time my five pounds was exhausted, the place was packed with families crowded around the arena, many on their knees crying with laughter at the huge flying bumblebee. Many people had appeared, running over the hills on hearing Bear's screams. I was bent double with laughter, even when I dragged Bear out of the side hoardings by his green and white spotted pop socks. Even the grumpy owner was trying to muffle his laughter and wiped away tears from his one good eye.

I helped carry my bewildered, disorientated and battered friend into what we later learnt was the roughest pub in Folkestone. It was full of very angry unemployed fishermen. As we entered, it was like one of those movies where everyone stops talking and stares menacingly at the innocent tourists who have stumbled accidentally into their lair. I use the term 'innocent' in its widest possible context.

"Don't do anything stupid, try and pretend to be normal," I said to Bear in a low voice, which was the most stupid thing I had ever said in my life.

"Right oh," said Bear, wobbling over to the juke-box.

Trying not to draw attention to myself, I decided to join in with the pub's sport of choice and put my money into the coin-slot to play pool. Bear dropped his coins into the juke-box. Just as I broke with the first pool shot, Jane Birkin came blaring out of the juke-box with the first of many simulated orgasmic cries in Serge Gainsborough's pop classic *Je t'aime*. I knew what H would have done, but from the looks we were getting I knew they were not simply going to let me disconnect the juke-box. No, this lot wanted a clean kill. We were only saved by the fact it took the locals three renditions of *Je t'aime* to decide how they were going to commit the act, which gave me enough time to escape out of the back and climb over a wall, dragging the still-befuddled Bear along.

We ran down the road with a mob from the pub running after us crying out "Kill the fucking poofs." Their cry was a bit strange as the song was about providing sexual ecstasy for the woman in your life. Bear and I hid behind a garden wall for a while until our pursuers gave up the search.

"You excelled yourself this time." I said to Bear. "I asked you to 'act normal' and the town forms a militia for the first time in over two hundred years."

"Well I learnt a lesson. In the next pub, I'll play safe and put on Cliff's Richard's *The Lord's Prayer*." Bear replied.

"Fancy another go on the Bucking Bronco?" I suggested, having found a couple of fivers in my pocket.

"Yer, I think I can get the hang of it now," replied Bear.

Our journeys of discovery and enlightenment led us to some of the more dangerous places in the world. The morning after an intensive night of drinking in Aberdeen, I proceeded to drive a car off a mountain road, along with my passengers H and Cockey. We found ourselves suspended on the edge of the cliff, similar to the coach at the end of the Michael Caine film *The Italian Job*. The one thing I remembered more than the near-death experience and the mobile crane towing our car back up the cliff onto land, was that it was the only time Cockey showed emotion.

"You bastard," said Cockey, looking directly at me as he got out of the car and slammed the door, nearly sending H and I to our deaths.

"He's got some serious issues," declared H.

I planned a little excursion to Hamburg, as I liked Germany, though my preference would have been to see Berlin again now that the Wall had been torn down. Surprisingly, H, Bear and Bourgeois decided to join me. They were very happy to engage with our foreign neighbours; even H, who always made that extra effort to enlighten our European neighbours on their deficiencies.

I reached Liverpool Street station first, followed by H and Bourgeois, but there was no sign of Bear. Suddenly Gorgeous George appeared. Sadly this was the last time we would see him; he died shortly afterwards, as I mentioned, from a heart attack.

"Any sign of my little boy?" he asked and I shook my head. "I thought he might be a little late. Gladys had to pop down the shops as she was short of custard doughnuts for his Wombles lunchbox." He then asked me in a sad, but in a very sincere way, "You will look after our little Bear won't you?"

At that point, I could see Bear bounding down the train platform in full regalia consisting of his favourite Welsh flag, his personally designed 'There's More Out Than In' T-shirt, of which he had fifteen in his wardrobe, and a nappy which looked like it was working in for a baby hippo. His costume was rounded off with a pair of twelve lace-holed boots. His travel bag, we later found out, contained a twenty-kilogram bag of baby talc which had to be applied liberally to his ball-sack.

"It's the size of my bollocks, you see, they keep rubbing off each other and a family pack of *Baby's Bottom Snow* stops me breaking out in a chaffing rash," Bear later explained to the entire bar on board the ship.

Apart from that, his travel bag contained a tube of toothpaste, no toothbrush of course, and the special food parcel prepared and packed by his mum and stuffed into his lunchbox, which had a picture of Great Uncle Bulgaria on the front.

Bear pushed past Gorgeous George and I, as his father shouted out, "Have a good time son."

"Bollocks!" his loving son replied, as he squeezed his hippo-sized arse through the train's door. Gorgeous George turned to me wearing a defeated smile.

"You will look after him, won't you?" he asked again.

I thought that even a team of gorilla-keepers would have their hands full with that blond lunatic. To emphasise my point, the metal arm of a British rail seat flew out of the train's open window across the platform. Bear had broken it off with a swift whack of his elbow; he always had trouble finding a comfortable position and had to adjust his seating accordingly.

"Yes George, of course I will look after your little Bear," I said with a sigh.

As the train set off, I looked out of the window and I saw Gorgeous George for the last time, waving us goodbye in his string-vest, holding his trousers up with his other hand.

As was customary on our weekend trips, one of us had to buy the beer for the journey and this time it was Bear's turn.

"Well here's your *Special Brew*," said Bear offering H, Bourgeois and myself a can each.

"I'm not drinking that loony juice," I said.

"But you told me to get twelve cans of the stuff," replied a bewildered Bear, which was true. "Well, I will just have to drink it over the next few days."

"You can't take booze on-board, you will just have to drink it before you get on the ship," I informed him.

"But I'll be off me bollocks if I drink twelve cans in two hours. Christ it may even kill me."

"Don't worry Bear, five minutes after the shock of your passing, we will be back to our old selves," I added, with H and Bourgeois nodding that they too expected to recover after a few moments of grace.

Bear then proceeded to finish off all twelve cans. By the time we reached the port of Harwich, he collapsed face down on the train platform as the train pulled in and the guard opened the door.

"Bear you're a bit drunk, if customs notice, they may not let you on-board." I said. "Best give me your passport, stand behind me and I will deal with it and get you through customs," I suggested to my friend.

"You're the best friend a man could have, I am so lucky to have you," declared a drunk and emotional Bear, now on the verge of tears as he handed me his passport.

A few minutes later, I turned to Bear, "On second thoughts, it might look suspicious if I hand your passport to the customs officer. It would be less conspicuous if you to do it."

Bear was again overcome with gratitude, "You're a great mate, you always look after me," at that point he walked straight into the door giving entrance to the customs building, knocking himself flat out on the floor. I dropped his passport on his head and jumped over him, along with H and Bourgeois, as we ran off in the direction of passport control.

From a discreet distance, having now got through customs, we watched Bear try to get to his feet and disentangle himself from his flag and adjust his nappy as he now headed towards the customs official. Bear handed over his passport,

"This is the ugliest photo I have ever seen," declared the official loudly for all to hear.

Completely unfazed, Bear replied, "Yeah, I always look like that."

At that point, the passport-controller showed Bear the picture of the pig from the film *Babe*, which I had cut out from an advert and stuck over his passport photo.

After thirty minutes of trying to explain why his passport contained a picture of an Australian pig, Bear was allowed through. Bear then exploded and give chase in our direction. On reaching our cabin, he decided to make himself even more at home than usual, firstly by kicking the sink off the wall. He then focused his interior design expertise on our bunks. He repeatedly jumped on the ends of them until they were all sloping at forty-five degree angles.

As we continued our voyage, Bear grew to like the idea of an interchangeable passport photo. So by the time we arrived at Hamburg, I glanced at Bear's passport to see he had a picture of the Mohican-haired lead singer from a punk band called Sig Sig Sputnik, sellotaped over his passport photo. Bear, unsurprisingly, was led away at German customs.

Bear was not detained by Hamburg customs for long; no one wanted to handle an English lunatic. When H, Bourgeois, Bear and I reached the centre of Hamburg, we spotted a street which was sealed off like a film set. But any resemblance to *Coronation Street* abruptly ended at the big sign stating in large letters 'NO LADIES'. Officially, the sign was to deter decent respectable women from accidentally entering the street, but there was no sign saying 'NO FUCK-UPS' so we entered. When the four of us entered the street on a bitterly cold winter's morning we found ourselves in an open air cobbled precinct, with prostitutes on either side getting ready to offer their wares in the windows.

Being about ten o'clock in the morning, the women in the windows were not expecting any customers and were relaxing reading the papers or doing their nails. One was even washing the windows, and didn't acknowledge the existence of three freezing pasty-faced Englishmen walking by. Then she spotted Bear waddling behind us in his bumble-bee jumper and a brown woolly Cossack hat, looking like the Sugar Puff Honey Monster. She started to bang on the windows, calling her colleagues to look at the outrageous spectacle on the street, which considering what they must have witnessed over the years was quite ironic. It must have appeared to them like a game of Russian roulette, in which one of them would end up having the legs broken off their bed in the three minutes it would take to entertain him.

Within a minute, all the previously half-asleep working girls were pressing themselves up against the cold windows to catch sight of the unusual character waddling down their cobbled street. It was soon obvious to the women in the street that none of us were interested in employing them. We were just curious, up for a dare and paying a visit to gawp at them. The tables had turned as Bear became the spectacle and the prostitutes in the windows were pointing and staring like excited teenagers at a pop concert. The irony was not lost on us, but Bear was oblivious and he just carried on combing his hair so that it retained its shape of an enlarged knob-end, despite the light rain. He nonchalantly continued along the street, listening to Cliff Richard's *Wired for Sound* on his cassette player headphones. His cassette player was the size of a home stereo powered by some form of crank-handle.

There was one other visitor to the street, a born-again Christian at the other end of the pedestrian precinct, who was denouncing the occupants of the street for their licentious behaviour. He was quickly manhandled out of the street by a policeman, but when as spotted the Bear he simply shouted in English, with a German accent, "Fuck me!"

I only have one ritual I adhere to whenever I travel. I go out of my way at airports to approach members of any obscure religious sects looking for new recruits. They're always surprised, but grateful to have someone volunteer personal contact details. Of course, I've never given them my details, I give them Bear's. On my return home, it is always good to hear that Bear has been receiving regular visitors, sometimes on a daily basis. A confused Bear met me at the station when I returned from one trip and told me that he was glad to get out of the house, as he had been filling in questionnaires all day for the 'Mormons', 'The Orange People' and 'The Daughters of Jesus.'

"Does everyone have to do this?" he asked.

"Yes, everyone fills in the forms," I said, "but after you've done that and they finish their tea, you drop your trousers and bare your arse to them until they are ready to go."

"Will do," replied Bear.

Most English supporters, whatever the sport, are perfectly well behaved when they watch their teams play on foreign soil, but a few get take the opportunity to do all the things they could never get away with at home. A perfect example of this occurred when H and I went to Amsterdam to see England play Holland in the 1994 qualifier for the World Cup. When we arrived at the train station we decided to have quick pint in a bar across the road. I went up to the barmaid, who seemed stoned due to the cloud of smoke in the pub, and asked for two beers. My request seemed to completely bewilder her; she untied her apron, put it on the bar and left the pub never to be seen again. I then turned around and noticed three English guys sitting around a bong and three ashtrays full of cigarette stubs and the tabs of spent *Rizla* papers.

"How long have they been here?" I asked the manager who now took over the bar.

"Three days, without a break," he replied.

"The whole town is drugged off its tits," observed H.

The city was certainly not as laid back as usual that particular weekend, as H and I quickly discovered. We hit the first bar in the city centre, and found television screens showing the running battles between the England and Dutch fans. Violence was breaking out, not just in Amsterdam, but also in Rotterdam where the match was being played and the Hook of Holland where England supporters were arriving by ferry. It was absolute bedlam. Captured English rioters were rounded up by the police in Rotterdam and put on trains to the Hook of Holland, where they joined the new arrivals and headed back into Amsterdam and got the next train to Rotterdam. It was perpetual motion, hooligan style.

H and I never had any interest in hooliganism so we decided to avoid the match altogether. We had no tickets anyway. We stayed in a small underground bar called De Burgs, near the town square. After watching the game on the television with the clearly ecstatic, but friendly locals, Michael our barman received a telephone call.

"Yar, Yar, Yar, Ah Yar!"

"What's happening," I asked Michael.

"Your countrymen are trying to level our city."

"Sorry Michael, but our country has been doing it for years, in the old days we dressed it up and called it colonialism."

"Fucking leftie," said H.

Michael's information was correct. When we went back upstairs, we found that some England fans had pushed a police car through the front window of the main police station. The majority of us, who had no interest in violence, went off to see the punk band the Buzzcocks play in a venue near the bar. Mixing with the local Dutch people that night did at least cheer us up somewhat after our dismal failure to qualify for the World Cup.

The following year, despite England's failure to qualify, H and I still headed to New York to see the World Cup. Ireland had qualified, under an English manager, Jack Charlton, and the Irish headed over in full force. The Americans had no understanding of football, or, as they call it, soccer. Indeed, they had no idea what trouble they were letting themselves in for by offering free booze on all trans-Atlantic flights to the tournament. The United Airlines cabin-crew on our flight could barely keep up with the demand and a drunken Irish passenger had to be restrained when he tried to open up the emergency door in mid-flight to go to the toilet.

H was suffering from some form of alcohol poisoning by the time we got to New York, so I headed off alone to see the famous Little Italy and experience the ambience of the Italy against Ireland game. Little Italy has now been swallowed up by New York's Chinatown, as most Italians have made their money and buggered off to New Jersey. The atmosphere in the restaurant I chose was electric, even though it was full of tourists and Mexican waiters. A BBC film crew entered and the female Scottish reporter started to talk in Italian to the customers. The reporter then came to me and spoke to me in Italian, but I informed her that I was English.

"What are you doing here?" she asked.

"I am enjoying the World Cup, even though due to a slight physical deformity, we are not in it."

"What physical deformity?" she asked.

"Having a huge arse called Graham Taylor as England manager," at which point the cameraman, who was the only other Englishman present, had to stop filming. He couldn't hold his camera steady for laughing. I thought no more of it, but a few days later I got a call from an ex-girlfriend in Scotland, who told me that my interview was being shown repeatedly on STV television. However, they had edited out my comment about the manager and implied that I had gone out to watch the World Cup, not realising that England hadn't qualified. Instead of laughing at my plight, she was not happy at being repeatedly ridiculed by her friends for having gone out with a "retarded *Sassenach.*"

Later on our first night in New York, H managed to get off his sick bed and joined me in a bar called Kennedy to watch the Brazil versus Germany game. At the same time, the worst cocktail barman in New York performed in front of us; drinks and their contents flew in all directions. Joe was a great old barman in the traditional friendly sense and he asked whether they we wanted anything else to go with our beers.

"Two waterproof jackets and matching goggles please." I replied.

I felt bad later, when he confided to H and me that he had Parkinson's disease. He was so loved by the customers that the owners dismissed the damage he caused every night by saying he was just getting clumsy in his old age. If a gorilla had put on an apron and took over the next shift, it could not have caused more damage. We kept Joe's secret, as we admired the old guy for trying to carry on working and you could not fault his cocktails on the rare occasion they made it into the glass. I then overheard one of the managers apologising to a customer who had received the contents of a cocktail over his white shirt. I realised that it was no secret to them that Joe was seriously ill. Such acts of kindness drew me closer to America, perhaps because of its brash image as a sink or swim society, kindness had a greater effect on me as I didn't expect it. Years later, I worked on a mental-health project in Manhattan and the kindness and support I experienced there over months, between the staff and the 'guests', I have rarely witnessed anywhere else in the world.

That evening with Joe, a good drinking session was had by all in the bar and I think I vaguely even remember a gorilla in an apron serving me chicken-wings. A few Americans came up to us and started to talk to us about the game on the TV screen.

"This is a sissy sport," said one.

After fully analysing this statement, H replied, "Bollocks."

I re-phrased my friend's comment, but the basic content was the same.

"A sissy game, well unlike Baseball and basketball, it is a contact sport and unlike American football we can go backwards as well as forwards, without changing the team."

"My ass!" replied Conrad, the biggest yank.

"Also," I added, "The World Cup involves every nation, while your World Series involves the US and one college side from the Canadian town of Pig's Knuckle Edmonton." I still think H put it better.

We received an invitation from Conrad and his friends for the following evening, "Guys, if you want a real game join us in the Red Lion in Bleecker Street to watch the basketball final. It's the New York Nicks against the Houston Rockets."

The offer was accepted by two English guys and the twenty drunken Irishmen behind us.

The following night, H and I went to the Red Lion to watch basketball with Conrad and his friends, surrounded by now double the previous night's quota of drunken Irishmen. Halfway through the game, we were suddenly watching the longest advert I have ever seen; bar Nicole Kidman promoting Chanel No 5. On the TV screen, a four-wheel drive car was being followed by twenty police cars, but that was all, no special effects, no semi-clad women and no bleached toothy smiles.

"I give up, this can't be an advert, have you made a substitution and he is being escorted to the game because of his coke habit?" I asked Conrad.

Conrad and his American companions were puzzled, but their bewilderment soon turned to embarrassment.

"I just got a call to say it's OJ Simpson. The police are in pursuit as he is wanted for the murder of his wife," explained Conrad.

"So they have stopped showing the second half of the biggest game of the year," I queried.

"Err! Yes," replied a subdued Conrad.

"As terrible as it is, if this was the FA Cup Final in England and Prince Philip had strangled the Queen Mother wearing a pair of false breasts, a rolling message would appear under the game, saying "QUEEN MOTHER CHOKED BY HIS ROYAL HIGHNESS PRINCE PHILIP IN BRA. EXTENDED NEWS AFTER THE MATCH ANALYSIS." I continued, "If they stopped televising the game, the Queen Mum would have a few TV producers for company."

At that point all the drunken Irish fans starting singing, "OJ, OJ, OJ, OJ, OOOOJ, OJ, OOOOJ ..." rather than the football chant of, "Ole! Ole! Ole," all to the bemusement and irritation of our American friends.

My travelling companions were not only H and Bear. I joined Chelsea Mark and his friend Morris to visit his parents in their villa in France one summer. That day Chelsea were playing the European football Super Cup, which was shown live that afternoon in a local bar. As we were the only Chelsea fans in town, it goes without saying that we were pretty drunk by the time the locals came out for their first drink later in the evening. Just when we thought the sleepy little town was closing for the night, we saw a sign for a disco and the three of us fell into it. Chelsea Mark stripped to his waist and threw his shirt across the dance floor in the manner of Travolta in *Saturday Night Fever*. The locals were surprised, as the music hadn't started and many were still eating a little supper of bread and cold meats with their parents and grandparents. It was a lively evening, but the next morning Chelsea Mark and I were hungover, much to the amusement of his mum and dad. Morris then appeared, swigging from a bottle of thirty-year-old scotch.

That afternoon Chelsea Mark's parents were invited to lunch with the town's Mayor, whom they knew well. The invitation was extended to their three English visitors, who rumour had it, had made a big impression at the local disco the night before. By the time we arrived at the Mayor's house, Morris had lost the power of coherent speech. We were introduced to the Mayor and her family who were all very pleasant, and the Mayor herself had made a traditional olive cake which was delicious. So delicious, that it seemed to trigger a response in Morris, who replied to every question for the rest of the day, "Excellent olive cake."

The questions put to Morris ranged from, "Will Britain adopt the euro?" to "Do you have any animals?" but nothing would deter Morris from his appreciation of the Mayor's olive cake. Chelsea Mark and I did not fare much better in the conversation stakes, but our hosts were the epitome of politeness and, with the help of Chelsea Mark's parents, all went well and our increasingly drunken state was ignored.

Morris decided to investigate the garden and we left him to his own devices as he disappeared to water a cherry blossom tree. When we heard a crash, we had no choice but to acknowledge that some disaster had occurred. As we turned our heads, we saw an example of how man's interaction with nature can have a destructive impact on the environment. Morris had passed out and landed face down on the children's rabbit hutch, flattening half of it. Marcel, the rabbit was unharmed, even though his home, indeed his world, had in one fell swoop been reduced to half its size. Indeed Marcel was now happily nibbled away on his new friend's bulbous red nose through the recently amended chicken wire partition. The family's kittens too were enjoying their new plaything and were excitedly jumping over the head of their comatose guest. No one knew what to say, but the Mayor returned with some more olive cake. Though we knew Morris to be quite partial to her baking, he was getting on so well with his new furry friends that it was a shame to disturb him.

Chapter 19 – *"Why Are There Not More Lesbians in the World?"*

What do women see in men? As I relate the tales of my male friends, I wonder why more women don't just give the local sperm bank a phone call when it's turkey basting time. When I look at what girlfriends have had to put with up from me, I am amazed that only two have tried to stab me over the years. One of them did try twice though but, like recording the real success of an Internet site, I go by single visits rather than repeated hits. Relationships have always been tricky things and if there is a formula to making them work, I have yet to discover it. One of the biggest problem is that women are complex, while men such as I are very simple creatures. In conversations between the sexes, men usually reach their limit when the main points are covered. Then we reach saturation point, turn off and just nod every now and then, in the vain hope that our silence won't be noticed while women continue to explore every possible eventuality.

This is not a 'lad's' story where women only have a functional role as child bearers, sexual partners or an audience to laugh at boys' tales of idiocy. Far from it, the women I call friends as well as rogues don't just hold their own with my male friends, they often streak past the boys without breaking into a sweat. If anything, it is the men that come off rather badly in this book and, if you narrow it down further, I come off worst of all.

The female rogues I know are not 'laddettes', a term commonly used in the tabloids, who are simply women trying to be men by drinking just as much, being as sexually shallow as many of my sex and turning violent when the England football team fails to win the World Cup every four years. Neither are they WAGS, another tabloid term, which I think means Women against Sobriety. The female rogues I know have more in their heads than the latest discounts at Harvey Nicks, fake tans and TV's *Pop Idol*; mind you, that doesn't mean that they don't enjoy those things too. I have had sexual relationships with very few women mentioned in this book, as the majority of my relationships are platonic, but either way I am lucky enough to call them friends.

If they share one common trait with the male rogues, it's that they will not succumb to peer pressure. They live by their moral code and do what they believe is right. Their challenge is even greater, as peer pressure is greater for women than it is for men. Family and friends expect more of women than they do of men and not just in the areas of dignity, style and family. I grew up in an environment where women were expected to put up with anything for the sake of their family, and divorce or separation was seen as a failure on the part of the woman rather than of the marriage.

When I worked in the park, I met a seemingly very demure blonde Irish girl called Maureen, who was a good friend as well as an ally in taking on Coldman. She worked part-time in the main office and was training to be a lawyer. While I just joined in the anarchy, Maureen passed her law exams before eventually leaving the park. That was not the limit of her ambitions though, and she made it clear that she would only use her legal skills to represent women who had been victims of violence. In that, she has succeeded and she has one of the highest conviction rates in prosecutions of rape and domestic violence cases in her field. Maureen is the first of the female benevolent rogues, possessing the core traits of moral courage, independence and, let's not forget, the ability to commit the most inane and stupid acts. OK, on this last point all she did was get on my motorbike, but bear in mind that H or Larry have never dared do that.

I really got to know Maureen when she was tasked by her law firm with teaching a group of kids from the East End about their rights under the law. This did not conflict with her ambitions to support female victims of crime, as this group had all been arrested at some time for possession of drugs. Her purpose was also to highlight to them that their continual law breaking would lead to an escalation of their criminal activities and, for some, increased drug dependency, potentially with their women and children becoming victims. After a few weeks of teaching this group, Maureen gave me a call and asked if I would talk to the class, as she was having trouble getting through to them.

"Why me?" I asked.

"I know you don't do drugs, but you're worse than all of them put together, so they may listen to one of their own." Maureen argued.

"Well, after such a gracious offer, dressed up in legal speak, how can I say no?"

I went along with Maureen to meet her group and, though they were boisterous, after a couple of hours talking to them everything seemed to go OK. Later, on the train back I related my positive view of how it went to Maureen. She agreed, but with a certain hesitation.

"Well yes, they did relate to you," she said, "incredibly some even took notes and I liked it when you said that a real man is one who channels his strength into changing society for the better, rather than lashing out at those weaker than yourself and that that was the greatest challenge of all for any young man."

"Maureen, you're wearing that little frown that I know so well, so come on what did I do wrong?" I asked.

"Your advice on the courts, which was not the message I was trying to get across."

"Which bit?" I asked.

"Well, it was it the part where you advised them that if they had to go to the courts, to opt for Crown rather than the magistrates court," Maureen frowned. "You then suggested Snalesbrook, as the jury is normally selected from Tilbury docks area where criminal activity is the norm rather than the exception. Hence, it is more likely they would get a 'not guilty' verdict followed by a nod and a wink from the jury."

"OK. Good point, well put," I said as I looked away from Maureen's embedded little frown, out of the train window at the drab landscape of East London that offered little hope to its youth.

One time, I had to pick Maureen up from the airport on my motorbike after one of her frequent visits to her family in Ireland. When I pulled up at the parking bay there she was, with her short blonde bobbed hair hanging just above the frilly collar of her white lace dress. Maureen looked like she had stepped out of the copy of *Vogue* protruding from her bag. It was a shame that the carriage that awaited her was my 750 cc Kawasaki Zephyr to take her home to her flat in Kilburn.

"You never said you were collecting me on your motorbike."

"You never asked," I replied with a big smile. I had grown to like that little frown of hers.

"How was Dublin?" I said, as I removed my bike helmet.

"Well I had a little bit of hassle getting through Dublin customs," she replied in her usual quiet, mannered voice. She continued, "Well you know I have six brothers? They wanted me to bring some contraceptives to Ireland, as they are banned by the Catholic Church."

"Yes, the backward bastards," I commented. For once, Maureen, who was a staunch Irish nationalist, did not bite, so disappointingly her cute little frown did not appear.

She continued her story, "Well I got stopped at customs and they opened my case to find a number of contraceptives inside."

"Superb," I replied.

Maureen ignored my remark and told me of the conversation she had had with the customs officer.

The main customs officer had asked her, "You realise you can only bring contraceptives into the country for your own personal use?"

"Yes," said Maureen.

"How many are here?" the customs officer asked.

"Three hundred," replied Maureen.

"For personal use?" he replied loudly to make sure his colleagues could hear. He had relished her embarrassment. "How long are you here for?"

"Oh, just the weekend," Maureen answered innocently. "The amazing thing was that he let me keep half of them."

"Do we need to stock up at the petrol station on the way home," I replied, trying to entice her little frown.

"Kindly fuck off," she replied in her beautiful soft voice and, to my delight, her little frown made a brief appearance.

What was even more ironic about this incident was that Maureen was gay. This was a secret that only a few knew and I was the only one in the park who had any idea.

One day she had left a message on my clocking out card that she wanted to meet me. We met in a pub in Camden Town and she was crying.

"She's left me," she informed me, through her tears.

"I am sorry, Maureen, but why has she left you?" I asked, knowing that she was referring to Ann, her longtime partner.

"She says she has found this guy," more sobbing "and though she says she does not love him and she still loves me," more crying then she continued, "she is leaving me for him."

"What did you say?" I asked sympathetically.

"I said I can give you more than he can, no one will ever love you as much as I do," her eyes were red.

"She said you can't give me what I need at this moment, you haven't got a …" Maureen could not speak the words.

"What just tell me, it's OK," I added as I held her hands.

"I haven't," followed by more sobs, "I haven't," more tears …

"What? Just say it, I won't make a joke, I'm not a complete idiot, you know, I have an understanding side."

Maureen finally managed to take a pause in her crying, "… got a cock."

I roared out laughing and took around ten minutes to control myself and refocus on

Maureen's angry face. She was frowning.

"I am sorry Maureen, it's just not what I expected," I finally answered as I tried to compose myself, but I failed and laughed uncontrollably once more.

Her girlfriend wanted a baby and for that she was prepared to settle down with a man. Times have changed, nowadays gay couples can do this through surrogacy or IVF, but at that time those options were not available. Poor Maureen, I think she wanted a child too, but with Ann.

Another girl friend, Tracy, I met many years ago in a club and immediately was struck with her direct manner. Tracy takes no nonsense from anyone, especially men. She really does use them and abuse them. I have never seen *Sex and the City*, but I am aware of the man-eating character named Samantha. Whereas the fictional character playfully flirts and then devours her men, Tracy has no time for such frivolity. She just grabs them by the balls, takes them home and throws them out in the middle of the night after, vampire-like, she has drained them of their fluids. Unlike the vampire she leaves no marks on her victim's neck, just a card from the local cab company between the cheeks of his arse.

I probably argue more with Tracy than any other woman I know, but our mutual love and respect is always there. Most of our arguments are based on her fearlessness. There have been times when she has disappeared for a night with two men she has only just met. She will not hear my protestations and I am frequently castigated by her, "I am a free spirit and I will do what I like."

Once when I was with Kathleen, an ex-girlfriend, Tracy showed us her holiday photos. The photos were of the usual number of Tracy's conquests during her two weeks in Bulgaria. Kathleen and I raised our eyebrows when we came to the picture of Tracy running along the beach with two young men, all of them were completely naked. Tracy apologized to Kathleen for the appendages shown saying "The water was cold that day."

Kathleen just smiled politely, while I looked out the window trying to suppress a grin at the actions of Tracy, a uniquely qualified rogue.

Sam is probably the only person I know who can handle any situation and to my male friends she is the ultimate rogue. To me, as I know how she has helped me resource each of my relief convoys, she is a benevolent one. In the twenty-five years I have known her, I have never seen her overwhelmed in any situation. As far as I know, no man or woman has ever got the better of her and I expect no one ever will. She was just as wild when we were at University. I met her when we were enrolling for courses and we immediately clicked like kindred spirits. We proceeded to the student bar for the rest of the day. The service that day was excellent, as the all-male bar staff focused on Sam, who has the kind of beauty which breathes a rare energy into men.

"Hey gorgeous, over here," she called out to the barmen. Service improved even more when she approached the bar and informed those in front of her, "Scatter, come on scatter, lady coming through."

In my second year at college, Sam popped into my flat, only for my girlfriend at the time, Kim, to completely freeze up. Sam said, "Hello" and didn't notice when Kim did not respond. When Sam had gone, Kim managed to retrieve her voice.

"That fucking woman scares the life out of me. Is she is a friend of yours?" She cried.

"Absolutely, I have no better friend, but she does have a gift for upsetting people. What has she done now?" Kim related her tale of the one and only time she had encountered Sam.

"When I started my first week, I was getting into a taxi after the Fresher's Ball," she shuddered, recalling the frightful events of their encounter. "Well, I heard this voice behind me shout out, 'How long have you been at college?' and without turning around, I replied, 'This is my first week'. Then someone grabbed my feet and pulled me out of the taxi, I landed face down on the tarmac. I looked up and saw this woman step over me and get into my taxi saying, 'Sorry love, but I have been here a year, so I take precedence', and she drove off with some bloke whom she appeared to have in a head lock."

I gave poor Kim a big hug and said, "Sorry about that, but that's Sam. I will protect you and I will make sure in future she focuses her aggressive nature on the college's women's rugby team instead."

My remark related to Sam's feud at the time with the captain of the women's rugby team who had made a pass at her in the team showers and received a swift, "Fuck off you fat hairy dyke!" for her troubles.

Sam dominated all of the men I ever saw her with, and I was worried that due to her strength and nature she would never find an equal; but she did. That man was Alan, who is very different to Sam, but he makes her happy and they have a wonderful family; I am honoured to be a godparent to one of their princesses.

When Sam and Alan decided to get married, the happy occasion resulted in a suitably outrageous wedding in the beautiful city of Bath. I had been to Sam's twenty-first birthday party in Bath a few years earlier and, after the party, had helped smuggle her into a nightclub, as she was barred from every venue in town. Cockey was with me for her party and Sam had told me afterward that her mother had interpreted his aloofness as a sign that he was gay.

H joined me when I was invited to Sam's wedding, as H had met Sam when he came up to see me at University and naturally they clicked as both are similarly direct and unapologetic about life. On the way into the wedding reception, we encountered Sam's mother and introduced H to her.

"I'm H and I'm not queer," he informed the mother of the bride and walked straight into the reception towards the bar. Sam's mother stood open-mouthed, but at least she would never get H confused with Cockey.

Sam's partner in crime was a girl called Anna. She is a friend but we often spar together, and years later, she is still barred from every club in town. She brought her new boyfriend, Lionel, to the wedding reception. Lionel put me straight immediately, by saying that he was French and the French pronunciation was 'Leon-nal' and as he put it, "The English, in their ignorance, always get it wrong."

I thought this would be good fun and got to work finding out a little bit more about Anna's new boyfriend who had such a high opinion of himself. Anna had briefed him about me and he kept coming over at different moments as the drink flowed throughout the evening. Each time he impressed on me his importance as the head of a publishing empire. I plied him with drinks on each of his visits and we stood each other, drink for drink. After ten drinks, it was time to find out a little bit more about Lionel.

"Publishing eh?" I asked. "I bet you worked hard, started at the bottom and worked your way to the top through superior intellect." Massaging someone's ego is always a good way to get someone to betray more about themselves than they intended and I rightly assessed that my irony would be lost on Lionel.

He opened up quicker than a hippopotamus at the dentist, "I came from a rough housing estate, like you," I looked at Anna whose tentative look alongside his comment on my roots, confirmed that she had briefed her boyfriend thoroughly.

"Ah a self-made man, what was your first job in publishing?" I asked.

Anna froze, but Lionel stabbed me on the chest with his finger and declared, "I was the one who put the red stars on penetration shots in porn magazines."

"I thought so, you started at the bottom and worked your way up" and with that comment my work was done: I went to congratulate Alan on marrying one my best friends. Meanwhile, Anna tried to drag Lionel away from H who had now picked up the conversation and was keen to learn more about the type of publications Lionel's company produced, in addition to negotiating a thirty per cent discount for himself.

I met a good friend of Alan's that night, who called herself FuFu. She was from

Northern Ireland, and though she tried to physically attack me on numerous occasions when we were going out together, it was a great few months.

I only ever introduce girlfriends to H and Bear after I have known them for some time, when I'm sure that they are strong enough to take the shock. But Fu-Fu had met H at the wedding, so when H suggested we all meet up for a drink back in London in a couple of weeks, I foolishly saw no harm in the idea. However, I forgot that as it was the end of season drink-up, that Bear would be present.

At the gathering a few weeks later, Bear walked in and after a while Fu-Fu turned to me and said, "Bear looks a total fuckwit." Well, Fu-Fu was from Northern Ireland so she never minced her words, "but he's a pussycat really once you get to know him." I thought it was still early, but I hoped she was right. Just then Fu-Fu reached for her lipstick which was in her handbag just under the stool where Bear was sitting.

As she bent down, I heard Bear say, "While you're down there love," as her head was now in line with his crotch.

Fu-Fu always wore a scarlet-coloured jacket, which should have told me something, and as she drew herself to her full six feet in heels, I saw that her the colour of her face now matched her jacket. What stopped her from smashing a glass on Bear's head is one of the great mysteries of life, but I did catch her arm as she went to punch him in the head. Bear was oblivious to the threat and just carried on annoying other people in the bar.

Only once did Fu-Fu and I discuss Northern Ireland, when we watched a television news report on the latest atrocity in her homeland. I asked her, "What would you do if civil war broke out in the North and English troops pulled out?"

"I'd do what I'm trained to do, join up with the Ulster Defence Association (UDA) and shoot as many Catholics as I could." As we lay on the sofa together I decided not to break such a romantic moment by mentioning my roots.

If H and Surf, more of him later, are my adopted brothers then Hannah and Gertrude are my adopted sisters. To call a male friend a brother is common parlance, but to call a friend of the opposite sex a sister should be equally acceptable. Can men and women be friends without a sexual motive? Yes, they can and that is how my relationship stands with Hannah and Gertrude, hence there are my 'sisters'; I love, protect and respect Hannah and Gertrude as if they were my family.

Hannah is from Australia and has no problem telling me straight if I am out of line. She frequently refers to me as an idiot, due to the scrapes I get into. We are close because we share the same values and beliefs. Though Hannah returned to Australia a few years ago to be with her family, our friendship is stronger than ever. A real friendship doesn't depend on seeing each other all the time, or need to be frequently replenished with hugs and laughter. True friends are always in your thoughts.

Gertrude, like Hannah, is a remarkably strong woman. She is beautiful and intelligent, which for a woman can be a difficult combination, as boyfriends are usually proud at first to be associated with such a beautiful woman. Then the insecurity kicks in. Gertrude is one of the few experts in the country who understands the potential in new energy sources, yet her bosses, both men and women are threatened by her drive and ability. It seems that both sexes would prefer her to be simply 'easy on the eye', as one director described his recruitment policy to me the first time we met; I do like it when someone makes it perfectly clear that we must be enemies right from the start. What is remarkable about Gertrude is her spirit and how she comes back even stronger after each knock and challenge. Never once have I seen her head bowed or her allowing herself to wallow in self-pity. She embraces life, but she is no one's fool.

I can give an example of Gertrude's enthusiasm. Whenever she calls me, I stop the nearest passer-by and ask them to hold my mobile phone for a minute. They absorb her opening lines.

"HELLO IT'S ME! GUESS WHAT HAPPENED TO ME TODAY? LIFE IS GOOD!"

Once the initial outburst is out of the way and Gertrude's tone relaxes to the level of the shout you use to hail a taxi, I retrieve my mobile phone. I always feel slightly guilty when I watch the people who have absorbed the first few seconds of Gertrude's dulcet tones, zigzag off in shock.

Perhaps I should define what I call friends, rogues and benevolent rogues. Hannah has no fears about rocking the boat if she has to, neither does Gertrude, but they don't do so lightly. They try to work with people and both adopt a sensible, sensitive and thoughtful approach to getting things done. They have never met, but both have worked in projects around the world offering what assistance they can in some very dangerous hotspots. Gertrude has worked in South American ghettos at extreme risk to herself and Hannah worked as a volunteer Red Cross first aider on match days at Chelsea's Stanford Bridge. They are 'benevolent', and they are kind, strong, characters, but they are not rogues. Bear and other characters you will hear of later such as Brains and Jockey, are rogues. Rogues trample on toes all around them, even putting in extra effort to reach those at a distance or those encased in steel toecaps. They never, for even the briefest of moments, take into account social niceties; they just do their own thing, often causing uproar and more commonly causing nausea. But, though Bear and Jockey are my friends, they have never done anything to warrant the label 'benevolent rogues'. Friends like Maureen, Rory and Sam are what I term benevolent rogues, because they cause mayhem but, like Hannah and Gertrude, they've helped others for no personal gain. All are still fundamentally rogues, and I have lost count of how often they have joined in a conversation with people they don't know, and don't agree with, using the most expressive word in the English dictionary, "Bollocks!"

You will not find the *Wizard of Oz* or *The Sound of Music* in my DVD collection, so not all of my relationships with women have been platonic. On very rare occasions, women have even made the first move. Alcohol is a wonderful thing, isn't it? When I used to drink in The Arch, a nurse from the nearby hospital always used to appear next to me and make it clear that it wasn't my brain she was after. Unfortunately for both of us, she was not my type. For a start, she smoked like a publican and was so thin that she couldn't walk across a cattle grate without fear of disappearing. One night, after a heavy session in The Arch, I couldn't find my keys and another nurse I knew said I could sleep on her floor for the night. That was fine until in the early hours I felt a woman slip under my quilt. I thought it was my hostess, until I got a whiff of ashtray, turned on the light and there was the nurse from The Arch. She was naked, her face was up against mine and her hand was moving below.

"Any chance of a kiss?" she said, taking her fag out of her mouth.

"What about six months in jail for sexual assault?" I replied, jumping up. I dressed quickly, tightened my trouser belt as tight as it would go and spent the rest of the night in the nearby café, The Glue Factory. The faded black and white pictures on the wall of various types of breakfasts, interspersed with winners of the Grand National, implied that the café sold food, but I thought it was best to stick to coffee.

As I say, for a woman to approach me was unusual, OK I admit it was only that once, so if you wish to be pedantic, the correct term is unique. Usually, my encounters were more along the lines of Patricia. She was a tall Irish woman with jet-black hair, who launched a verbal attack on me the first time we met. She turned to me in a bar and accused me of looking at her in a sexual way, which I denied.

"I didn't even notice you until just now," I said, "when you turned and accused me of looking at your breasts. I am not a breast man; mind you're beautiful when you're screaming profanities."

That made her even angrier. "There is more to women than their looks!" she screamed.

"I agree, so let's talk and see if there is more to you than a desire to set fire to my head."

From that moment, we started going out. I did not realise that Patricia was a fervent Irish nationalist, until H came around to my flat a few months later and I introduced them. Patricia refused to speak to him. Now H is not the most ardent conversationalist at the best of times, but even he nods his head when introduced to someone. Later, I asked her why she refused to talk to H and she replied, "I don't speak to enemy occupiers of my country."

How someone as intelligent as Patricia could be so bigoted, was beyond my comprehension. I had seen the aftermath of a 'nationalist bomb' while I was working in the park and then as now I failed to see how the dismembered limbs of young army bandsmen furthered the cause of a nation's liberty. Patricia was wonderful in so many ways, but when her prejudice against anything English became visible, all we ever did was argue. Patricia, I was to learn was angry with everyone, but I never got to know her well enough to learn why. The strange thing was that if she was arguing and her opponent stopped, or started to agree with her, then she was disappointed. The argument was the main thing, rather than the result.

The relationship could not survive and the last time we saw each other was when I joined her family for a Karaoke night in her local pub. Her younger cousin Helen had Down's syndrome; I prefer 'had' to 'suffered from', as the kid was so happy that she didn't appear to be suffering. I gave Helen a hug whenever I saw her; she was the only woman I have ever known who never took 'no' for an answer. Helen always wore me down until I gave into her demands and that night she got me up dancing. It made her laugh as I danced liked an epileptic on speed. She loved singing Karaoke songs, but that night one drunk tried to impress his mates, by shouting abuse at Helen while she was singing. I decided to shut him up, as I saw Helen falter mid-song as his abuse grew louder. I threw him out of the pub. Everyone quickly went back to enjoying the night, but Patricia went ballistic.

"How dare you patronise Helen? You had no right to act for her or my family," she yelled.

"What about freedom of expression for the poor stripper you knocked out with a pint-glass in your dad's pub last week?"

"Fuck off, she degraded women," Patricia yelled back.

"She needed six stitches, I know that." I said.

My last memory of Patricia was of her trying to beat up the one Karaoke judge who did not vote for Helen, even though Helen had won. I said my goodbyes and gave Helen a hug; I knew it was the last time I would see her. I hope that Patricia became a little less angry over the years.

I have a long catalogue of failed attempts at trying to chat up women. I have been told to 'get fucked' more times than I care to remember. One of the most memorable brush-offs was when I offered to buy a girl a drink and she replied, "Sorry, I only drink my own piss." I missed the point entirely, thinking that she was just like that weird old British actress who had recently said she drank urine as it was beneficial to her health. I offered to get her a glass, but her look made me realise that her response was along the lines of the 'piss off' variety. As I got the message, I moved on, though I laughed a lot.

My bumpy journey along the highway of relationships continued and in The Arch, I met another beautiful Irish girl called Nula. I was lucky in this encounter, I wasn't found squashed flat like the rabbit that stopped to wonder what the bright lights heading towards him were. The night Nula and I met, there was the usual type of band in The Arch, a bunch of guys on electric guitars, but slightly unusual in that the sole woman in the band was playing the triangle with a lollipop stick. Now, under new management, The Arch house policy was only to book bands that had a woman in them and the week before I had seen a woman play the washboard with her foot. The manager explained to me the reasons behind The Arch's new equal opportunities policy.

"The customers love a bit of skirt. You can't expect them to be looking at just a load of hairy-arsed *eijits* all fucken night."

The band played the usual set, which consisted of murdering popular rock numbers, especially pop anthems by U2. At one stage, the lead singer burst forth with a rendition of a "that popular Irish tune" as he introduced it; he then crucified Bruce Springsteen's *Born in the USA* in an unintelligible Irish brogue.

We ended up back at Nula's place.

"You need to know that there will be no going the whole way tonight," Nula informed me.

"So you get your kicks by watching a grown man cry?" I replied and smiled.

"Actually, we will never go the whole way," she continued.

"I'm serious now, I really am going to burst out crying," I replied with considerable more feeling. I then enquired why.

"Is it one of the usual reasons that women give?" I asked, "Such as "I'm a good catholic girl", "My girlfriend's a bouncer" or the one thing all heterosexuals fear more than anything, "My nuts will get in the way."

"It's the first." Nula replied. "I have traditional values. I'm keeping my virginity for that special day."

"If it's Tottenham winning the league, you better invest in a cat and a vibrating rabbit," I said. Nula laughed and then for a few months we spent some fairly innocent time together.

A few weeks after that first night, I noticed for the first time a photo on Nula's dressing table. It was a picture of a large hairy man, sporting a wild beard, one eyebrow stretched across his entire forehead and he wore the smile of a goat-fiddler dressed for his weekly appearance in court.

"Was this taken after his first cave-painting exhibition?" I commented.

"Fuck off. He's my favourite uncle, Bernie."

It suddenly occurred to me that I knew him, "Christ it's not Bernie Docherty, who has just been jailed for twenty-five years for blowing up a police station in Belfast."

"Yes, that's him, isn't he wonderful?"

"Another in a long line of terrorist bombers, well, as you said, you have traditional values," she smiled as my comment went over her head.

Christ, not another blonde deluded Irish nationalist I said to myself. As a young man in north London it was difficult to stand by one's principles, especially when those that challenged them were as beautiful as Patricia and Nula.

Those months that Nula and I went out together were interesting; I challenged her and her friends' belief that terrorism was somehow not terrorism, if you supported the *cause*. It made no sense to me, particularly when Nula and her friends professed they wanted a peaceful solution. I have never had an affiliation with any political faction, but over the years it has led to irreconcilable differences with women who have delusional and sometimes dangerous beliefs, such as Liberal Democrats, terrorist sympathisers and Arsenal supporters.

Chapter 20 – *Death by Stiletto*

Pubs are the engine room of England. Any occasion draws us to the pub, sporting events, birthdays, redundancy, even funerals. You only have to compare different countries' cultural influences, particularly their soap operas, to discover where the heart of their community lies. The American TV series *Friends* and *Frazier* were based around coffee shops such as Central Perk. As for Australians, though they are not averse to the odd drink, the characters in their most popular TV series, *Neighbours*, meet in the breakfast diner. In England, we have rival soaps such as *Coronation Street* and *Eastenders*, but the community focus is the pub, the Rovers Return and the Queen Vic. Pubs are also the main places where men and women meet, at least in my experience.

I met Colette in a bar and she was the first woman I lived with. She was French and could speak excellent English, but my ability with languages was so bad she would only speak to me in French purely as a matter of principle. She had a violent temper. I did not complain, as I loved the passion and the excitement of not knowing what would happen next. On Sunday mornings, we would regularly set off to nearby Camberwell Market to buy a new set of crockery, after she had smashed every piece of kitchenware we owned the night before. It was usually my fault but, even in her wildest rage the verbal abuse was always in her native tongue. *Merde* was Colette's favourite retort, but she would occasionally helpfully add, in English, 'head' to the end of it, so that I understood the crux of her message. Our relationship was based simply on the electricity we created; it was enough to keep us together for over six months.

On the subject of commitment, I have never been married. There are those that think it is because I have no respect for the institution of marriage, but it's the opposite; I have the utmost respect for those who do make a lifelong commitment to one another. My view is that when you do get married, you sign up in the belief that you've found the person whom you will grow old with, who you will always love and protect. I have met many a woman that I would have been honoured to marry. However, I have been involved in a number of voluntary missions abroad over the last twenty years and I felt it was not right to commit to a family when I had a desire to be elsewhere in the world. Perhaps I never met the right person, for if I had, I would have given up that side of my life to be with them.

Many years ago, all the elder members of my family got together to discuss the careers and marital status of the younger members of the family. Nearly all my cousins were either married with kids or engaged and then the conversation got around to me. Biddy shook her head as numerous excuses were offered by various members of the family, ranging from "He has not met the right woman yet," to "He is probably putting his career first for now." Uncle Mick was still alive then and commented, "Maybe he's bent?"

I heard that Mad Kathrina tore into him later, not to protect my character, but rather for having the gall to open his mouth. Good old Uncle Mick I thought, God bless him. I hope he finally has some peace.

Another of my numerous failings is that I am a very basic, simple man. This does not mean that I believe that women have to do the cooking, the cleaning and raise the kids. Far from it, I am happy to do all of these things as I am self-sufficient and most of the time I live and travel alone. But I am out of sync with the modern view of what a man should be. I am not in touch with my feminine side, in fact, I would say that we are not even on speaking terms. When I first heard of a 'metro-sexual male', I thought it was the technical term for those perverts on the tube who try to touch up women in crowded carriages.

I have seen many relationships where the woman is so close to being her partner's mother that the relationship verges on the oedipal. Many men have no idea how to do the simplest repair around the house or how to make their own decisions on anything outside it. Some husbands, with the right moisturiser, could be mistaken for the eldest son in their own family, as they are impotent as a parent. This is not necessarily wrong, as such relationships can lead to both persons fulfilling their individual needs, particularly if the woman is the 'mother earth' type. Not for me though, I have never needed mothering. Hell I wouldn't even let Biddy do it.

I do love a good rogue wedding and I have been a best man three times, firstly, for Loftus and Suzy with both of their children in tow. Secondly, for my friend Chris, who married a girl he had only known for three weeks. That arrangement was simply lust and it was similar to that of Colette and me, with the exception that his bride could not speak one word of English. Looking back, I think it was illegal. And thirdly H, to a lovely woman called Shelley, whom Bear, to H's anger, calls Shelley Oblige. Naturally, utterance of those words earns Bear a good punch to the side of the head.

Another pressure on my relationships is my friends. I would have to admit that, for many of my girlfriends, being with me must have been hell, partly because at some point they had to meet my friends.

I went out with a girl called Dana, who was great fun, very smart and a fantastic laugh. Plus I seem to vaguely remember that she also had forty-two inch breasts. Dana was very eager to meet H and Bear, so I asked her if she wanted to join us and go to the college in the park for their summer ball. Dana naively accepted. As we walked into the main student bar, Dana grabbed my arm very tightly.

"Oh my God, keep away from those two lunatics at the bar," she said.

"Well spotted, say hello to H and Bear."

After the introductions and H getting the drinks in, we sat down. Bear had purchased a big plate of apple-pie and ice cream to go with his cocktail of Tia Maria and Babycham. Over the years, Bear's taste in alcohol may have varied, but he was always consistent in having whatever concoction he ordered served up in a half-pint glass. Dana was wary of both of these 'circus-freaks' as she later termed them, but bravely decided to engage Bear in polite conversation.

"What have you been up to today?" she asked.

"Watching the Millwall game."

Bear described the game in his high-pitched voice, attracting the attention of the entire bar. As he grew more animated, relating the finer points of the match, he knocked over the entire contents of his plate on to Dana's lap with pie and ice cream landing upside down.

"Oh fuck no, me pastry!" he screamed.

Bear began to frantically scoop up his dessert with his spoon from Dana's crotch area. Dana looked at me with her mouth open in disbelief.

"Don't worry I think he likes you," I comforted her. "Just pat him on the head, but don't make any sudden movements."

A few moments later, a few students from the college rugby club squared up to H as he headed to the toilet. It appeared they did not approve of outsiders moving in on their 'territory', as they described the female students. They seemed to think that a stylish and sophisticated woman like Dana was from the college and could not be an associate of ours.

"Why don't you all leave quietly, before we make you," said the chief Neanderthal.

I turned to Dana and asked her to make a quick exit and meet me back at my place. Once Dana was safely out of the way I joined my friend.

Then Bear entered the fray. Bear hated violence; his approach was to play the clown and he proceeded to act out rather than describe the highlights of the Millwall match. The twelve students who had squared up to us realised that they had failed to intimidate us into leaving. Instead, they now had to watch a blond man-mountain jump up in front of them, head an invisible ball and slide across the floor to save the same ball from entering an invisible goal. It made me think of a kind of spot the ball game, if it became a national tour.

Bear's plan worked; the Neanderthals grew highly embarrassed at being part of this ridiculous spectacle in front of their fellow students and they started to disburse. The chief Neanderthal realised that his numbers had dwindled and that we were three to his five. He did not like the odds, so he too disappeared into the crowd.

"Good one, Bear," I said.

"Don't get me wrong, if it had kicked off, I would have gone in fists flying into you," replied Bear.

"Once again your kind thoughts are quite overwhelming," I said.

"Well I hate fighting but if I have to I might as well be on the winning side, or I could get hurt. Gladys would not be happy if anything happened to her little boy," continued Bear.

"Of course, your welfare is my greatest concern, so please feel free to turn on your mates at any time," I replied.

"Ah thanks, it's my looks you see, I won't always be at peak as I am now, and it's hard to believe now but one day I may start to lose them."

H did not take part in the post-confrontation banter. Instead, he just affectionately punched Bear on the side of the head and went back to his lager.

"Christ, H, what was that for?" Bear cried out.

"Supporting Millwall," replied H.

Dana hadn't left, she had just taken cover amongst the crowd. She came back to join us, giggling at our "circus performance."

Unfortunately, another obstacle preventing me from forming a long-term relationship was my lack of social skills. Lisa, an ex-girlfriend, telephoned me out of the blue one day to tell me that she was a lesbian.

"You're a strap-on? Excellent, but why tell me?" I asked.

"Oh great, thank you for your sensitivity as per usual, you arsehole," she replied. Nevertheless she ploughed on, "I am telling you, as I need your help with something."

"No problem. I am happy to help, what is it you need doing?" I asked, trying to be serious.

"Look you're the only person I know who would have the front to do this. I need you to…," she couldn't finish her sentence.

I was wondering what Lisa was going to say that was so troubling, as she had no trouble telling her ex-boyfriend she was gay.

She began again, "It's my girlfriend, she is quite adventurous in bed and on my trip to London next weekend she wants me to purchase some things."

I burst out laughing.

"Let me get this right. You're too embarrassed, so you thought you would ask me to get your lesbian toys for you?" I laughed.

"Look, you're the most immature man I know, but I trust you and you're the only person I know who would do it. I know you'll crack schoolboy jokes at my expense every chance you get. But will you help me, please?"

"Of course I will and once you're on the train home, you can put it all behind you, so to speak," I roared out laughing again. I could tell that Lisa was already beginning to regret asking for my help.

The following weekend, I met Lisa at Charing Cross station and we began our tour of the sex-shops around the Soho area. While Lisa sat in nearby coffee shops trying to look inconspicuous, I went into the shops with her shopping list. It was not easy for Lisa to remain inconspicuous, as I came out of each shop waving various items in the air shouting out, "Is it big enough?" or "Do you want this with a belt?" Usually, this would not cause offence when buying items of clothing for a friend, but as I was waving a butt-plug that would startle an elephant or a foot-long strap-on Black Mamba Eye-Popper, this perturbed some passersby and certainly bothered Lisa who tried to bury her blushing cheeks in her newspaper.

When I finally sat down with Lisa at the end of my shopping expedition, she said, "This is the most embarrassing day of my life."

"Well it's not over yet," I said, and I produced a huge double-ended dildo from a bag and added "be careful with this brute" as I slammed it on our table, terrifying the life out of our waitress.

Lisa dropped her head into her hands. Meanwhile I looked around smiling, still swinging the 'double-diamond love-pipe' in the air, hitting the light fixture above our table.

I saw Lisa off at the station later that evening with her two bags of goodies and, though she was exhausted from having been ridiculed by me all day, she raised a wry smile.

"You're one of a kind." She said and added, with a sigh of relief. "Thank God," and gave me a goodbye kiss.

"Well it wouldn't be a trip to London without a bit of shopping for the ladies." I said. "Be happy."

It seems impossible, but I once met a woman with an even wickeder sense of humour than my own. Her name was Stephanie and when I met her parents for the first time she introduced me saying, "He's completely mental." I'm not saying that Stephanie was wrong, but I thought that at least I should try to construct some kind of defence.

I turned to her, "I would have preferred 'Hi, this is my boyfriend,' and by the way what right have you to say that?" I retorted feigning outrage.

"I am a psychiatrist at St. Bart's Hospital," Stephanie reminded me.

"May I remind you to respect doctor and patient confidentiality," I protested. Stephanie's parents just looked at each other, no doubt of the joint opinion that there was no need to save for their daughter's wedding anytime soon and they could go ahead with their summer cruise as planned.

Stephanie was typical of the female rogues I've known, as she could put any man in their place. When both Chelsea Mark and H met her for the first time, each was attracted to her and both made advances.

Stephanie stopped both in their tracks with, "Sorry guys, probably best not, you see

I've got problems with me snatch."

My two friends were put firmly in their place, but acknowledged their better by supplying Stephanie with free drink for the rest of the night to show their appreciation of her superiority in the war of the sexes. I joined them and made a toast, "To Stephanie. Death by stiletto."

My love affair with America continued when I met a woman who had not the slightest concern about the life I led, but actually embraced it. I was drinking in a student union bar in the University College London with H and I started chatting to a beautiful American girl called Kerry. She was very proper and, though she matched me drink for drink, I knew I was not going to get anywhere with her that night. She did agree to meet me the next day though.

The following day, H popped around to my flat as we had agreed to go together to the football. He reminded me about the beautiful, feisty New Yorker, who meeting us at the station and joining us for the Chelsea game that afternoon. I laughed, as I thought it was unlikely that she would be there. I am not normally that lucky and also, with the amount of drink she had consumed, I thought she would probably be throwing up for the rest of the afternoon. To my pleasant surprise, Kerry was at the station waiting for me and she looked even more stunning than I remembered, while I looked like I had been dragged through a hedge backwards.

On our way to the game, Kerry told us that there were three things she and her American college friends, who were visiting our troublesome little island for the summer, had been warned not to do on their visit to England. Firstly, do not go to a football match. Secondly, do not go to Brixton. Thirdly, do not go to an Irish bar. These guidelines were even printed and laminated on a card, which was given to each student to keep on their person at all times during their stay. Naturally, Kerry being an adventurous girl and given my tendency to break rules, we did all three that very day. Kerry epitomized many of the qualities of her country. She was naïve to some extent, brash, patriotic and she had a lust for new experiences. She was also not afraid to break away from her more staid companions.

Kerry was one of the few girlfriends I took to meet Biddy. Kerry had insisted on meeting her as she was fourth generation Irish and she had never actually sat down and had a conversation with anyone who had been born in Ireland.

"On your own head be it," I said, though I knew that I would come off the worse in the encounter.

Biddy took an immediate liking to Kerry, so much so that after a few straight vodkas she offered Kerry some words of advice.

"You're a lovely girl. You could do so much better than him, you know."

Kerry thought this was hilarious, as so many mothers of her past boyfriends had thought no one was good enough for their son.

"Biddy's a bit different to most mothers," I said, "Just like Lucretia Borgia and Boadicea before her."

When I went with H to the World Cup, as I have mentioned, it was a few years later and I met Kerry for the last time. We agreed to meet in a plush bar in Manhattan's Soho district. When I saw her walking down the spiral staircase into the bar, she was dressed in an executive outfit, but her feminine beauty shone through and she was as beautiful as when I had first met her in the student bar in London.

"Hi, still fucking up the world?" was her opening line.

"No need, as you lot are doing a great job without me."

She laughed, but I could see that she was playing it cool, even though she was starting the evening off with pints of Brooklyn Ale.

After a few rounds, Kerry relaxed and started to laugh. She told me that she was moving from New York and putting her wild days behind her, as she had been offered a job with a film studio in California. She then told me that she was getting married the following month. I understood why she was keeping her distance. We were being served by a particularly unfriendly barman who kept grunting and slamming our drinks on the bar, but we just carried on talking and ignored him. That is until Kerry ordered another round and he turned to me and said, "She's had enough. A real lady doesn't drink more than two beers."

Kerry never took crap from anyone; she was trying to keep her composure but I could still see that familiar fire in her eyes. I looked at her and thought, "that's the Kerry I know" and though she did not say anything, it was good to see she hadn't changed.

However, I was happy to strike up a dialogue with the barman, "Fuck off," I told him.

At that point, the whole end of the bar burst into applause, including the other barmen. It seemed that our barman had been sacked earlier by the manager, but refused to leave until the end of his shift as he wanted to be as disruptive as possible. Following our brief discussion, he stormed off and Kerry and I drank free beers for the rest of the evening, supplied by a grateful management.

Kerry turned to me and said, "I miss you. We never had a dull moment in London."

H was still recovering from alcohol poisoning, but he was delighted to see Kerry when we fell through the doorway of our hotel room a few hours later. He gave her one final hug and then she said her goodbyes and I took Kerry to Penn Station to get the last train home.

"Ask me to marry you and I will be with you forever," were her last words to me.

I was stunned, but I was too young and too wild to get married. Also, I was shocked, as marriage to anyone was not something I had ever contemplated. So that was that, we never saw each other or spoke again. I hope she is happy and that if she has raised daughters they have a laminated copy of things not to do if ever they go to London.

Kathleen, whom I have already mentioned, was a wonderful woman, who really tried to put up with me and my 'Kids', a collective term she coined for my friends. God bless her, she even asked if I wanted to go with her to a friend's wedding in Kenya.

"Kathleen, surely you want to retain some of your friends?" I asked.

Despite this, she still wanted me to go with her and we set off for Africa. Her friend

Marion was marrying the farm manager of a national park in the Masai-Mara.

We entered her friend's fiancé's beautiful house in the middle of the picturesque safari park and walked straight into a domestic dispute between Marion's fiancé and his black cook.

"My coat is missing and it has to be you," our host was arguing.

"I did not take your coat sir," James the cook, kept saying.

Eventually, the coat was found hanging on the back of a door, just where Marion's fiancé had left it.

Later I asked, "Why did you think it was him?"

"You can't trust any of them," he declared.

"Cooks?" I asked.

He did not find this funny, "No, blacks!"

War was openly declared between Marion's fiancé and myself, while poor Kathleen looked on, worried and hoping that I would accept the situation and be a respectful guest. I couldn't.

"How long has James been your cook?" I asked Marion's fiancé.

"Twenty years."

"Well he's either been very good at not getting caught over two decades or you thought he had decided to risk everything to wear a hunting jacket in forty-degrees of heat." Kathleen was not happy, particularly as I think I might have used the term "You prick."

Later that evening, a tent was erected for Kathleen and I in the garden. I apologised to Kathleen for ruining her trip, but I would not apologise for what I had said. I did not get on with any of the white residents I met, except one kindly old man. It was just the case that every white inhabitant I encountered would at some stage declare, "I'm not a racist, but…"

The bride's sister, Kay, had a boyfriend called Danny, who was also in our party. He was from Newcastle and he was a rugby player, so though we came from working class backgrounds we were very different. Like me, he did not give a damn for the etiquette of our colonial friends, which made us strange allies. Kay's parents hated him; they wanted their younger daughter to marry into wealth, as her sister was now doing. Danny was a hardworking man, but he had no money and little education.

Over dinner on the first night, Marion asked Danny how he met her sister Kay.

"She was throwing up in the toilets where I was a bouncer," he replied, "so I helped her up and took her home."

Now the dinner conversation was at the crossroads. Kay's family could end the conversation there and not discuss their daughter's fondness for cheap alcohol, or they could focus on positive side, in that Danny helped their daughter when she was vulnerable. Unfortunately, Marion's family chose neither.

"It is disgusting that you would let a young lady get into such a state," said Kay's father to Danny, as if he were solely to blame.

I later found out from Kay that Danny had been a real gentleman and he had cleaned her up and taken her home when she was practically unconscious. The club's owners had sacked him the next day for leaving the door and Marion's parents had him arrested. He bore them no grudge, but they took every opportunity to denigrate him. To Kay's credit, she stood up for her man. They were completely in love with each other and not even her parents were going to break them up. I admired them both and I knew that, as they were strong enough to fight for each other, that their love would last a lifetime.

The wedding took place by the lake in the middle of the Masai-Mara and on the big day it was beautifully adorned by a flock of flaming pink flamingoes. It was a gorgeous day in the most idyllic setting, the only problem was that the vicar, who had been paid, then refused to carry out the service unless he received an additional bribe. Ever though the bribe had to be paid, the bigots were in a frenzy of self-righteousness proclaiming that all blacks were corrupt. One of these was a Member of Parliament who was embroiled in the expenses scandal back in

England some years later.

At the reception, I was at the buffet loading up my plate with a cooked leg of Thompson gazelle, when an elongated chinless member of the local white gentry approached me. This Throwback was the son of the local lord, the kindly old man I mentioned. I had met the Lord earlier in the week. He lived in a big house on a hill and played with his train set which ran the length and breadth of his forty-two-room house. I liked the eccentric eighty-five-year-old man, who was the only person I'd met who was civil to the local black workers. They returned that respect. The same could not be said for the son, who stood toe to toe with me in his blue blazer, white pressed starched flannels and riding boots. He also slapped a ridding-crop repeatedly against his thigh like some public school master getting ready to mount some future member of parliament in the toilets of Eton College.

"What did you think of my father?" the Throwback demanded.

"I liked him," I said, "he is a good man, generous and kind, unlike many I have met here."

"You upstart," boomed the Throwback.

"Upstart. Which Bertie Wooster novel did you fall out of?"

"He's an old fool, whom the natives take full advantage of," he continued.

"He seemed happy enough to me."

"This country needs to be dragged into the civilized world."

"Oh dear, but what will happen to you?" I asked.

He ignored my responses, as if I were one of his abused servants and carried on firing accusations and questions at me without any interest in my answers.

"Do you know who I am?" demanded the Throwback.

"No, but I would be happy to ask someone and let you know."

"I am Lord Fimblewayne Massingbird, sole heir to the fifth Lord Massingbird and future owner of all of his sizeable estates in Kenya."

The rest of the herd at the reception was looking over at the Throwback and me butting heads. Not that it would have made any difference to me, but I was later told that he had shot dead a trespasser on his land the previous year and the night before the wedding he had been arrested for wounding a local prostitute in a brothel, when he drunkenly fired his revolver at a dart board.

"I will give you a twenty minute start, before I track you down and shoot you," he declared.

"OK, I will be ten yards further up the buffet by then." I replied.

That was not the end of it and, though we were physically separated by guests for the rest of the evening, the Throwback later tried to run me over later in his Range Rover. As he drove at me, he nearly killed the groom's little boy from a previous relationship, as both of us had escaped the drunken throng inside to play football in the garden. I pushed the kid out of the vehicle's path and then I jumped onto the car's side, grabbed the wheel and steered it into a brick wall. The police were not called, as the Throwback was deemed untouchable. A few years later, events repeated themselves and he is now in jail awaiting trial following shooting another trespasser on his property. The prosecution may have a case this time as the deceased worked on the lord's property. I wouldn't hold my breath though.

On the Kings Road one beautiful sunny day, I bumped into Amelia, whom I hadn't seen since those days when we put trucks on the road to Romania. We were pleased to see each other and there appeared to be no animosity, or so I thought, so we went for a coffee. She had not changed at all apart from having a baby in a pushchair beside her. I whacked my mind trying to work out dates, wondering if the beautiful little girl was mine. Amelia informed me that since we broke up, she had met a man, who was sensible, caring and who offered her the security to raise a family. Well that told me. Good for Amelia, she has a big heart, but she wasn't afraid to go for the jugular when she needed to.

"He sounds great, I am very happy for you, do I know him?" I asked her.

She mentioned a name of an old rock star; I thought that it could not be the same man. Amelia confirmed that it was.

I smiled, "So let me get this right, an ageing rock star, from a culture of drink, drugs and rock and roll. Well, if he is more stable than me, then I must be truly gone in the head."

"Correct," she declared without even a pause, "even if he shot himself out of a cannon into a brick wall three times a day without a helmet, he still would be saner than you." Amelia's impassive face then broke into a smile and she collapsed into a full hearty laugh, which made her beautiful baby girl smile too. I believe the three of them really are very happy and in so many ways Amelia, as always, was right about me.

Such post-mortems at the end of a relationship are not unusual for me. Sometimes they even happen when things are going well, or I think are going well until the "Right, there are a couple of things we need to talk about," conversation.

A few years ago, I was going out with an Italian girl, Alexandra, and we were having lunch.

"A couple of things we need to talk about," she began.

Eyes down and look in as the bingo caller used to say; I tried to fake a combination of interest and contriteness dependent on which menu of my many shortcomings she was about to read from. The first thirty minutes covered the usual areas and, to be honest, it was a completely accurate assessment of my numerous failings, as a man and as a partner. She covered the standard deficiencies, lacking in culture, arrogance, ignorance, independence, poor humour, lack of fashion sense and, added to them, her disapproval of my past adventures, some of which you know of.

Nothing unusual there, was my thought, sex wasn't mentioned and that was a plus, but the rest was pretty much par for the course. Unfortunately, she then moved on to my personal failings as a human being, which was unusual, as this was usually blended into the standard 'lack of relationship skills' conversation. I was impressed that Alexandra took the time to split them apart; it made it far easier for my simple little uncluttered brain to take it in. However, two hours of criticism was a bit much I thought, but the coffee was on tap and Alexandra seemed happy, which was the main thing.

After three hours, I thought surely my time on the psychiatrist's couch must be at an end, but she had merely paused to take a sip of coffee and now suitably refreshed, she returned to my weaknesses as a boyfriend.

"One of your biggest failings, Righten, is that you have lived too much." she continued. "It's very hard to deal with in a new relationship and it will be hard for a woman to grow with you."

I could see the sense in that and I wanted to say that they were so many adventures still to come and to share with someone, but my subconscious told me 'Keep your mouth shut and just nod.'

Before I could say anything though, Alexandra completely threw me.

"However, I have decided that we will go on holiday together to Italy," she said, "and you can meet my family, as I feel we are now at the stage to move on."

How we had moved from the "crude philistine" and "you're a little in love with death" of two hours earlier, to the "meet my family" was beyond me. Maybe her parents yearned for the old days when daughters married the slow-witted gladiatorial type?

I leant forward, gave her a kiss, smiled and said, "I am sure your parents are lovely people, but I can't do this to them," I paid the bill and said goodbye. I thought that was a nicer thing to say than, "No thanks, I have no ambition to be in an unmarked grave at the bottom of a Sicilian mountain, thank you."

It was good to see, and it was so often the case when I met ex-girlfriends like Amelia and Kathleen, that they were so much happier when we were no longer together. Amelia and others had married good men, others had secured good jobs and some even made enough money that they could afford to put out a contract out on me. OK, I exaggerate, as I only know of one ex-girlfriend who seriously considered hiring a contract killer.

Though my values were minimal back then, I consciously didn't get involved with married women, as I was not going to break up any families. That value went beyond the social sphere and on humanitarian missions I never took anyone with me who was married or had children. I did not want to cause stress to any families left behind. I have seen the distress I have caused girlfriends when I was away for weeks and was not be able to get in contact; mind you some would say I couldn't communicate when I was home. I have no desire to extend the stress to the families of those who have offered to help me over the years. Also, I know that it risked the core objective of getting aid to wherever it was needed, for if anything happened to my co-drivers I would have to take them home immediately. It was always better to work with loners like myself, as we had no dependents. It's very selfish of me, but it is one rule I strongly adhere to in that with all my convoys, 'sad cases only need apply'.

Everyone has a natural defence mechanism and it takes a lot to break through mine, but we all carry scars and some of my deepest are the result of meeting a woman called Jane, whom I saw for the first time in a club in Hampstead. When I first saw her, she was so natural, so free of pretence that I went straight over to her; I had to say hello. She laughed a lot that evening, but made it clear that she had a young son and was not interested in any kind of relationship with anyone. I soon found out why. The father of her son, Jamie, was recorded as 'absent without leave' from the army, and in addition, he was a violent head-case.

It was unusual for me to get close to anyone, let alone to get involved with a woman who had a kid. However, Jane was quite a woman and over the following months, I found myself drawn closer to her. Jane was a nurse who also did a little part-time fashion modelling to earn extra money, which puzzled me as her overheads did not seem to be much and I knew it hurt her to be away from Jamie any longer than she had to. Over time, her defences along with mine began to come down and against our better judgement we became quite close. Sometimes Jamie and I would join Jane on set for one of her shoots and we would occasional look at each other and smile as we were both very proud to be with such a beautiful and accomplished woman. Each night though, my thoughts would return to Jamie's father; he was out there somewhere. I knew a new guy would not be welcome on the scene and that trouble was just around the corner. Unbeknownst to me, it was far closer than that.

Jane had a habit of getting up and going downstairs very early each morning. I would hear her opening and closing the back door for some purpose that she had never explained. Then, very early one morning, I heard heavy footsteps bounding down the metal staircase leading to the basement. Thinking that someone was breaking in, I jumped out of bed but Jane grabbed my arm just as I saw a big guy coming back up from the basement carrying a black bag.

"Please don't, its Grant, Jamie's dad."

"What is going on Jane, tell me?" I demanded.

She broke down crying and told me that he dropped his washing into the basement twice a week. She washed and ironed his clothes and he picked up his clean clothes and all the money she could afford to give him on his next visit.

"Why?" I asked, kneeling down to wipe her tears.

"He doesn't care for me or Jamie, but I'm scared of him," she sobbed. "He hurt me before and once he put me in hospital and Jamie was taken into care. I am terrified he will hurt us if I don't do what he wants or give him enough money."

Jamie appeared and ran to his mum, he was crying too. Christ, no wonder she didn't want a relationship, I thought. I stood back and watched as they clung desperately to each other, trying to contain my anger at Grant, trying not to add to their fear. A few days later I got up early and Jane asked why.

"Don't worry I will sort his washing out." I said.

Jane was terrified and pleaded, "No, please don't do anything."

"Leave it to me, I will help. Don't worry, I won't do anything stupid," I lied.

When I washed Grant's latest deposit of clothes I realised how he had evaded the military police for so long; his clothes were of the New Romantic style, the popular fashion at the time. It was clear from the stains on his shirt collars that he wore that deep pancake makeup that was part of their image, a brilliant disguise. He didn't even bother to remove lipstick from clothes or condom wrappers from his pockets. After I dried and then ironed his clothes I left them in a bag outside on the basement step. I then went back upstairs to Jane and we wrapped ourselves around each other.

Then all hell erupted outside, there was a loud banging on the door downstairs followed by a man screaming "You fucking whore."

"My God it's Grant, what have you done?" Jane asked, now completely terrified.

"I don't know have you paid the milkman?" I said.

"Don't, he will kill you!" shouted Jane as I raced downstairs.

"It's Ok, just look after Jamie. I must have forgotten to press his collar."

As I opened the door, he was about to land a punch on what he expected to be Jane, but stopped in his tracks at the sight of me.

"Where is that fucking bitch? She burnt a big hole with the iron in every fucking thing," he screamed.

"That was me I'm afraid, sorry about that," he wasn't expecting this, neither was he expecting the flying pan which I now landed right across the side of his head. I knew he was a big man and would undoubtedly be full of rage and pumped up with adrenaline. I knew it would take all my strength and guile to stop him getting to Jane and Jamie. We crashed about in the basement for what seemed an eternity. Then the MPs whom I had tipped off, arrived later than I expected, but welcome nonetheless and Grant was bundled off in a military police van.

The aftermath of all this was that Grant was sent to a prison for a minimum of two years, during which time he died of a drug overdose. Jane, rightly, never forgave me for betraying her trust and putting her and Jamie in danger. I knew that what I had done meant that she would never trust me again and it would be the end of us. Jane did not know that I had a further plan and I went to see Grant in jail.

I did not know that he was going to die soon. He was very willing to see me, if only to make it clear that he would kill me as soon as he got out. I slipped him my address through a gap in the wire partition and said that I, and a few others, would be delighted to reshape his head with a crowbar, if he ever went near Jane and Jamie again. I told him the names of the others I mentioned, each of whom had a well-earned reputation for violence and funnily enough, they were all inmates of that same prison. Grant then smashed his chair against the wired-glass partition that separated us and thrust his fingers though the now fractured glass, with only the wire mesh laced with fragments of glass separating us. I sat back and watched his blood stream down the broken glass grate partition as the wire cut through his fingers, as he screamed and spat at me, as he repeatedly smashed his face against the broken wired-glass until the prison wardens restrained him. I had no doubt he would kill me when he got out, but I prayed he would spare Jane, though I doubted it, and Jamie.

I will always regret what I did to Jane and Jamie. It was stupid and if Grant had got past me that morning, I have no doubt he would have killed Jane and perhaps even Jamie. Unfortunately, the story did not end there. Over the following months, Jane met a good man, Peter, a man she could trust, not a lunatic like Grant and me. Jamie though, missed me and kept running away from home to find me. When he turned up on my doorstep, I would take him back to his mum. It was destroying Jane, as Jamie blamed her for us splitting up. I suppose I had grown close to Jamie too; I had often taken him to the park to play football, or to the zoo, which was his favourite and we just did the things his dad never did.

To stop Jamie running away from home, Jane and Peter left London and moved to Bedford. Within a few hours of their move, I got a call from a distraught Jane to say that Jamie had run off again. I too had moved, so I got on my motorbike and went to all the parks in the area where they used to live and where we used to play football. Then Jane called me to say that he had been found by the mother of one of his friends, at a football pitch near where I lived. I found him with the concerned mother. He was soaking wet, dirty and scared.

"My mum hates me, come back home," he said.

At that point, I had to do probably the hardest thing I have ever done in my life I said, "Jamie I don't care about you, I don't care about your mum and that is why I left you both."

Jamie was stunned and I felt that I was ripping the little loving heart out of him. I continued and in doing so, I ripped my older cynical one to shreds as well, "Don't you ever run away again, or I will hit you so fucking hard."

Jane and Peter pulled up in their car and Jamie, now terrified of me, ran to his mum. Jane could not work out what had happened, but that was the first time Jamie had gone to her since we broke up. Jane and Peter did not say anything they just looked at me and nodded. Jamie clung to his devoted mum and he turned to look at me for the last time, with that same frightened look he had given his father on the morning we fought.

After they had gone, I sat on a bench in the rain. I could never have hurt Jamie or Jane. That was the hardest thing I had ever done. I felt sick at the words and threats I had uttered to that poor little kid. God knows what harm my threats have caused over the years and I knew he would always hate me. I try to look back and think that what I did was the right thing to do, but I look at the results, one betrayed mother, one dead father and one terrified little boy. I think of Biddy and of how she took control of the threat that my dad became. She had used her wits and, as a result, I escaped unscathed. I had confronted violence with violence and put the innocent at unnecessary risk. If that is what happens when I try to do the right thing, imagine the harm I cause when I set out to make trouble?

I decided after that never to get involved with women with families. I thought that if I stuck with single women, any harm caused would only involve mature adults who knew what they were doing. My mistake in this approach was that I judged myself as mature.

A few weeks later I met a girl in Shoreditch one Saturday night and we ended up spending the night together. We went our separate ways after having breakfast in her plush hotel the following morning. It was a great night, we both had fun and I thought nothing more of it. Later that day, I headed off to Rome for the rest of the week as Chelsea was playing a Champions League match. When I got back on the Friday, I opened my e-mails to read the following from the girl I left with a big smile on her face:

Monday:

"Hi that was a great night on Saturday and I know we did not arrange to meet again, but I'm back in London next weekend if I can buy you a drink? I hope you don't mind, but I think we were great together."

Tuesday:

"Are you ignoring me, I'm not used to being ignored."

Wednesday:

"You fucking bastard, you just used me, I feel so degraded and humiliated. You shit."

Messages of a similar theme continued throughout that day and into the evening.

Thursday:

"I have told my brothers how you used me. You're fucking dead," twenty or so messages in a similar vein followed that.

Friday morning:

"You bastard, I have deleted all your details, you're scum and you're not worthy of me. Fuck off!"

So after twenty minutes of catching up on my week's e-mails, I found out that I had been dumped from an intense and, at times, passionate relationship, that I hadn't even know I was in. Saying that, I liked her approach, I gave her a call immediately and we ended up going out for a couple of months.

I decided after Jane that I was not going to get involved in relationships for a while, but only a fool believes his head rules his heart. It was my last Christmas at University, when I heard that a girl who I had an affair with the previous summer, Rachael, had been knocked down and killed by a hit and run driver. I also learnt from her sister Gillian, who had called to tell me this tragic news, that Rachael had been carrying a child. Up to that point I had hardly thought of Rachael, as our relationship had been a summer fling, but I hit the college bar hard that night. Rachael had separated from her husband, who used to knock her about. I didn't know about that at the time, but maybe there is something in my subconscious that attracts me to such women. Perhaps I want to help as there was so little I could do to protect my mum when I was a kid.

I knew that Rachael's ex-husband drove an articulated truck and the police informed me that it was such a vehicle that had killed Rachael. I lost it for the first time since school and I went to London to find him, in the belief that he had murdered her. Maybe again, if I was on a psychiatrist's couch, he would say that this was as a result of never bringing Loftus' killers to justice, or simply that I had some kind of macho vengeance complex, so just pay the receptionist on the way out and don't come back.

When I found Rachael's ex-husband, I found an emaciated drug addict who could barely tie a tourniquet around his arm to inject heroin, let alone turn the wheel of a truck. He didn't know or, for that matter, care that Rachael was dead. I closed what was left of the door I had kicked in and left the addict to what little life he had left. I don't know what I would have done if I had found him to be the villain that I envisaged. Rachael is buried in Scotland under a tombstone that reads ... well best to keep some secrets and leave her and her unborn child in peace.

I decided to find out more about Rachael's life and contacted Gillian, she agreed to meet me on Christmas Eve and tell me more about the woman who I had spent the summer with, yet barely knew.

"A bit late aren't you?" Gillian said to me. "Pity you didn't take the time when she was alive," she was angry with me and who could blame her? She continued, "She was a good looking girl, but it was a cross to bear as she kept attracting the wrong guys."

I knew Gillian saw little difference between me and her brother-in law. After a while she opened up and told me more about Rachael.

"She was happy, I think last summer, her bastard of a husband was out of the way and she told me that you were new and exciting and you made her laugh for the first time in a long while. I'll give you that"

I took no comfort from her words.

"She told me you wouldn't hang around, guys like you don't."

"Sorry, I didn't know it was going to end like this," Gillian cared little for what I thought, but sat back deep in her own thoughts.

After a while Gillian continued, "Did you know she was pregnant?" she asked.

"I had no idea, until you called and told me."

"Me neither," said Gillian, which surprised me.

"She never opened up to anyone, our dad was a shit too," again, I knew that her dad and I were on the same list.

She continued, "There was another guy after you, so I have no idea whose kid it was."

"We always used protection," I added.

"They split, a bit like guys like you," she spat.

Then we both sat in silence and then Gillian asked, "Do you think Rachael knew that she was pregnant?"

"I have no idea," I just wondered that if she did, did she welcome the news and see it as a fresh start? Or was it another unexpected kick that life gave her? I hoped she did know and that it brought her some happiness, but I guess Rachael was one of those people that no one, not even her sister, ever really knew.

LOVING ROGUES

"You despise me don't you?"

"If I gave you any thought I probably would."

Humphrey Bogart and Peter Lorrie

Casablanca

Chapter 21 – *"It's Alright For You, I Have To Live In Here!"*

On the subject of love, it will come as no surprise to discover that Bear's love life was not what most people would call normal. There have been a few dismal attempts of Bear trying to secure a mate and, at one stage, he actually did have a girlfriend. It was back in those very early days when Bear was a young man running around the park, desperately trying to turn it in to a desert oasis. It started when H and I were sitting in the hut reading the papers, when we suddenly heard a high-pitched explosion.

"JULIANNA, JULIAAAA, IT'S JULIANNA HERE!"

"What the fuck was that?" said H, startled by the assault on his eardrums.

"JULIANNA, JULIANNA HERE!" we heard again, as if a tsunami were announcing its arrival.

We were then confronted by a pretty little blonde who was working in the park over the summer, in between her university studies. H nicknamed her Gunshot.

At that point, we did not know that Bear and Gunshot were soon going to be an item. A few weeks later when H and I walked into the Wellington pub on Tottenham Court Road, we discovered the two of them together, plus twenty of her fellow psychology student friends. The reason she had such a large number of chaperones was because Bear was filling out psychology questionnaires for all of them.

The Bear joined H and me at the bar and whispered to us, "Who told them about how I feel about my mum?"

I asked him if it was love and he replied, "Yes, that is if love requires you to fill twenty questionnaires a day covering 'Have you ever masturbated on hearing of the death of a relative?' or 'Fantasised about sex with a small rodent?'"

It was clear that Bear had fallen in love, because he was worried that one day he might lose her. That was proof enough, as Bear never thought of the future in those days.

"Luckily all my answers so far are yes, but what happens if one time I have a question, where I have to say no? Do you think she will lose interest in me?"

"Don't worry Bear, if it's about you and masturbation there is no such question. Your only fear is that you won't be able to tick the relevant box because the men in white coats have tied your sleeves behind your back," I replied, to soothe my friend's anxieties.

It was short and painful for Bear as Gunshot left him as soon as it was time to go back to college. I guess her research was over.

They say you can't keep a good man down; Bear was soon hopelessly in love again. Hopelessly being the optimal word, for this was the time when he had been sacked by Coldman and somehow found work in a little public park in central London. Next to Bear's little park was a little white church and next to that was another little park. There worked a little fat spiky-haired girl in dungarees known as Portly Punk.

Bear had fallen for her and when I asked him why he replied, "Well just look at us, we have so much in common."

"Ah, so she is a lover of the cake trolley too," I suggested, "what else attracted you to her?"

"Well, I guess it was the way she smokes her pipe, while playing pool." I saw one problem in this budding romance of his, a sticking point that Bear was also soon to discover.

A few days later, Bear summoned up the courage to leave his little barren wilderness, wobble past the white picket fence of the little church and enter Portly Punk's beautifully kept and petal populated environment. The scene reminds me of those Swiss clocks where little male and female figurines come out of their little porches to greet each other when the clock strikes the hour. It's a sweet, but flawed image, as in this scenario the rotund male would struggle to squeeze his backside out of his small hut door, while the female was frantically nailing her door shut. Bear went up to Portly Punk and presented her with a bunch of daffodils.

"You fat bastard, I only planted them last month," she responded with fury. She looked at the overturned soil at the front gate and at the mangled floral tribute in Bear's hand.

"It's the thought that counts and sometimes you have got to seize the moment," responded the love-struck Bear.

"Get stuffed fatso," replied the object of his affections.

"But I love you!" persisted Bear.

"I am gay!" replied Portly Punk.

"I'm happy too!" he replied, just as she kicked him in his privates and he collapsed on top of her freshly-planted marigold display holding his now aching 'love-conkers'. Even as he was blacking out, his instinctive hatred for all forms of plant life drew him in their direction.

The next day Bear entered his little barren park only to discover that Portly Punk's girlfriend had burnt his little hut to the ground. As a result, Bourgeois phoned me to say Bear was very depressed: he was sitting on his little stool in the rain, surrounded by smouldering embers and we should go cheer him up. Never one to neglect my good friend in his hour of need, I agreed.

Bourgeois had arrived at the park ahead of me and was engaged in conversation with Bear, whose arms were now waiving furiously, as he voiced his woes to his friend. I realised that Bear could not see me, so to announce my arrival, I picked up a crushed conker, not one of his, but one from the tree that had blown in from the neighbouring park, and lobbed it at his head. It missed him by an inch, but he still didn't spot that I was there. Then I picked up part of a dead branch and threw that at him. It caught him on the temple and levelled him out on the rose bed behind him. The difference from his collapse the previous day, was that it was a flower bed in name only; his fall was cushioned only by the stalks of roses he had deadheaded with a chainsaw on the day of his arrival and the abundance of weeds in between. Bourgeois did his usual when Bear was in trouble and ran off to the pub. Not much going on here I thought, so I hailed Bourgeois and joined him.

A year later I received a call from Bear early one Sunday morning, "Eighty-four pounds." he said.

"Well, I thought your daily food allowance was more than that," I replied, still half-asleep.

"No, I just got twelve women for eighty-four pounds," he screamed excitedly down the telephone.

"I'm glad you told me, now give me your location and I will have the police there in no time."

"I was on the underground eating a Poundsaver's extra-value family-sized cheesecake made with Azerbaijan cheddar, when I spotted an advert for 'Datehole'."

"I think you will find, that it is called 'Datemate,' but don't let the facts get in the way." I said. "Bear you will find that it provides lonely tunes such as your good self, with the opportunity to be introduced to women, who then decide if they want the relationship to go further."

A long pause followed as Bear considered this and finally said, "Well that's me fucked then."

A few days later on New Year's Day, I received another call from Bear, "I'm depressed. I'm going to kill myself."

"Good luck, bye" I replied.

"No, I'm serious."

"So am I. Look do I have to hold your hand for everything? A stun gun to the head will work, just like they do to bullocks, or just wear an West Ham shirt to a Chelsea game, either will do the job."

"I'm in a phone-box in the middle of Trafalgar Square, so I'll meet you in the park in the middle of the square, you can't miss me, as I'm wearing my bumble-bee top, I'll be crying my eyes out and the punk-rockers will be throwing birdseed at me again."

When I arrived, there was the depressed bumblebee sitting dejectedly in the middle of the park in the square,

"What future is there for me? I want to meet someone and have little Bears of my own."

"Your best option, as you have the jumper, is to stick a TV aerial on your blond mop, float above the flower beds and start making honey." I said.

"Fuck off," muttered the depressed Bumblebee, as once again at the punk-rockers started throwing bird-seed at him again.

"Ok, what happened?" I asked.

"Well I am suing Datehole; they said they can't find me anyone. I told them I will take anything, fat, green, handicapped, coffin-dodgers, anything, I am desperate."

"What happened to the twelve prospects?" I asked.

"Well I sent in my application, plus a picture of Richard Gere, but eleven of them walked off when they saw me in the flesh."

"That is surprising, I thought they would run, but at least one of them didn't walk off," I added.

"She had a car," said Bear as he took another mouthful out of his Marks and Spencer meal-for-four trifle. "Look the bastards even put it to paper," he passed me a letter on officially headed notepaper.

"*Dear Sir,*

It is with deep regret that for the first time in our company's history....,"

well you can guess the rest.

"Well, H always did say you do have the face that launched a thousand shits." I laughed.

"Is that it? Is that all the advice and sympathy I get? Quoting that fucking psycho?" said the dejected Bear.

"OK, here goes and I mean this from the heart."

Bear looked up.

"Here is a number of someone who will listen to your plight and is an expert on these matters," I then scribbled Cockey's number on a scrap of paper and handed it to the depressed bumblebee.

"Excellent, share my emotional turmoil with that fucking android, I might as well pour my heart out to the statue of Eros up the road. What's your other plan; meet up with H so he can punch me in the head? Yes, that will sort me out, of course why didn't I think of that?" as he rolled up my scrap of paper and threw it at the pigeons feasting on seed.

Then Bear said something which showed how vulnerable and depressed he really was.

"I miss Gorgeous George. I know it's a few years since he dropped dead, but I miss my old man," I relented and I sat down next to my old friend, as I too, missed his dad and thought it best to talk Bear though his grief.

"What are your memories of your old man?" I asked.

"Well he had a fantastic sense of humour and he used to make me laugh out loud."

"For instance?" I asked, trying in all seriousness to bring him out of his depressed state by getting him to talk about happier times.

"Well once I nicked chips off his plate and quick as a flash he said, 'You bastard, you nicked my chips,' he had such a quick wit my old man."

Hardly up there with Shakespeare's comedies, I thought, but what do you expect from someone whom I had only known laugh out loud once? And that had been when he spotted a pub called The Cock.

"My Dad's death really hit me hard," he continued, "I wish we had talked more, but when we did talk, he often took little things the wrong way."

"Bear if you had the chance again, would you act any different with Gorgeous George?"

Then after a moment of reflection he added, "You mean like when I sold his disability scooter to go on the piss?"

As always, the counterbalance to Bear was Chelsea Mark. Chelsea Mark was very successful in attracting women, but, as with Bear, it always ended up in tears although without the birdseed. I had known Chelsea Mark for a quite a few years when, as we were getting drunk in his parents' house one evening, when he said slurring, "Have I showed you the wedding photos?"

He pressed a framed picture into my hand, but I was finding it hard to focus. I spotted Chelsea Mark's dad who looked just like his son, only with a perm.

"This must be his first marriage," I said, "your mum is much shorter."

"No, it's my wedding," he confessed, as he had failed to mention it before.

His reason for marriage was unusual even for Chelsea Mark. As a fully signed-up member of EMU, he hated Americans in the belief that, as the leading superpower, they were the cause of all the problems in the world. He then met Marjory, an American student studying in London.

"I expected all yanks to be ogres, so when I found her to be really friendly and nice with a big pair of breasts, I was so overwhelmed that I married her." he said.

"How long did it last?" I asked.

"Three months, but we had a break in between."

We then returned to the serious business of breaking into his father's best rum.

Chelsea Mark was cynical after his experience and I later asked him if he would remarry. "Fuck that married lark, it would be easier if I met some woman who hates me and just gave her all my money," he replied.

Later that night, we were in a bar and he turned to me and pointed his fingers to his head, "It's alright for you." he said. "I have to fucking live in here!" Chelsea Mark irritates everyone he meets, including me, but I wouldn't change the rogue for the world.

.

Chapter 22 – *The University of Hard Knocks*

I have tried to tell the stories of the rogues I have met using my life as the structure, but sometimes I have devoted a chapter to focus on a theme. The previous chapter told of rogues in love, a cul-de-sac for most. It is time to resume my fitful journey. After I left the park, I went to a university that had the highest pregnancy and suicide rate per head of any educational establishment in the country. Fortunately, the two were not related. Once again, as with St Vesuvius, my educational institute was groundbreaking in its extracurricular achievements. Being a northern university it was also the first time I had lived in that part of the country. Up to that point my knowledge of the North was based on images from the television. The TV series that made the greatest impact was Alan Bleasdale's *Boys from the Black Stuff*. For comedy value it was Warren Clarke's short but hilarious portrayal of the Northern Industrialist in *Blackadder*, who expounded northern wisdom as "I love my wife more than I do my best pig" and "I would rather put my John Thomas in the hands of a madman with a blunt pair of shears than give my daughter to any man." I was in the minority of the University's new intake, being 'a soft-southern bastard' and, more specifically, 'that flash London bastard.' I was in my element, loving this honest and sharply observed humour of my new northern friends, even though I was usually on the receiving end.

My first year, there was a lot of drinking and I made a few serious enemies. I settled in very nicely. These enemies consisted of a military group of right-wing nutjobs; the self-termed the Officer Training Taskforce (OTT). For the next three years, I would clash with this militia of upper-class halfwits on regular occasions. Complementing these were the Young Conservatives, who had their own share of dysfunctional droids. If a student was a member of both, it often led to invitations to speak on Christian gatherings in America's Deep South, indeed their claim to fame was that one member, who had changed his name to Silas Stephens as he liked the initials, gave a lecture on his belief in creationism.

These droids had opinions on everything, as the world was divided into them and everyone else – there were no grey areas. A grey area would just confuse things for them, as it would have warranted some kind of reasoned discussion, and God forbid, perhaps even reflection. Their simplistic method of assessment also meant that they had a forthright opinion on everything; no subject was off limits. I remember one union debate, where one of the droids opposed a motion that called for the removal of the tax on tampons. The droid stood on a platform arguing that the tax should not just remain; it should be increased to boost our military might in the continued battle against communism. He ended his speech with the argument that tampons were a luxury item and that women still had the option, as in the past, of using twigs wrapped with leaves. The leader of the Women's Society went ballistic and challenged him by asking why no government tax was levied on men's razors, which were not classed as luxury items.

"Those are not luxury items," he replied "as western men have to be clean shaven to counter terrorism; for it is essential instrument in the defence of our national security to differentiate us from sandal-wearing bearded militants." Surprisingly, he had a girlfriend, but I put her with that small group of female lunatics who regularly sent proposals of marriage to murderers and rapists in prison.

As I mentioned, it was at University that I first encountered Sam, but I also made two other lifelong friends during my time there; Terry 'Fuckwit' and Jonathan Waffle. Terry was from Southport, just outside of Liverpool in the north-west of the country, whereas Waffle originated from Hull in the north-east. Hull, at the time, was the most violent city in the UK, including even Belfast, which was being bombed to buggery by various paramilitaries. This went someway to explaining why Waffle sat in our first history tutorial with a bandage wrapped a bit too tightly around his head. Someone had broken a bottle over his head when he was having a quite drink and reading his newspaper during 'Happy Hour'. Waffle moved with sudden jerky movements, so with hair sticking up in the middle of his bandage, he looked like a farmyard rooster and reminded me of the Looney Tunes cartoon character, Foghorn-Leghorn.

Our tutor opened with the question, "Can we outline any subtle differences between the North and the South during the American Civil War and why the North eventually defeated the South?"

I pondered on the force of the civil rights argument which divided support more in the south than it did in the north; the superior military might of the Northern armies and the leadership qualities of Abraham Lincoln and his generals. All of this was well known, so I wracked my brain for more 'subtle' reasons. Fortunately, when it came to the obscure, Waffle stated that the reason that the North beat the South was due to the greater size of their horses. The conclusion at the end of his thirty-minute monologue was that the Southern forces transported fewer provisions and munitions as their horses could not carry as much as their Northern counterparts, so they were ill-fed and out-gunned. Added to this, their horses' legs were shorter and they covered less distance than their Northern counterparts, so the defeat of the South was inevitable.

After our tutorial ended, I went to congratulate my fellow student, "Sticking with your theme, that was the biggest load of horseshit I have ever heard."

"You're missing the subtle objective of my argument, Johnny boy."

"Which is?" I asked.

"The tutor believed it."

My first year was one of high personal achievement, for example I was elected President of the Atheist Society. Mind you this was *in absentia*, as I had not run for the post; I was not a member of the Society; nor did I know of its existence until I was informed I was its President. I found out when I was having free Sunday dinner with the Catholic Society, which provided a roast for its members. The priest broke the news to me of my success and immediately challenged me on my appointment, accusing me of hypocrisy for being a member of both societies.

"Not true, as I am not a member of either," I corrected him.

"What? So why are you here each Sunday?" he demanded.

"Same reason as ninety percent of the people here, we are too hung-over to cook a proper meal on a Sunday. By the way have you ever thought of providing lunches on Saturdays, as Friday is a big night out too?"

I was sent a letter later in the week informing me that 'I was excommunicated.' I never did find out who nominated me as President of the Atheist Society, but I was soon deselected by the members when it was discovered that I was also a regular at lunches held by the Jewish and Muslim societies.

Our lecturers were exceptional. Not from an educational viewpoint, but from an entertainment perspective. Professor Bent tutored on modern African history and I saw photos of him alongside the leaders of many African states, as he had advised many of them during their fight for independence from their colonial masters. Academically and intellectually he had no peer, but when it came to the mundane tasks of life he couldn't find his arse with both hands. When I attended his first lecture, he arrived an hour late, his trousers dripping wet, apologising profusely.

"I am sorry that one is late," he said, "but I locked myself out of my house this morning and had to get back in through my kitchen window. At which point I fell into the sink where my underwear was soaking after I had a little accident during a dinner with President of The United Republic of Tanzania."

As he turned to write on the blackboard, he displayed his bare backside to the audience, he had torn the arse out of his pants. To the relief of the class, the split in his trousers was covered by his shirt which thankfully was always hanging out over his belt. Each time he had to stretch to write at the top of the blackboard, the whole class would look at their shoes. Professor Bent did not make our next class; he had been arrested for vagrancy while campaigning for the local Social Democratic Party outside Woolworths in the town centre. While Professor Bent was absent, Professor Harrington stepped in to take his class for the morning.

Like Professor Bent, his introduction was also unusual, "Many of you may have heard that last night I was drunk in the student bar," he said, "and that later I was seen chasing after a male fresher. This is entirely true, so let us now proceed with the lecture."

In my second year, I actually had a steady girlfriend, Karen, a Scottish woman who is now one of the country's foremost surgeons and who has helped me over the years secure specialist medical equipment for my aid missions abroad. We first got together when she noticed a photo I had cut out of the newspaper and placed on my wall.

"What are you doing with a picture of a can of Shark Repellent above your bed?" she asked. "And what does the minute writing at the bottom say?"

I told her to read it for herself and she moved close to the wall as she attempted to read the small words I had scribbled in blue pen at the base of the notice. Karen proceeded to read the words very slowly, as the print was so small it was very difficult to read, "IF YOU CAN READ THIS, IT'S TOO LATE!" She then burst out laughing. They say that women often go for humour rather than looks, thank God that in my case that occasionally turned out to be true.

My third year was similar to the first, in that if anything could happen, it did. I found myself allocated to the last flat on campus and Terry took one of the other three rooms available. The other rooms were taken by Neville from Newport and Mackenzie, a Glaswegian. Putting a Welshman and a Scotsman with a flash soft southerner and a straight-talking northerner was like cramming 'nitro' into a small tinder-box packed with gunpowder and phosphorus.

The Welshman, Neville, used to watch the Australian soap *Neighbours* twice a day because he couldn't fathom the plot from just a single viewing. His hygiene was questionable; indeed it was regularly questioned in the student magazine, so we were happy when he somehow got a girlfriend, called Janet, who was able to lead him to the launderette.

The Scotsman and I clashed the most. If there was a bad news story, it was somehow always England's fault. Of course the Scotsman was usually right, but we had a stand-up blazing row in the student Union bar one Friday night, when he was championing the secession of Scotland from the Union. That wasn't why the row erupted; I had no problem with Scotland going its own way, I just thought it was a bit premature. Maybe it was the way I phrased it, "Evolution before devolution," that might explain, though not excuse, why he then resorted to some pretty abusive language for the following eight months.

They say you never really know someone until you live with them. Sadly, it was only when it was too late and we had moved in that we discovered Terry's passion for *Sesame Street* and in particular his devotion to one of the less-well-known characters, Captain Vegetable. Unlike Kermit and Miss Piggy, the character was pretty one-dimensional and simply repeated his name to song, plus a few words, "Captain Vegetable, Captain Vegetable, with my carrot in my hand. Captain Vegetable..." Terry would often sing along over breakfast. Occasionally, Terry's singing would be interrupted by his musical Tetley Tea teacup, which had a light-reactive base. It played the Tetley tune whenever he lifted it up. I relate this combination of tunes only because Terry would occasionally and in all seriousness, knock on my door and comment on my music, saying "Turn that shit off."

Mackenzie laid down the rules of the house on our first week, when he stood in the middle of the room and read aloud from a list he had drawn up.

"Right you three bastards," he began.

That was not a good start, as it was just me cooking dinner for Karen, who would be over later. Neville was in his room trying to work out how to turn on his bedside lamp and Terry was in Southport. Though I was the sole member of the audience, I decided it was only polite to hear out my fellow flat-mate; plus he was always good entertainment especially after staggering back from an afternoon drinking session. There was nothing unusual in his being drunk on a Monday afternoon, but I think he had good reason on this occasion. As I remember, the woman he had being trying to seduce for over a month had just beaten me in becoming President of the Lesbian Society. I should mention that I had gone a bit power crazy and campaigned to be president of numerous societies including the Student Union, under the alias of Penelope-Perrier Porsche.

Mackenzie stood on a chair to read out the first of the rule of the house, "Terry is to stop singing any more stupid fucking kids' tunes."

"Does that include the Scottish national anthem?" I asked from the floor.

"Fuck off you English twat," Mackenzie continued, "Neville's to learn how to turn the fucken television on himself and stop waking me up in the middle of the fucken night to do it for him."

"What about his girlfriend's vibrator?" I asked, "You don't expect him to switch it on himself?"

"Get to fuck," he replied.

I was next, "Your London wankers, I had enough of them last Christmas, they are not fucking coming up here anymore and taking the piss out of me, particularly H."

"It's H's way of saying he likes you. Also, if I pass your message on to him you may hurt his feelings. He has a sensitive nature, so there is a good chance he will come up here, tear your head off and shit down your little Scottish neck."

Mackenzie fell off the chair and immediately fell asleep under the television. I decided to not disturb him, but when Karen arrived and sat down for dinner she kept asking if he was OK and why he had the TV remote in his hand.

"I put it there for when Neville comes in and wants to watch the evening episode of *Neighbours*." I said.

The next day, Mackenzie found that the list of house rules in his top pocket had been replaced with another piece of paper with only the words DEVOLUTION BEFORE EVOLUTION as a title. Naturally, he could not remember any of his house rules, so I thought it was not worth going through them again for the others, except for one that I added. That was that the ashtray of butts must be emptied daily, which would cut off Mackenzie's sole supply of tobacco when he had no money at the end of each month. This rule was adopted by a vote of three, with one abstention, as Mackenzie was down the pub.

For the rest of the week and into the weekend, Mackenzie was in a foul mood, cursing me at every opportunity. That wasn't helped when H and Bourgeois burst through the door, paying a surprise weekend visit. I knew they were coming; the surprise was that I forgot to tell Mackenzie. H threw his weekend bag on the table, sending Mackenzie's breakfast flying into the air and he then slapped Mackenzie on his balding head while giving him a hearty welcome.

"Alright Sweaty," he said, which was rhyming slang, derived from Sweaty Sock, meaning Jock.

"See I told you he liked you," I said, and went back to reading my paper.

H and Bourgeois had already sunk a few lagers on the train, so after throwing their sleeping bags into Mackenzie's room they were setting off to the local pub.

"Fancy a pint?" asked H.

Terry and I declined, but Neville joined them and Mackenzie was already out of the door. When Terry and I joined the four of them later, Bourgeois and Mackenzie were already barred, even though it was only three o'clock in the afternoon.

"What happened?" I asked H.

"Mackenzie called the bar manager an English bastard, and you know how sensitive the Welsh are."

"And Bourgeois?" added Terry.

"He's drunk and when he dropped his pint and the glass smashed, the Welshman came over and screamed at Bourgeois, "This place isn't filled with glasses you know!'"

"And he barred Bourgeois for that?" asked Terry.

"Not exactly, Bourgeois replied, 'This is a pub, isn't it?' so the manager threw him out and Bourgeois fell over Mackenzie lying on the ground outside."

"Well it's always good to see everyone is getting on," I added.

"What about him?" I asked H, nodding towards Neville.

"No idea. He shut down two hours ago, I think his brain blew a fuse when he started to read what was written on the beer mat," said H, as he ordered three more pints from the aggrieved manager.

We never received them as we too were barred, when H added to his order, "Do you ever smile you miserable Welsh bastard?"

So we picked up our two colleagues outside and set off to the student union bar. A friend of mine, Vanessa, was already holding court at the bar. She was looking forward to meeting my London friends. H liked Vanessa and he had asked me about her the last time he had come up for the weekend.

"She speaks three languages, loves horses and gets horny eating marshmallows," I had informed him.

H went over to Vanessa and offered her a marshmallow from the bag he happened to have in his pocket.

"H, I take it you have been talking to Righten? There is more to me than just sex you know, I also speak a number of languages and I have my own stables," replied Vanessa.

"Wombat never mentioned," responded H, and knowing him, he wasn't lying. It was just that all that had registered from my description was "horny" and "marshmallows".

Vanessa and I were laughing, but H was bemused as his introduction had seemed perfectly proper to my good friend.

The party later that night back at our flat was a little raucous. Sam came up to me and said the Head of Halls, was outside the flat; I said I would deal with it.

"It's too late." Sam said.

"Why?" I asked.

"H is already talking to him."

I sobered up instantly, as indeed did everyone in the room, even Mackenzie lifted his head briefly from under the television table and Neville moved an eyebrow, which indicated that something had seriously disturbed his brain cell. By the time I got there, the Head of Halls had gone, so I looked over the banister to the landing at the base of our staircase.

"Thank God, no body. What did you do with it, H?" I asked.

"Nothing, I explained the situation and he left," replied H.

Somehow, I felt that I was only getting a selective version of their meeting. Sure enough the next day, the Head of Halls walked straight into our flat unannounced. That morning, Vanessa and a few of her friends were cooking breakfast for us; or "taking pity on those poor retarded delinquents", as she put it. H had already put *Debbie does Dallas 2* into the video-player, and provided a brief introduction to the film.

"There's a bit too much plot," said H, "but it has much better production values that the first one. Plus, her tits look bigger."

Vanessa and her friends didn't bat an eyelid, but the Head of Halls burst through our door only to stop instantly as he saw the porn on the TV. Rudely, he completely ignored Mackenzie who was still lying unconscious beneath it.

I greeted him, "Pull up a seat and we can put on another sausage if you like?"

He ignored me and then he somehow managed to turn himself away from the TV where Debbie appeared to be skiing with a penis in each hand. He then saw H and he immediately backed out the door.

Bourgeois turned to me, "I believe you have recorded the largest amount in fines levied on a student in the history of the university."

"True, thirty-five fines, even more than Mackenzie down there, who had three fights with members of the rugby club in one evening and has only accumulated seven. He's a disgrace to the flat and he is lucky I'm still talking 'at' him."

"Neville, how many do you have?" Bourgeois asked.

Neville appeared to be having trouble getting a Rubik's cube out of its box. "I don't know," he said, "I never open my mail."

On Monday morning, I was ordered to the Dean of Students' office and was immediately challenged by the Head of Halls,

"Your psychotic friend nearly fucking killed me," he said.

"Err, can you be more specific? All my friends are psychotic to some extent," I replied.

The outraged Head of Halls turned to the Dean, who tried to calm down her colleague, but she failed.

"His friend said that unless I fucked off, he would rip my face off and use it for a lampshade."

I turned to the Dean, "Well he would say that, wouldn't he?"

The Head of Halls exploded again, "Are you saying I am a liar?"

"No, I am saying that sounds just like something H would say," I replied.

"Christ, let's just agree on a fifty pound fine this time and by the way, is there any chance of you paying any of the previous ones?" the Dean pleaded.

"You will have to talk to Neville; he deals with all my mail." I said.

Over the three years I was at university, I met many of Waffle and Terry's friends who were easily on a par with my collection of life's fuck-ups. One of Waffle's mates was 'Fish-shop Bill'; he was not blessed with much luck. When I went up to Hull for a New Year's party, Fish-shop Bill was being teased by the others over the tattoo he had got the night before.

His misfortune was that he had needed to get really drunk to have his football team's logo tattooed on his arm. That explained why he hadn't noticed that the guy who applied the tattoo had also been in the same bar drinking all day. Hence Pawn-shop Bill ended up with HULL CITTY tattoo.

Later that night, we went to the local pub in the little town of Swampland and Pawn-shop Bill was very reluctant to go in.

"He in the shithouse again," Waffle informed me.

Pawn-shop Bill looked very uneasy.

Waffle continued, "He needed money to go drinking last week, so he decided to make a copy of his dad's video porn collection and sell the copies down the pub. All went well until the next day, when all the guys down the pub who bought a copy wanted their money back and were set to kill him."

"Bad copies?" I asked.

"You could say that, he did not have a video copying machine," then Waffle turned to his subdued friend, "So what did you use?"

"My mum's cam-recorder," Pawn-shop Bill replied dejectedly.

"Yes, the quality would be suspect," I added.

"Actually, the picture quality was pretty good," added Waffle.

"Err, so what was the problem?" I asked.

At that point Waffle's mate, Boverby, leant forward somewhat agitated, "Well, when you buy a snide porn-video and the missus is out and you settle down with your pants around your ankles, you don't expect to see the reflection of your mate in the film with his pants around his ankles as well."

"Look, it was good quality porn, so I got carried away and I hadn't been getting any from my girlfriend," Pawn-shop Bill replied.

"Even less now, as she's the barmaid in the pub he sold the copies in. She dumped him when she found out that he sold her dad a copy. Actually the whole town knows about it," Waffle informed me.

Pawn-shop Bill added, "My dad's not too happy either. My mum's not talking to him, she didn't know about his stash of porn."

I love Waffle's family. I have never been at another breakfast table where there has been so much warmth and where I have laughed so much. Waffle's dad turned to me at the breakfast table and talked proudly of his family, "I am so lucky that none of my children have a talent or they would be at home playing the fucking recorder, all day."

"You're right there, father, I couldn't blow a Member of Parliament," replied Waffle.

Waffle's mother, Vivian, joined us at the table, "The hardest part is that my sons never clean their rooms. Well, I know every mother has to deal with that, but I can't believe the amount of porn they get through. I can't even clean under their beds without being confronted with a big hairy fanny staring me in the face," she said with deep irritation. Waffle's dad looked at him, with the kind of pride that only a father has. There was constant banter, delivered with honesty that is rare in families. I loved it.

Terry was not to be outdone by his northern neighbour in the silly as arsehole stakes, as he had his own assortment of eccentrics on the other side of the Pennines. On a visit to Terry's hometown, Southport, I met his friends. One of them, Bennett, known commonly as Benefits, seemed to have a grudge about something.

"What's up?" I asked him.

"These bastards have fucked my life up." he said.

The others sitting around the table in the pub, looked in every direction but his. I then learnt that Benefit's girlfriend went missing a few years ago. The rumour, started by his friends, was that he had murdered her. Despite his repeated protestations of innocence, the rumour severely curtailed any potential relationships for him, as when he was seen chatting to a woman one of mates would interrupt, "He murdered his girlfriend, you know."

He had some reason to think that his friends had ruined his life.

"You're a bloody strange one, though," said Terry accusingly to Benefits, "you trapped my girlfriend on the sofa at that party last night and bored the arse off her with a conversation about how to tell a car by its indicator."

VD, Terry's best friend, interjected, "You're a fucking weirdo, so my money was on you that you did kill her."

"Bollocks, I'm supposed to be your mate, you should have given me the benefit of the doubt. Something else could have happened to her, you know." demanded Benefits.

"Only that she took her own life, because you bored the tits off her," added VD.

I should add here that, eventually, Benefits' original girlfriend did turn up after having travelled the globe. At that point, Terry decided to move the conversation on to a friend's wedding which they had all attended the previous weekend. "An excellent day, shame though that Dan got a seat at the top table." said Terry.

"Yes, when he got up and did that robotic dance and then started singing that was a surprise," said VD.

"Yes, I couldn't hear what he was singing, as they put all the fuckwits on the one table at the back," said Terry who was also sitting at the table. "A bit weird they put the grandmother with us and seated her next to me," said Terry.

"They originally put her next to me, but she moved across to you. She said she wasn't sitting next to a murderer," said Benefits bitterly.

"Fucking right, that is why I'm sitting opposite you now," added VD.

"I didn't fucking kill my girlfriend!" screamed Benefits.

Terry continued, "When Dan started singing, granny next to me kept straining on her hearing trumpet and asking me about the lyrics. 'Did he say Shithouse?' she asked me at one stage, as we both looked up to see Dan dancing, jumping from one leg to the other singing some inaudible ditty." Terry continued, "I strained my ear a bit more to work out what he was wailing on about, but even then all I got was the chorus of 'Dan, Dan, the sanitary man, I pressed the handle and my new arrival went down the pan."

"Bog-cleaner, is our Dan," Benefits explained to me.

"And an accomplished songwriter too; he wrote that himself you know," added VD.

"Yes, that was his first public performance too," added Benefits.

"His mum and dad weren't happy," said Terry.

"His sister, the bride, wasn't too pleased either," said VD.

"Yes, you would have expected Dan to wait until the speeches were over," added Terry.

As for our last term at University, everyone knew that it was going to end in tears, everyone, that is, except the occupants of our flat. On the last Saturday before our final exams, we were raided by the police. This was due to Neville's moped catching fire in our flat, which then went flying out of the window. The police were delighted to seize an opportunity to raid the most notorious flat on campus. Terry and I were a bit bemused, but I think so too were Neville and Mackenzie. We quickly agreed not to say anything, mind you, we had no idea what was going on anyway. When the Detective Inspector walked into our flat, he asked us if we wanted to take a seat.

"Who the fuck do you think you are to come into our flat and tell us we can sit down if we like?" Mackenzie exploded.

We were all promptly arrested. If Scotland ever has a seat on the UN Council, I will immediately get to work building a nuclear fallout shelter. Mackenzie and Neville could not assist the police with their enquiries, as everything from the night before was a drunken blur and Terry and I had not been present.

Mackenzie was never so happy as when we were arrested, finally achieved the notoriety he craved. We would be the talk of the campus. However, in the police station, there were no cells left and he was locked up in the cleaner's' broom cupboard, along with her mop and bucket while Terry, Neville and I got a cell to ourselves. We were released the following morning, and with the court case pending, any right-minded people would have focused on their final exams. We headed to the pub, but forgot about our friend who was not released until discovered by the cleaner on the following Monday morning.

Some of the female students gave Waffle a colouring book in preparation for his final exams, and he was delighted to cram in some final crayoning.

"Facts only restrict the imagination" he later told me.

The week of my finals was chaos. The wheels were stolen off my car by my old friends in the OTT, so when they ganged up on me in the student union I was delighted to fight back. I also had the aftermath of the moped incident and the police kept taking me in for questioning during the night and dropping me off late in the morning for my examinations. Surprisingly, I ended up with a good degree despite the efforts of the local constabulary.

On the final day of our exams, Waffle came out from the exam slightly perplexed.

"What's wrong?" I asked.

"Well, I was a bit surprised by some of the questions." he said.

"Yes, it's the throw of the dice, sometimes your specialist field comes up, and sometimes it doesn't." I said.

"I know, but I thought the question on the reasons for the fall of the Ottoman Empire was tough."

"What?" I exclaimed.

"I think I handled it well and I gave an answer on the basis that, with a name like Otto, their leader was not going to inspire the troops. A bit like why we nearly lost the Second World War when Neville was prime minister, but we were saved by having a leader with a good solid name like Winston leading the nation."

"You idiot, our paper was titled 'Twentieth-Century American Politics', you sat on the wrong side of the hall and did the medieval history paper." I said.

Sure enough, within the hour his mistake was noticed and Waffle was called into the senior tutor's office. All of us waited outside for the University Board to decide if his four years were all for nothing.

"What did they say?" I asked, when he came out.

"I can re-sit the paper," he cheerfully replied.

"Brilliant, but how did you wrangle that?" I continued.

"I told them it was your fault."

"OK, whatever works," I said.

"Yes, I told them that you told me where to sit and they just went, 'that explains it.'"

I had to admit that Waffle knew his audience. He knew that where there was a problem, the University Board would believe that I was at the centre of it.

We celebrated that night and Waffle moved in on a very attractive blonde swigging from a two-litre bottle of sweet cider.

"You're a classy lady, what's your name?" he asked.

"It's Annabel," she replied, taking another swig.

"Ah, Annabel, that was my father's name."

Her rugby-playing boyfriend appeared later and caught them together, which resulted in Waffle leaping out of the bathroom and landing in the fish pool, just as Terry urinated into it. The next morning Terry and Waffle were discovered in a field, tripping on magic mushrooms, while dancing a tango for the pleasure of a bemused cow.

Though the court case over the burying moped was dropped, I did not attend the graduation a few months later; I could not abide all the pomp that went with it. Terry left university and, soon after, entered the world of psychiatry.

"Staff or client?" I asked him, when I heard of his career choice.

"Staff, so I can violently abuse you when you're admitted," replied Terry.

Terry is very good at what he does and over the years he has worked with patients who suffer with severe mental disabilities. I admire Terry a lot. Incredibly, Waffle moved into the same field of psychiatric nursing and, like Terry, he has excelled. These two benevolent rogues are now experts in their field, helping those with some of the most severe health issues in the country. Yet, given the right circumstances, it wouldn't take much to reunite the duo to perform some old dance routines for Daisy and her grass-grazing friends.

I don't understand websites that offer you the opportunity to get back in touch with friends from school or college. If you had wanted to remain friends you would have stayed in touch with them in the first place. Having said that Karen once phoned me after we left university to express her annoyance about the fact that I had got married without telling her. It was news to me too. She said that she looked up Friends Revisited and, under Sam's entry, it said that we were married with two little kids and lived above a Costalot Coffee shop with a guinea pig called Beckham.

Chapter 23 – *A Game and Too Many Halves*

I mentioned that my football team was Chelsea, but don't worry, I won't regale you with tales of the club, but rather of the camaraderie that comes with any sport. The team you support is the first conscious decision you ever have to make. When you're a child you have to pin your colours to one team, even if you are like the French football national team – you don't have an interest in the game or know its rules.

Your parents have already determined your religion, you may renounce it later as I did, but it's likely that you'll have been mentally or physically scarred after being nearly drowned in a baptismal font or having had your foreskin removed by a hairy man in a frock and a skullcap. Discovering the love of your life and even deciding upon marriage all happens later in life and, even then, it is by no means till death do us part. You and only you, as a child, select your life-long passion, and whether it's football, baseball or cricket, you are fettered to it forever through the highs and more numerous lows. Even when you sign a player you despise or you are led by a manager who could not manage a fart, let alone your team, you stick with your decision. Your passion will disappoint you, betray you, it may even fade a little as the years whittle away, but if you cheat on it and change your team, you will be forever be mistrusted by those around you, and rightly so.

Chelsea is for the uninitiated, known as the 'Blues' to supporters and as 'mindless lunatics' to the media and the general populace. Chelsea's origins are in London and their ground is called Stanford Bridge and, for the following reasons, they are hated by many and often justifiably. In the seventies, they had one of the most violent followings in the country; some of their followers were also associated with racism, while most were seen as flash, arrogant Londoners. Even when the club had a hotel built next to the ground, the sign above its reception was written in Latin, which when roughly translated read, "Rome wasn't built on democracy it was built on destroying its enemies." In the end of year's report of the club's finances, it stated that the hotel was operating at a loss. The chairman of our club at that time, known as Captain Birdseye due to his snow-white beard, was renowned for making enemies: he didn't even like Chelsea supporters and had applied for permission to erect electric fences in order to barbecue us if we fell against them.

Since those days, the club has changed for the better. In the seventies and eighties, racist violence was common on the terraces, but then fortunately things changed. We had one of the first black professional football managers, the brilliant Ruud Gullit who originally came to us as a player. We had a new billionaire Russian owner and then the greatest manager on God's earth or his earth as he might say, José Mourinho, the self-proclaimed 'Special one'. So you see that we had all avenues covered on the 'We hate you' front: foreign intake, a billionaire owner, an arrogant egotistical manager and, let us not forget, vocal fans who were not afraid to voice their opinions, like H, Chelsea Mark and Amelia. Our critics were spoilt for choice.

As a kid, I felt like I was in the middle of a child custody battle. My father's family were Tottenham supporters and Biddy's family were all Arsenal fans. There is a locally held belief that when my dad fell off the scaffolding, there was a sighting of a Mrs Doyle-like figure in an Arsenal top making the tea on the platform of the scaffolding that day. I asked Biddy about this rumour years later.

"Where's your evidence?" she replied and carried on drinking her tea, never once removing her eyes from the B-list celebrity actress eating a live beetle on a TV reality shown.

Each side of the family tried to lure me into their camp, very much the way the Catholic Church had tried and, like them, everyone failed. When you're a kid everyone wants you to carry on their traditions. There is a picture of me somewhere in a pushchair wearing a Tottenham scarf and holding an Arsenal rattle; today such behaviour would rightly be registered as child abuse. I was dragged along to mind-numbing, boring games at both grounds and plied with sweets to entice me into their folds. Naturally, the view of this ten-year-old was 'fuck 'em' and I remember my Uncle Paddy telling the local priest of my bad language after he asked me if I would like to be an Arsenal fan,

"He told us roughly the same thing," replied the priest, shaking his head.

Then I discovered my passion and was drawn to the roguish nature of West London's Chelsea. Players like Peter Osgood and Alan Hudson, epitomised their 'Jack the Lad' flamboyancy, and were much more exciting than the dour players in the two North London sides. Their antics were legendary. When Chelsea won the FA Cup in 1970, journalists had to slip a fiver into the hand of one Chelsea player to get on the bus and take photos of the team on their victory parade. Peter Osgood, our renowned striker, was once castigated by the manager Dave Sexton for not trying hard enough in training when the team was running laps at Lingfield racecourse. At the next training session, Osgood beat everyone else to the finish line, though he did so astride a horse. Even the manager smiled. I first saw Osgood on television when he scored against Tottenham and he then ran up to their supporters and gave them a V-sign. It was different and I loved these talented rogues. It was at that moment that I decided to become a Chelsea supporter. I made my decision and to hell with the opinions of family and school friends; I was a self-made Chelsea fan. Then the team was split up and there followed twenty-five years of crap.

You learn a lot by being constantly abused by supporters of more successful teams. You become like tempered steel, forged under intense barracking and an occasional battering. If I went against the grain and supported a non-local team, then my good friend Tommy Spitfire did the same, but more of him later. Naturally Bear went beyond just going against the grain; he cleared the forest. Often he would turn up to meet us in a pub wearing the colours of some obscure team. H and I met Bear later on the day that his sister had bought him a Norwich shirt rather than the Brazilian one he wanted.

"It's good of you to turn up in at least wearing a shirt from an English team," I said, "though I think our game against Norwich is?" I pretended to check my invisible diary, "Never, as we are not in the same division."

"Well you know it's my birthday today?" he informed his best friends.

"I didn't and I don't care," replied H.

As usual Bear gave H a side look and continued, "I can't believe my sister bought this health and safety vest, instead of the Brazilian national football shirt. And, she made me promise to wear this every birthday."

"Yes, what else could you possibly want? Other than the national kit of a team you could not find on a map," I added.

"It's in Africa, university boy," replied Bear.

"When did your sister last see, when you were at school perhaps?" I asked. "You're ten stone too big for it. You should wear it every birthday, so we can watch it ride above your increasing girth each year. When you're forty it will look like a neck wrap," I observed.

Bear was very depressed by the shame of wearing something uncontroversial. Then H suggested that one day his boob-tube may well fit him.

"Really, when?" said Bear, perking up at this news.

"A few hours after Cockey has been gradually feeding your limbs into a ham slicer," suggested H.

In the eighties, our London rivals Spurs and Arsenal won more trophies, while Chelsea was relegated in the 1986/87 season. It did not matter though, as a hardcore fan's loyalty is not dependent on continued success and it now meant that we played lower division teams and enjoyed new adventures in foreign lands such a Grimsby. I remember Grimsby for being knocked to the ground by a rival fan holding an inflatable fish.

In 1970, the most violent FA Cup game in history was played when Chelsea clashed with Leeds, which we won in the replay. Since then, there is no love lost between the rival fans. Our club unveiled one of the first electronic video screens in the country on the same day that Leeds came to London to play us. The screen was very large and basic compared to what you see now; it displayed something that looked like a blue Mr Man which jumped up and down when either team scored. In case we did not gage the significance of the dance, it also had the word 'GOAL' flashing underneath. I laughed at this until we played Cardiff: they needed such a prompt, as it had been so long since their team had scored that some fans failed to register it. Leeds were already two goals down by halftime and their fans had had their fill of Mr Blue's dancing. When we put the third goal in, the Leeds fans then engaged in what is now called 'audience participation'. Some of their fans grabbed a girder that had been left behind by the construction crew building our new East Stand and used it as a battering ram. They charged headlong into our new state-of-the-art video screen. Maybe it had nothing to do with the game and they just lashed out at this new frightening technology, which was yet to arrive in the north. This similar to stories you hear of newly discovered tribes smashing cameras in terror that they would steal their soul.

Tottenham was another game you wouldn't want to take your mum to, unless of course it was Biddy who feared nothing. Sullivan, a staunch Tottenham fan, agreed to meet Chelsea Mark, H, myself and Kathleen, who was attending her first football game, in a nearby pub after the match. Sullivan chose the pub, as it had wheelchair access. We nearly all needed wheelchair access afterwards, as it turned out to be the meeting place for Chelsea's hardcore hooligan crew and it was right alongside the Tottenham supporters' main access road to the ground. Within a few minutes, the pub was surrounded by Tottenham fans trying to break through a police cordon, while throwing bricks and anything else through the pub's windows.

Kathleen was terrified and couldn't believe what was happening.

"This is nothing, wait until the police dog comes through the main window and trust me he won't be wearing a red cape with a big yellow 'S' emblazoned on it either," I informed her.

When the pub's manager told everyone that all Chelsea fans had to leave the pub and make their own way through the braying mob, Chelsea Mark declared himself a Tottenham fan, to the disgust of Sullivan and H. Only Chelsea Mark could unite fanatical Tottenham and Chelsea fans at such a moment.

"Oh Christ, we are all dead," said the bar manager, as more fighting broke out and the police lost control of the situation.

Then, in a surreal moment, and to the surprise of the embattled police and the bar manager, all the Chelsea fans started singing Don McLean's song *American Pie.* They roared their lungs out when they got to the line, 'This will be the day that I die, this will be the day that I die.' This really didn't placate the baying Tottenham fans who now really wanted to make our wish come true.

Going up to Liverpool for a game was another situation where you had to keep your head down. Unfortunately, I was not very good at it. Before one FA Cup game, I was having a pre-match game drink with fellow Chelsea supporters H, Wurzel and Churchill in a bar near Liverpool's Anfield ground and trying to keep a low profile. Then some locals entered the bar wearing black permed wigs, matching fake black moustaches and red shell suits. After a few pints, I decided I had to go over to them and congratulate them on playing up to the stereotypical image of Liverpudlians.

"What the fok do you mean? We're the foken *Magnum PI* Appreciation Society," the biggest scouser replied, they then all held up their Private Investigator badges featuring pictures of beaming Tom Selleck.

"We're dead," said H, as I returned to the bar.

"Ah fuck it, it wasn't much of a future anyway," declared Wurzel.

As I have mentioned, I had no interest in hooliganism and never swung a fist at anyone because of the colour of their scarf. No, for me the appeal was the camaraderie and I loved the banter, particularly after a game. This involved people with absolutely nothing in common. They may not have come from the same area; the same school and in some cases not even support the same team. Chelsea Mark, H and I were all Chelsea supporters, Bourgeois hated sport of any kind and Bear changed team more often that he changed his underpants. We all met after one game and we were discussing the obscene wages paid to the players. H, a free marketer, had no problem with it and he announced that he too expected soon to earn similar amounts.

"Protection racket, no doubt? OK, I'll buy you a pint if you stop fucking hitting me," added Bear.

"No, I have sent off for a book from the *Daily Mail* called *The Secret of How to be a Millionaire*," explained H.

I asked H what the book's secret was.

"The secret is to write a book on something you know."

"How can it be a secret, if all you have to do is order it through a national newspaper? Anyway you haven't received it yet, and you know the secret," responded Bear.

I had to admit that Bear had asked a perfectly reasonable question and deserved a sensible answer. He didn't get it; H just punched him in the side of his head.

"If you keep doing that, one day I will end up fucking brain damaged," Bear complained. No one made a comment, for that 'one day' had well and truly gone.

As Bear never learnt from experience, he continued to press his good friend on his new enterprise. "So the bloke who wrote the book is a millionaire, so I wonder why he needs to write a book?" enquired Bear who then answered his own question, "Ah, I get he's probably a millionaire by getting mugs like you, H, to buy his book," Bear started laughing again, now seeming hell bent on getting another punch in the head from H. He succeeded. Bear picked himself off the floor, sat back in his chair and then spoke of his master plan to secure fame and fortune.

"I want to go on *Who Wants to be a Millionaire*? How far do you think I would get?" asked Bear.

"Well, if you manage to sit on the chair without breaking it, that would be a result," I answered.

"Well, I might get a question right," said Bear, excited by the idea.

"Only if they ask you what your name is, you went fifty-fifty and the other answer was Nelson Mandela," I added.

H still thought Bear would get it wrong and Bear sat back and then said, "Yeah, I'm no good under pressure."

Chelsea Mark was drunk, but he too was excited by H's millionaire idea, "It makes sense to me to write about something you're an expert on, what do you do, H?"

"He's a landscape gardener. That is what the reading public needs, another fucking book on laying turf. A bestseller if ever I heard one. When was the last time you saw a book on gardening?" asked Bear, laughing his head off and only stopping when H punched him again.

Chelsea Mark told us that he was subsidising his studies by stealing academic books for college students, which might be worth a book.

"What happens if they catch you?" I asked.

"They have done, but I only steal from left-wing bookshops, so they never turn you over to the police," he replied.

"You're a true socialist," said H.

"So your idea is to write a book about stealing books, which people who want to steal books will buy? Perhaps you won't make money directly from the book, but maybe people will pay you to steal your book for them?" I suggested.

"I don't understand any of this," confessed Bear, shaking his head.

Bear then turned to H angrily, "You bastard, I told you I would end up fucking brain-dead if you kept walloping me." You don't need me to describe how H responded to that.

Grabbing a coffee at halftime was always a good chance to catch up on world events. H was having a beer with Wurzel, Churchill and Cuthbert. Cuthbert was usually miserable, but today even more than usual. The conversation took off in its usual random way.

"Wurzel was punched in the head by some Arsenal fans during an England World Cup qualifier," said H.

"They took offence to my wearing a Scotland top," explained Wurzel.

"Good old Wurzel, wearing the shirt of a team who wasn't even playing." I continued. "Of course when the day comes that Scotland does qualify, you will be a drooling vegetable following an assortment of blows to your nut."

"Even if they qualify, the sweaty socks will be home before the postcards. By the way, in case your dad didn't mention it you're not even Scottish," added H.

"Get to fuck, you pair of wankers," said Wurzel, sounding almost like a native of his adopted land.

"You're a very angry man, a few more whacks might help, in the same way one kicks an old television to get the reception back," I informed my good friend.

Cuthbert just stood there, contributing nothing to the conversation, not even breaking into a smile at our usual idiotic banter.

"I hear, Wurzel, that you ended up having sex with Doris, the barmaid from the Pig and Crackling, in Warshalton station car park last Saturday night," said H.

Apparently, I was the only one who did not know this. Warshalton in South London is the mirror image of Trumpton, in North London. Everyone knows one another.

"She's a game girl that Doris, I only asked for a packet of cheese and onion crisps at closing time. A game girl indeed," added Wurzel and all present agreed that she was indeed a game girl, though Cuthbert said nothing.

"Yep, I hear the station manager had to screw the 'Welcome to Warshalton' sign back up on the wall and God knows where the litter basket underneath it disappeared to?" noted H.

This was followed by a kind of contemplative silence, H turned to Cuthbert. "Your cousin, isn't she?"

Cuthbert gave a dejected nod.

Warshalton was the home of H, Wurzel, Cuthbert and Churchill and many other Chelsea fans. It is a strange place, a bit like *The Stepford Wives*, though only H Cuthbert and Churchill have mustered enough social skills to attain a wife, which gives you a rough idea of the social capabilities of the others. You know you are approaching Warshalton, by the increasing number of Union Jack flags hanging out of residential buildings. The number of men carrying William Hill's betting slip blue pens behind their ears also rises. No one ever talks about anything shown on television, except for the Gene Hunt character in the TV series *Ashes to Ashes,* which is based on the Jack Regan character in the *Sweeney* of thirty years earlier. Both characters represent values of a bygone age, when a mugger was given a fist in the face rather than a council house.

I enjoy my visits to Warshalton, though the values of my friends there are not those of a liberal lefty like myself. I respect the spirit of my friends in Warshalton and also in Trumpton who reject political correctness and judge everything on its merits without reference to a pre-determined script issued by the politicians or the media. Warshalton, like Trumpton, is home to many black and Asian families and yet I have never witnessed any hostility on my visits. The rules are simply; English is the common language, you work for a living and if you denigrate British troops fighting abroad you will be on the police's missing persons' list by the end of the week.

When I first went there, the local pub, the Bomber Harris, was having a quiz.

"You have a brain, so you're in our team," H informed me.

I have never attempted a pub quiz, but I agreed to give it a go. The questions were all on popular culture, from twenty years earlier, such as, 'Why did *Kojak* suck a lollipop?' and 'What did Wyatt Earp have in his trousers that was six-inches long?' Unfortunately, much to H's annoyance, our team came last with me having provided all the wrong answers. My answers, to those particular questions were 'To break his smoking habit' and 'a six-shooter', but the 'official' answers to nearly all of the questions usually included the word 'cock'. The women's team won handsomely and carried off the prize of twelve bottles of *Hooch* and a frozen duck. I am sure it was merely a coincidence that the local pond was comparatively depleted of wild fowl when considered with the other ponds I passed on my motorbike on the way down to Warshalton. Of course, I knew that the quiz was a set up and that the answers given were a joke at my expense, mind you it would be good if one day someone would confirm that that was the case.

After the porno-quiz, we joined Wurzel at the bar.

"My daughter has just turned sixteen and she is on the Internet now," Wurzel informed us proudly.

"What's her website?" asked H excitedly, believing that a new porn website had just entered the market.

"Err, I mean I bought her a computer for her birthday," responded a bewildered Wurzel.

"Well, let us know if anything develops," said H.

"Err ... I will," replied the confused father.

I caught H's eye, as I looked away, tears breaking freely and rolling down my face on to the bar. Good old H – whatever planet he was on, it was yet to be visited by civilised man.

Gareth, another Chelsea follower and friend of Wurzel, turned up and updated his friends on the progress of his court case against a lap-dancing club.

I turned to H, "Is it another cloning card scam, when someone innocently receives an enormous bill, even though they have never been there?"

"No," replied H.

Gareth's case was that he had been overcharged and that he had only spent a thousand pounds rather than the itemised two-thousand pounds at 'The Thai Ping-Pong Club'. I admire such rogues; caring little about what people thought, he was prepared to stand in the defendant's box to argue his case while the local press took notes in the gallery.

"It was the principle of it," said Gareth, who was not short of a few bob and had made his money through various dodgy scams.

One scam involved the sale of hardcore porn. He owned a company in Holland, but the videos he mailed contained content no more hardcore than an episode of *Songs of Praise*. When disgruntled customers wanted their money back, he sent them cheques specially printed in Amsterdam, with a picture of a woman sucking what appeared to be a bollard, printed across each cheque. Out of roughly fifty-thousand pounds worth of orders, only five cheques, totalling a few hundred pounds, were ever cashed to secure a refund. Wurzel believed he knew three of those who went to the bank to cash the cheques, they were all from Warshalton and I think the others were from Trumpton. H cross-referenced the 'Infamous Five' and agreed, as one of them was his neighbour Mr Patel.

Plank was another resident of Warshalton and one of those who had been squashed behind the door with Wurzel, when God entered to distribute brains and good fortune. His misadventures were legendary, particularly when he went to Chelsea away games. He worked in a newspaper shop and he had trouble getting all the school kids out at the end of the day. This wasn't helped when someone added a 0 in permanent ink to the number on the sign 'NO MORE THAN 1 CHILDREN IN THE SHOP AT A TIME'. As a result, he often missed the train to away games and had to drive to the game on his own. It was a rule that when you travelled to any game, you tried to blend in and not make yourself a target. Wurzel interpreted this rule slightly differently. His approach was to keep his head up, declaring to all and sundry that he was a Chelsea fan and provide commentary on the weaknesses of the rival supporters' team, their future prospects and general appearance. Hence the sudden, but frequent arrival of some metal object to his bonnet.

Plank though, would try to be unobtrusive, but would always suffer some calamity similar to his friend Wurzel. He went to a game in Glasgow and returned to find his car window had been smashed. He realised that it was best not to hang around, so he drove all the way back to South London and it was said that the fire brigade had to use a crowbar to pry his fingers from the steering wheel, when he pulled up outside the Bomber Harris. Even with the heater on, it must have been cold driving in January through the snow in sub-zero temperatures. We later learnt that Plank had left his Chelsea programme on the dashboard of his car; identifying himself to Glasgow's locals as a rival supporter and a Londoner at that. Plank said he had learnt from the experience. You can therefore imagine his distress when he returned after watching Chelsea play Manchester United in Old Trafford, only to find that his van had not only suffered the same fate, but, in addition, it was covered in abusive graffiti.

"I don't understand it, how did they know I was not one of them?" Plank said shaking his head, and he continued, "I even parked in the Manchester United car park."

"You probably are one of them, but that is a different issue," I added.

Wurzel believed he had worked out the answer, "I may be wrong but even in a troop of northern monkeys, there might have one that could read."

"What has that got to do with it?" said Plank who then turned to me. "And by the way I just understood what you meant, so you can fuck off and all, Righten."

"Well your vehicle does have *The South London Newspaper* blazoned in foot-high letters on both sides of the van," noted Wurzel

Plank nodded, realising the slight fault in his camouflage and once again, he declared that he would learn from the experience. He never did.

Plank had a number of other problems, mainly with the Inland Revenue, so he fled the country. At his leaving do, he was blind drunk by the end of the evening, as indeed we all were, including Wurzel, Churchill, H and I. H called on me to say a few kind words on Plank's behalf, as I was one of the few at that point in the evening who could still speak. I stood on the stairs, asked everyone to toast Plank and delivered a eulogy on behalf of my good friend.

"'A toast to a keep-fit fanatic, a lifelong Tottenham supporter,'" at this point Plank peered down at his well-nurtured beer-gut hanging over his Chelsea shorts, "and a standard bearer for our multi-cultural society, especially, in his charitable works for the local gypsy population despite the animosity of all present here." I stood down amidst a deafening silence.

I have never seen a mob in such a quandary, bemused by my words and wondering whether to give Plank a goodbye hug or lynch him, which I believe is still legal in certain parts of Warshalton. I did say it was a eulogy.

I wouldn't say that Plank's move to Spain was a tremendous success. He is limited in terms of job opportunities, which may be due to the language. At his farewell party, I asked him how his Spanish was; he replied, slightly puzzled, "Why do you ask?" He found two jobs, one as a window cleaner specialising in ground floors, as he suffered from vertigo. The other job was selling cardigans on the beach; again, the moneymaking opportunities don't leap out at me. This may explain why, when we all went to see him in Spain, as Chelsea were playing a European Cup qualifier near where he lived, someone commented that he smelt of dog-biscuits.

On our way to an FA Cup game against Manchester United, we decided to spend the night in Blackpool. We all peered out of our minibus as we pulled up outside our hotel.

"Looks like their halfway through building it," observed H.

"Hard to say whether they're halfway through building it or demolishing it," I replied.

We went in, thinking perhaps that its real attraction was its decor, perhaps even its ambience. I was wrong on both counts, as the wallpaper was hanging off the walls, the furniture was smashed to pieces and it looked like there was a dead dog in the corner.

"H, I'm confused again, as this is how we leave a place, not how we enter it. Bear would love it here," I said.

"Is it just you?" the little manager of the establishment asked me, straining his eyes, as he looked me up and down.

"Err ... No," and then I pointed at the other sixteen standing behind me. He proceeded to go along the line looking each of us up and down, straining his eyes even further. I later found out from his wife that he could not see beyond his glasses, in fact I think he could not see as far as his glasses.

Later I had to come downstairs, as I had received an urgent message from work to send them some papers. I went up to our short-sighted, bewildered manager.

"Hello," I said.

"Who said that?" replied the manager.

"Me," this could take a while, I thought.

"Me who?" he replied.

"I'm the Chinese lady in room five," I continued, "Do you have a fax machine?"

"We don't tolerate that kind of behaviour here," he angrily replied.

Well there was only one thing for it and that was to head towards the huge pub directly opposite, leaving our manager to pondering what disgusting acts one could commit with a 'fax machine'.

The bar was an interesting place; it had a huge banner across its windows declaring, WORLD CUP FOOTBALL FINAL LIVE HERE. The World Cup occurs every four years; I could not work out whether it was an advanced piece of promotion for the World Cup in two years' time or if the banner was still up from two years earlier. Upon entering, it was clear that it was based on the final that England had won in 1966, thirty years earlier. As I stood in the middle of the bar, which could only be described as a shithole, Scottie, one of our number who was never backward about chatting up women, came over to H and me.

"Fuck this, I don't know where to begin, they're all mutants," he said, just as a wheelchair ran over his foot and the woman pushing it shouted at Scottie.

"Get the fuck out of the way and let my mum through," she said, and the mother and daughter combo headed to the bar. The mother gave Scottie's privates a quick squeeze on the way by.

The bar was memorable, in being one of the worst I had ever been to. If the fighting and flying glasses was not dangerous enough, an old woman was going around the tables, lifting up the heads of drunken men and sticking her tongue down their throats. Piles of vomit were starting to appear at the tables of the abused gentleman and one threw up over the food counter.

I turned to H and said, "It can only get worse. We would have to be out of our minds to stay here."

"Same again?" asked H.

"Absolutely," I replied.

We would often travel abroad to watch our team play international games, but sadly not everyone wanted to extend the hand of friendship to our foreign neighbours. When Chelsea played a European tie in Prague, H, had no great love of Germany and refused to get out of our van on Aryan soil. I remember when H was offered some porn videos when we worked in the park and he refused a German one, not on moral grounds, but because 'he hated to see the bastards enjoying themselves.' H refused to leave the vehicle even when we stopped at a German service station to take a toilet break and get something to eat. After an hour of taking our time over coffee and some delicious apple-strudel, we watched H urinating out the open door of the vehicle. We all acknowledged his commitment, but it was tempered with anxiety at sharing a vehicle with one of England's top lunatics.

When we reached the stadium on the outskirts of Prague, we entered the main bar of the only hotel in town. Bottles of pilsner flowed freely until somebody shouted out that the coach to the game had arrived outside. We all ran from the bar and I was the first to open the back door of the coach. It was an old battered transit van; I fell over a wheelbarrow in the back and the bags of cement exploded as more of us piled in. With around twenty of us sandwiched together in the back of the vehicle, I had to admit it was pretty small but I was glad we had official transport to get us to the ground in time for kick-off. We were seasoned travellers, and reconciled ourselves to the fact that in little European outposts you couldn't expect too much of the hosts. The driver finally pulled up on a building site and ran off into the distance. Only then did it dawn on us that we had jumped into the poor man's van. He had nothing at all to do with the game and was probably just making his way home as he encountered us at a stop sign. No wonder he ran for the hills on finding a mob of Chelsea fans covered in cement powder staring at him through bloodshot-eyes. It was a long walk to the ground and, of course, it was raining.

When Chelsea had an away game against Majorca, a 'projectile of vomit' (Chelsea Mark's collective term for Chelsea fans) headed for Heathrow. One member of our entourage was Michael, who in all the time I knew him, never smiled. Yet, he did the most ridiculous things and would have me crying with laughter at his deadpan expression.

When we got to the main town, the first bar we entered had a drinking ritual during happy hour of someone putting on a motorbike helmet and then the manager hitting them with a cricket bat. On hearing this, Michael put the helmet on and rushed head-first into the side of the bar, wedging himself in it. This came as a complete shock to the manager who confessed he had never seen anything like this in twenty years of running the bar. He couldn't stop laughing, free drinks were served all afternoon and I later heard that Michael's adaptation became part of the bar's nightly entertainment.

When we excavated a dazed Michael out of the hole in the side of the bar, he did not smile or even acknowledge what had happened. He headed straight over to the rotund little barmaid, Monica. She was standing at the end of the bar with her mouth still open in shock. Michael started chatting to Monica as if nothing had happened and they went out together for the rest of our week there.

I bumped into Michael and Monica a few days later, in Majorca's very small natural history museum just before the start of evening's game.

"You enjoy natural history Monica?" I asked.

"I lick the big dinnersaws," she replied, looking at the structure of an aggressive looking mini-T-Rex.

Monica turned to Michael and pointing at the reconstructed dinosaur bones, asked him, "The monster has a huge mouth, if he existed now do you thing he would eat me whole?"

"No, he would spit that out," replied Michael without a hint of a smile, which she responded to with a big grin, pleased that Michael had assured her she would be safe in the event of such an unlikely encounter. As always on these occasions, I turned away with tears streaming down my face.

Michael was just one of those characters who occasionally comes into your life and leaves just as quickly. I haven't seen him in over ten years, but I would have no doubt the tears would start to stream down my face once more, if I saw his hangdog expression.

On these trips to Europe, Dan, a true blue Norwegian Chelsea fan living in Spain would meet up with us. Chelsea, despite its reputation in the seventies, has thankfully shaken off its xenophobic reputational and has a real cosmopolitan following, particularly at international games. Dan was six-foot-two with long blond hair and looked like he was a roadie for a heavy rock group, such as his favourite band Black Sabbath. On the plane after we lost to Stavanger FC, the stewardess came up to Paul who was a good friend of Dan's.

"Are you a Chelsea fan?" she asked.

"Yes," Paul replied

"Ah! Shame you lost to that little team, well I am a Man United fan myself," she said with an insincere smile. "If there is anything you need, please don't hesitate to ask."

"Well," said Paul, pointing towards Dan. "You can cheer me up, by going up to the big blond fellow over there; he is the guitarist from Status Quo. Would you mind asking him for his autograph please?" Paul looked at me as she tottered away and said, "This will cheer me up."

As the stewardess made her way up to Dan, it occurred to me that as a Norwegian who supported a foreign team, which had just been beaten by a small Norwegian team, he might be a little grumpy. My suspicion was confirmed when he informed the stewardess that firstly, he couldn't play the guitar, secondly, that Status Quo was soft-rock shit and thirdly she could therefore fuck her Man United-supporting, Surrey arse in various ways. He might not have been of English stock, but he certainly adopted the culture.

When Chelsea reached a Cup Winners final, we set off in Chelsea Mark's car for Stockholm and I agreed to share the driving with him, while H sat in the back. After three days, we reached the Swedish border, and the guards seemed delighted to see us.

"Are you English hooligans?" the border guard asked checking our passports. We all let out a collective Neanderthal growl. They laughed and waved us through.

When we reached the city centre, we set off immediately to the pub, having no time to savour the delights of our hotel even though it offered free snacks and hardcore porn. We joined up with Wurzel and Churchill outside the Cheers Bar right in the middle of the city centre. We were having a beer outside on the street in front of the bar, when I noticed an old man in a pink tracksuit crossing the road.

"Have you seen the old gay fellow coming towards us?" I said to Wurzel.

"Err, that's my father," replied Wurzel.

Indeed it was. Wurzel's father Ken, an interesting rogue like his son, was always trying to subsidise his travelling costs by buying contraband abroad and selling it back in England. Nothing unusual in that, if you are talking about a few packets of cigarettes tucked into his jacket on the plane home. However, those ten turtles he brought back following a game in Indonesia were slightly out of the ordinary. Ken did tell me that he received strange looks from the air stewardess when he asked for extra nibbles and then, as he put it, "dropped pretzels into my undergarments."

The atmosphere outside the pub was relaxed and we engaged freely with the locals. However, someone in Swedish police headquarters must have panicked at our growing number and, the next thing we knew, about ten riot police vehicles pulled up in front of the Cheers Bar. Armed riot police, with helmets, visors, truncheons and riot shields emerged and a police cordon was formed in front of us. None of us were the slightest bit bothered, until we noticed that many of the smaller curvy riot cops had long blonde hair sticking out from below their riot helmets.

"By Odin's hairy arse, they're women in fetish gear," declared Wurzel.

"I think you will find they are police, but I get your point," I said as we followed Scottie over to the cordon.

"Sally forth! What! What!" uttered Wurzel, in the voice of the archetypal English character, Terry Thomas.

"Ding-dong, bang on," replied H in similar voice, as the entire English contingent was now engaged in conversing with the police women.

Even the policemen seemed disarmed by our non-aggressive approach and some even took a drink from our beer cans before we headed off to the game. Whatever training the border guards and riot police have in Sweden, it is a far better approach than salivating police dogs and tear-gas, as there was no trouble reported anywhere. Even when Wurzel discovered that his Swedish goddess, upon taking off her helmet, was a male midget with highlights.

"He's a chip off the old block," I said to Ken, who fortunately had not heard my early comment about him being gay. We watched his son as he realised his mistake, and explained to the little policeman, "I'm butter side up you know," indicating his sexual preference.

"I'm sure you are, but which side is which?" I said to Wurzel later only to be met by another bewildered look.

We won the final and had a few beers in Stockholm that night, just to be sociable. The next day all of us, despite the euphoria of winning our first trophy in years, were shattered through suffering from a combination of sleep deprivation and drink. Chelsea Mark seemed to be the worst for wear and appeared to be having a mental nervous breakdown. I first noticed when he informed me that he didn't know how to drive and therefore he couldn't take a turn driving the car on our return journey. I was slightly suspicious, as he had shared the driving with me on our way to Stockholm and we were driving his car. Later, when I pulled into a petrol station, he came up to me and asked what the object was in his hand.

"It's an apple," I replied.

"Where do they come from?" he asked, looking in awe at the round object in front of him.

"Looks like you're doing all the driving back to England," interrupted H, as he didn't have a licence then.

Chelsea Mark did not utter one word on our journey back, apart from screaming when H attempted to strangle him for trying to light a cigarette behind the petrol station's fuel tanker while I was refuelling the car.

When we were back in England and only a couple of miles away from his home, after having dropped H off in Warshalton, Chelsea Mark finally turned to me and spoke his first words in three days.

"You're riding the clutch a bit too hard," he said, after seventy-two hours of me doing all the driving.

Naturally, I replied with a torrent of abuse, but he just looked at me with a big stupid smile. 'What's was the point?' I thought, as my old friend was obviously away somewhere; he continued to look at me with a smile on his glazed and vacant face.

Whereas Chelsea Mark's mental breakdown was a temporary one, I think Wurzel was a fully paid Looney Tune and could easily hold his own in conversation with Yosemite Sam, Elmer Fudd and Daffy Duck. As a result, he would often get some thought into his head and wander off from the group. On many occasions, this made him an easy target for local yobs on the lookout for a stray rival supporter wandering the streets on his own. Having his head cracked opened on a number of occasions only added to his wandering nature. I decided to stay close to Wurzel as he is a good man and we didn't want anything to happen to him, unless of course it was of a non-violent nature and made him the subject of extreme ridicule.

A delegation of Chelsea fans paid a flying visit to see Plank in Spain, our non-Spanish speaking tax exile. We were drinking at a table in the bar Plank worked in and after a few pints Wurzel sat up abruptly.

"Christ! Where is my owl?" he asked.

"What owl?" I asked.

"I bought an owl," he said.

"When?" I asked.

"When Chelsea played Galatasaray in Istanbul, I think," he replied even more agitated.

"Wurzel that was twenty-three years ago. Why ask now?" Churchill interjected.

"Well, I only just thought of it," replied Wurzel.

It was how his mind worked; Wurzel; he is not malicious, he just blurts out what he thinks, or more likely imagines, is in front of him.

Later, he added, "I think it could have been a chicken, but I'm no David Attenborough."

I went to the toilet, more to regain my sanity than to relieve myself. However, when I returned, I discovered Wurzel standing at the bar.

"Your peanuts are a bit rank, Mr Righten," he told me.

"Well they would be, if it wasn't that they're the pips from the olives I have just eating," I informed him.

On reaching the hotel Plank had booked for us, as he was sleeping under a fuel tank in a garage, we signed in at reception while Wurzel was distracted by the lobby's aquarium, which contained a few lobsters and a rock. When it was Wurzel's roommate Barclay's turn to sign in, he couldn't find his pen which had fallen off his ear. In an attempt to help, Wurzel grabbed one of the lobsters out of the tank and handed to his roommate, at which point the poor startled receptionist nearly jumped out of the window.

For the next few days, whenever Wurzel walked through reception, the abused lobster immediately scuttled towards the glass wall of his tank; he was bug-eyed and kept banging his pinchers on the glass. He was clearly seeking revenge for being assaulted in front of his soon-to-be-boiled colleagues.

For our away trips, we always put Wurzel and Barclay together when a hotel was involved, simply for the comedy value. They were complete technophobes and couldn't even wind up the hotel clock in their bedroom. On that first night in Rome, they required assistance to operate the standard two-tap shower. Later that week in another hotel, I noticed that the shower had six knobs and three dials. It looked like the control panel for Apollo Ten, so I knew Wurzel and Barclay were in trouble. Sure enough, Wurzel was in reception an hour later, wearing only a pair of shorts and demanding assistance to programme his shower. No member of staff would dare enter the Englishmen's room, so I had to do it. While I disabled the shower controls, I then noticed on the way out that someone had squeezed toothpaste on to the television's remote control.

The next day, I was having breakfast with H and Churchill across the road, when Wurzel made an appearance on the front balcony of the hotel. Again all he had on was his shorts and, though I could not hear everything, the words, "fuck", "shower" and "sweaty bollocks," could be heard across the Piazza. Locals looked up at the Best of British bellowing out obscenities and waving his arms in the air, in what looked to be a cross between John Cleese at his most animated and Benito Mussolini as his feet left the pavement when the locals hung him from a lamppost. Wurzel and Barclay joined us a few minutes later; both looked pretty rough and unshaven, having being without hot water for a few days. We were crying with laughter.

"Look, what do you expect when you put the two biggest bookends together in one room?' declared Wurzel angrily.

"Quite right, you tell em Wurzel!" said Barclay, supporting his friend's argument.

I mentioned to H that I would often keep an eye on Wurzel when he was abroad.

"You mean to make sure something happens to him?" laughed H.

"What do you mean?"

"Well remember in Genoa, when we all got soaking wet in the rain and he wanted to go off and buy some dry clothes and you went with him?"

"Yes and your point is?" I asked.

"Well, you brought him back wearing a latex bondage outfit."

"It was waterproof," I added in my defence.

Wurzel then interrupted, "And what about the time, when we were drinking late and I left my phone on the bar?"

"Yes. You never thanked me for picking it up and returning it to you."

"Thanked you? You texted my ex-girlfriend at three in the morning, saying, 'BUM SEX NOW.'"

"Well, it's your fault," I said.

"How the bollocks can it be my fault?" Wurzel countered.

"Well, when she phoned you the next morning, you apologised. So it must be your fault otherwise why apologise?" I reasoned.

"Well, I was so drunk, that I couldn't remember what had happened, and it sounded just like the kind of thing I would do," and he then added, "and another thing, you also texted the same message to some Chelsea lunatics as well. It was fucking embarrassing when I walked into the pub the following Saturday before the game and everyone shouted, "Hello Sailor.""

"As I said, it's your fault," I continued.

"But I didn't text anyone, how the fuck can it be my fault?" he argued.

"Look, I'm your friend and who picks your friends for you?"

"I do," said Wurzel.

"You see, it's your decision to have me as your friend, and only you can make that decision, no one else. You're a grown man and you have to take responsibility for your actions."

"Well, he has a point Wurzel," replied H.

"Yes, you're quite right, harsh, but fair," conceded Wurzel.

On one of our trips, we stopped off for a quiet drink in Amsterdam on our way back from a game in Berlin and ended up staying there for two days. Some of our group kept disappearing, as one put it, 'to seek a further understanding of Dutch culture', which meant to go with the working girls. H, Wurzel, Barclay and I spent an evening playing pool, getting pissed and listening to the Clash on the juke-box in the Excalibur bar.

One of our group Kelvin, returned from 'looking up an old friend' as Wurzel so delicately put it. He said to me, "I understand H is in a relationship and, in the case of Wurzel and the others, any decent prostitute would stick her head in the oven if they knocked on the door," he paused, "but you're single, so how come you're not sampling the local delights?"

I turned to him and repeated the words had I once heard from my good friend Sullivan, "Not for me, you see I like women."

I am not prudish and I didn't mean to judge him, – each to their own, I say, but I was with Sullivan on this. To have sex with a woman for money, without any emotional contact, would be a sterile exercise. Even if I did, I know I would feel terrible afterwards for having used another human being in such a way. I need some kind of chemistry, even if it's just sharing a laugh.

You learn a lot as a fan travelling abroad, particularly that you are stripped of your rights. Our Minister for Sport at the time branded all British fans scum. That included the poor English guy who was staying with his wife in a small central hotel in Dusseldorf, who popped out to get a packet of fags and went into the bar downstairs which was full of England fans after a game. He was then escorted on to a plane by the Dusseldorf police along with everyone else in the bar and sent back to England.

I remember meeting a young couple on a ferry to Belgium when I was on my way to see Chelsea play Bruges. I parked my motorbike on the ferry's loading bay and headed up to the restaurant, which was deserted except for a young English couple. I got talking to Matt and Charlotte and discovered that, unlike every other English person heading to Belgium that day by train, plane and car they were not going to the game. Instead they were spending a day shopping in Belgium and buying each other wedding gifts. I congratulated them, but I was a bit worried and warned them that the boat would be packed with Chelsea fans returning home after the game. I suggested they should try to get an earlier ferry if they could.

Matt then said, "Is it true what they say about Chelsea fans, being dangerous and all?"

"That's all nonsense, we are nice people really. Just don't go near us when we're eating," both Matt and Charlotte were still in their teens and they looked genuinely shocked. "Don't worry, just joking," I added, and both breathed a sigh of relief.

When we docked, I rode my Zephyr motorbike (I lovingly called Blue Nellie as I am a sad case) up to the Belgium customs control. I then saw that Matt and Charlotte had been stopped by the police. I got off my bike and went over. The border police told me that because the couple had British passports, but no match tickets they would not be allowed to disembark. It was blatantly obvious that they were not ticketless fans trying to bunk into the game, but the guards refused to let them through. I then refused to move my bike until they were released and I read the border police the 'riot act' ranging from the Geneva Convention to the Magna Carter. A few other late travelling Chelsea fans joined the argument and the border guards started to get worried and eventually waved the innocent couple through. The delay meant that I was going to miss the first half, but I need not have worried, for on entering Bruges it appeared no one was getting in for kick-off.

As I drove into the town centre, the police were aiming water cannon in the direction of the Chelsea fans on the main street. It was pandemonium. I pulled up, thinking that I would not be going anywhere for a while, when all of a sudden one of the riot cops spotted me and ordered the water cannon to be turned off. He directed me to continue on my journey. Then the full force of the police in riot gear rushed into the Chelsea fans, raining their batons down on English heads left, right and centre. I was at a loss as to why the police changed their tactics to wave me on, but I'm not one to miss an opportunity; I started up my engine and carried on my journey.

Motorbikes don't have number plates on the front, so the riot cop must have thought that I was a local, trying to get to the match and so he ordered the riot squad to clear a path for me. The Chelsea fans were cheering as I motored through with a big smile on my face, which bewildered the riot police until they saw my British number plate. It was too late to stop me. I drove into the stadium car park and parked up in the directors' car space. I was one of the first to take my seat for the game, in a still empty stadium.

The story did not end there, as the same thing happened on the way back. That time though, I blended in with the cars of the local fans leaving the ground. I did stop when I got past the same riot cop who was in charge of launching repeated attacks on my fellow supporters. I gave him a few toots of the horn and a wave just as I sped off, to the amusement of my fellow supporters. I saw one familiar face get hit full in the face by the jet of water from the cannon, but I knew that it would be water off a duck's back to Wurzel.

Back on the ferry, I decided to get some food only to be confronted by a very large angry woman; the ship's cook. She was berating the Belgium passengers who were objecting to the meagre portions she was lobbing on to unwashed plates. I wondered what was going to happen when she encountered some of the 'Orribles', an endearing term for the small number of Chelsea fans who were famous for their lack of social skills.

I met the young couple again, who unfortunately hadn't managed to get an earlier ferry. Charlotte was upset; the cook had splattered gravy all over her new dress and then told her to keep moving or she would have them both thrown off the ferry.

I bought some drinks for the three of us and tried to cheer up the young couple after their horrendous day trip. I said to them that they should sit back and wait for the cabaret to begin, as more Chelsea fans were starting to arrive on-board. We did not have to wait long. The cook threatened anyone with her ladle whenever they questioned the slop deposited on their trays or frowned at being splattered by the bilge masquerading as gravy. Then the first of the 'Orribles' fell into the queue. I had never talked to MacNasty, but I knew him by sight and knew that he fully warranted his reputation and name. The cook dropped a small amount of what looked like vomit on MacNasty's plate. He was so drunk he could barely focus, but his diction was impeccable and he spoke each word slowly and with just the right amount of stress on each syllable.

"You ... can ... stick ... this ... right ... back ... up ... your ... arsehole ... you ... fat ... ugly ... bull-fucker."

The whole restaurant erupted as the cook hit the alarm bell and the crew, plainclothes policeman and more 'Orribles' now boarding the vessel started trading punches. It brought back memories of my trip with the Finns all those years before. I got up to leave and advised the young couple cowering in the corner to join me. I assured them that they were safe with me. Charlotte, who barely said a word earlier on the voyage, now made me laugh.

"You were right about the eating bit," she said in a very low voice.

The soon-to-be-married young couple finally broke into the first grins of the day, as we viewed MacNasty doing a drunken tango with the mortified cook.

True football fans never support their club because they want the kudos of being linked with continued success, a point even more relevant to Arsenal fans these days. We were there more for the fun of it and defeat was never the end of the day for us. When playing Liverpool up at Anfield we were three nil down at half time.

Our response was, "We're going to win the league and now you're going to believe us," to the complete bemusement of the rival fans.

We continued singing even when Liverpool reached goals four and five.

We did finally make it to the finals of the most treasured of all trophies, the Champions League Cup in Moscow, which we lost. Afterwards I was drinking at a bar not far from the Kremlin with some lifelong Chelsea fans. Each had a different perspective on life. On my left was Cuthbert and on my right Kieran.

Cuthbert's mobile phone rang, "Yes love, we lost. Ah well, we gave it a good go though. How're the kids?" the conversation disappeared into the ether, mingling with all the other conversations going on the bar, until I heard Cuthbert again over the din, "Oh love, while you're on, can you video the film tonight? Yes Channel Five. Yes, that's the one – *Brokeback Mountain*."

Well that was the start of the ridicule ...

"Oh fuck, you heard that," replied Cuthbert.

"No worries Cuthbert," I reassured him and I then added, "You should get the Director's Cut; I hear it has extra cock."

"Fuck off, it's a good film," replied Cuthbert.

I then turned to Kieran, whom I only ever met at European games, and said, "Well could it get any worse than this? All these years of toil and heartbreak to travel around the globe to this Moscow shithole and then we lose everything on one kick." Our captain, John Terry, had missed the final penalty. "Then to hear one of your mates asks his missus to tape a gay love story for him, while he's out here. I bet she wonders now, why he takes her to see *Priscilla Queen of the Desert* every birthday."

Kieran let out a long deep sigh and said, "Oh yes, it can get much worse."

"Really, how?" I asked with real interest as to how the despondence could get worse.

"Well," he said, "the wife won't be happy when she finds out that I had to sell her car to get here."

Chapter 24 – *Magpies and Custard*

After university, it was time to get back to reality. I again flitted around from one job to another. When I add up all the jobs I have had, the sum comes to forty-three and number thirty-five was as a heavy goods vehicle driver at a recycling centre. That was when a combination of Amelia, my new job and the rogues I'd met started me on a new path: delivering medical aid to troubled areas across the world.

I had not worked with a team that was as mad as the new crowd was since Arkham. Three characters stood out, the boss Nobbie and my co-workers, Bertie and Angus, who was known as the 'Colonel'. My latest nickname was the 'Graduate'. Bertie and I started on the same day and we were introduced to Nobbie, a rotund ginger-haired gentleman with his shirt permanently hanging out over his trousers. Nobbie was distressed, which, we were to learn, was his normal state.

One morning Nobbie was having a team meeting, when all of a sudden he stopped in mid-sentence, raced to the telephone and called his wife.

"Are you wearing any knickers?" Nobbie demanded to know.

"He could at least have gone through the niceties and said good morning," was my comment to Bertie who was sitting next to me.

When Nobbie's wife replied that she was, he responded, "Then whose I am I wearing?"

"I think one has put one's dick in the custard," whispered Bertie, as he nodded knowingly to me.

Bertie would often use such unusual euphemisms. When I first met him, he was coming out of the interview room as I was going in.

He commented to me as I passed, "They wanted to know the ins and outs of a magpie's arsehole." Where these expressions came from was a mystery, and I decided that it was best that they remained so.

We all liked Nobbie and he was very good at his job. He was so totally committed to it that he never took a holiday. When he eventually decided to take his wife away for the weekend, we all chipped in some money to buy some supplies for their trip. Bertie took charge and told us he had got them some surprise gifts for the long journey to Cornwall.

"Will it be a surprise, Bertie?" I asked, but having no doubt it would be.

"It will be when his wife opens the glove compartment to get some sweets and the dildos and vibrators tumble out," he said.

"I hope you included a shovel, so she can bury his body at the next service station," I added.

Surprises were not reserved just for the staff, but were often produced for the public too. A member of the public walked into the office later that week with a recycling question.

"What do you do with blue bottles?" he asked.

"I normally hit them on the head with a rolled up newspaper," replied Bertie.

At that point I laughed my tea out of my nose, a feat I had not been able to accomplish since school. "Thank you very much, that is very helpful," said the poor man, backing out the door as quickly as he could.

I too could have been up before a tribunal when another gentleman that week opened the door of our tiny hut and asked, "Excuse me, do you have anywhere I can leave a wheelchair?"

"You can stick it in the scrap metal container," I replied, genuinely trying to be helpful. Unfortunately, I did not realise that he meant to keep it out of harm's way, as his elderly mother was sitting in it. The old lady, understandably, gripped the chair's armrests in panic.

Bertie, as well as the Colonel, became not just my friends but also benevolent rogues. Both had the requisite requirements of gallows humour, a personal moral compass, which showed when they helped me put my first medical convoys together, plus a healthy disrespect for social niceties. Bertie's humour and relaxed view of the world reminded me of my childhood hero, Ronnie Barker's Fletcher. The only difference was that Bertie was even more obsessed with sex, than any of the sex-starved inmates of the fictional Slade Prison.

We had a nasty piece of work in the gang at the time, whom I named Wing-nut. At our first meeting, he informed Bertie and me that he didn't like blacks, women or homosexuals.

"Well that has me told, you bitch," said Bertie who then put on a completely over-the-top camp act whenever he encountered Wing-nut.

"There's no need to bang on about your prejudices, I hate bigots with big ears, but you don't hear me going on about it. Well, apart from now Wing-nut," I informed him.

Wing-nut would come into the hut each morning and ask the same question, "Where are the keys to my van?"

Each morning, my reply would be the same, "They're in it."

He would then return ten minutes later, trying to find the right keys and would eventually find them in the key-cabinet on the hook below where I had written 'WING-NUT'.

He didn't like me much, even though I would always welcome his appearance with a big smile and a wink and I'd blow him a kiss. We all played up to his racism and homophobia. Wing-nut never realised that Lover was actually Bertie's surname; we used it in the same way that Rowan Atkinson used "Morning Darling" to greet his nemesis Captain Darling in *Blackadder goes Forth*. Each morning I would arrive and say "Morning Lover," to Bertie, who would reply with lines like "You're late; were you chained to a tree on Hampstead Heath again?" He'd then put his fingers on each of his nipples and playfully circle them while slowly licking his lips. Wing-nut could have beaten the shit out of all of us together, but I think he was terrified that he would catch our "gay disease" as he once confided to the Colonel. The Colonel was a master of the put-down and once told Wing-nut of his time during the war, when he was on naval manoeuvres hunting German U-boats.

"Is must have been great to kill Germans, but Hitler was right about the Jews though." Wing-nut said.

Unperturbed by Wing-nut's remark, the Colonel continued, "Ah, but not as much fun as taking turns in the barrel, as you didn't know whose trouser-snake would drop through the hole in the side next."

Wing-nut's mouth fell open and his tea dribbled back out of it. The Colonel was no more a homosexual, than Bertie or I, but like us, he had no time for Wing-nut's bigotry. The Colonel always went for the jugular, as he explained to me one day, "I didn't take prisoners during the war. I don't see why I should now."

Wing-nut was stalking his estranged wife at the time and would tell the Colonel, but never Bertie and me, how he would follow her everywhere. One such night, she had gone bowling on a girls' night out. Wing-nut took a bowling-ball out of his bag and proudly displayed his stalking prop for all to see.

"This cost me three hundred pounds, but it's money well spent to keep an eye on that bitch." Wing-nut declared.

"Wow, you paid three hundred pounds for that? I wonder how much it would be worth without holes in it?" I said. I smiled and winked at him and blew him a kiss in my usual way.

Bertie then entered the hut and performed his standard greeting when he met Wing-nut. He dropped his pants, circled his nipples with his fingers and licked his lips while looking seductively at Wing nut.

"Ooh! Wing-nut, what I could do to you!" he simpered.

A terrified Wing-nut ran out of the hut and that was the last I ever saw of him. We then got a call from Nobbie to say that Wing-nut had resigned.

"You were all piss-taking faggots and he was afraid of getting bummed." Nobbie said.

I guess it could be said that Wing-nut was the victim of homosexual bullying by heterosexuals. Mind you, it would take some time for the case to be heard by a tribunal, just for the lawyers to agree on the precise phrasing of the charge.

When we heard the news that Wing-nut had thrown in the towel, I turned to Bertie and said, "Now he's gone, if you ever drop your pants in my presence again, I will set fire to your walnuts."

"Oh you're incorrigible, you would do that for me?" said Bertie, dropping his trousers, circling his nipples with his fingers and slowly running his tongue over his lips once again.

The Colonel carried on reading his *Financial Times*, but I could detect that wry knowing smile of his. Though we were different in background, education and lifestyle, though Bertie and the Colonel were both grandfathers whereas I couldn't keep a relationship together, we were as one when we opposed people like Wing-nut who would have to trade their bigotry elsewhere.

On my first week working in the centre, I was trying to repair one of the machines, when I heard, "What the fuck are you doing?"

I straightened up and looked directly at the tall, wiry guy who was now sizing me up. "Limbering up to lamp you around the head with this," I replied nodding to the large monkey-wrench I was holding in my hand.

The man, who was called Francis, was now squaring up to me and now that we faced each other, only a few inches apart he said, "OK, I'll hold it and you hit it with a sledgehammer."

Women find it difficult to understand such testosterone-charged incidents and it's hard to explain the male psyche, but men often test each other; it's a primeval instinct to see how far we will go, it helps define our character and marks out our boundaries. By squaring up to Francis, I had secured his respect and, from that moment on, we never again exchanged an angry word.

Francis was employed by the local council and his job was to empty out the contents of the flats of people who had recently died. His nickname was Dirty Harry, because anyone who was assigned to him rarely got to the end of the week without either being hospitalised or resigning from the job because they couldn't keep up with his work rate. He was a phenomenal worker. He never stopped to take a drink or even a sandwich throughout the day; quite simply he was so full of energy that he constantly had to be doing something. I never even saw him sit down.

One week the council attempted to pair Francis up with a co-worker and assigned him a huge lumbering hulk in his early twenties, who looked like he would actually last the day. He didn't. Francis came into our yard, as he did every day to work through his official lunch break. He told me that he had come out of his first house clearance that morning to find that the new guy had already disappeared. Later on, the police discovered his work colleague lying unconscious next to where Francis had parked his vehicle on their first job. A fridge with a big dent in it lay beside him, one that Francis had thrown out of the third-floor window of the flat when he was clearing it out.

I liked Francis, he did his own thing and he didn't give a damn about what anyone thought, but I was a little surprised to hear he had a family.

"You have a woman in your life!" I asked, wide-eyed with surprise.

"Oh she buggered off yesterday, so I'm stuck with my Wendoline," he said as he summoned his daughter into our office.

At that point, a sweet little girl of about twelve came into our office. Her eyes were red from crying and she was still sniffling back tears.

"Are you OK, little girl?" asked the Colonel.

"Oh, she is OK, she's just upset because I found a spider in the bath and threw it out of the window," explained Francis.

"Don't fret, the spider will survive," said the Colonel kindly.

"It was," she tried to get the words through her sniffles, "my pet tarantula," she blurted out, but then started her sniffles again.

"She bought some fish and the only glass tank I had had her pet tarantula in it, so I put her pet spider in the bath," continued Francis, very matter of fact.

"How did it end up flying out the window?" I asked.

"Well her mum wanted a bath and I hadn't warned her about the insect when she went to get in, so she had a fit. I just picked up the thing and threw it out of the window. Women, they're never happy, now her mother has fucked off and this one's been crying all morning," added Francis, shaking his head at Wendoline.

I was not the least surprised to hear that Frank would pick up an 'insect' such as a deadly tarantula with his bare hand and just lob it out of the window on to some unsuspecting passer-by. It probably landed on his unfortunate work colleague, who would have been staggering home rubbing his bandaged head.

At that point, the Colonel, having listening to Francis' tale of a man at one with nature, burst out laughing.

"You laugh again and I will stick a knife through your bollocks," uttered little Wendoline. She wiped away tears, while waving what appeared to be a flick-knife in her hand.

"She's definitely daddy's little girl," I said to Francis, and that was the first time I ever saw him smile, no doubt with a sense of pride, as he turned to look at the knife that his little girl was threatening us with.

If someone was down with a suspicious illness or went red-faced at the sight of an attractive woman, the rest of us went in for the kill. An injured Thompson Gazelle had a better chance of safely navigating its way across the African Plains, than someone had of emerging unscathed after exhibiting weakness in our presence. Beaker, our very straight press officer from the head office, who was always in a state of distress whenever she had to deal with us, hence her nickname taken from the terrified lab-assistant in the *Muppets*, came to the yard. She was looking for a Mr Doyle, one of our drivers, as the personal details he gave were discovered to be those of someone who had died ten years earlier. His name drew a blank; real names meant nothing to us, we inhabited the world of Scouse Bob and Bollock-chops. Beaker was always uneasy when having to deal with Bertie, as he had that kind of look that somehow he viewed everyone in a state of undress. She expressed her nervousness by contorting her legs around each other, like climbing ivy, while standing as if she was trying to ward off an accident, hence her nickname.

Beaker then described Mr Doyle's job from the file. Again, we could offer her no help, as we had no idea what anyone's actual job description was, we just got on with any jobs that needed doing.

"Hmmn, what does he look like?" I asked.

"He is a grey-haired gentleman," answered Beaker.

"That could be anyone of the coffin dodgers around here," I said, recording the presence of Bertie and the Colonel. The Colonel nodded back an acknowledgement; Bertie, though, silently mouthed the words, "I love you and you treat me like shit."

I continued to try to assist Beaker, "Any unusual features?"

After she offered more bland description, finally she implied that Mr Doyle was partial to a tipple.

"His face is a little bit red and blotchy," she said, in a low voice.

"Ah," said Bertie turning to us, "You know Doyle; he's the one with a face like a baboon's arse." Doyle walked in just then, and if you had a picture of him and another of a monkey's backside, you would swear they were twins.

Not everyone was a rogue and amongst our number we counted Leonardo, a very decent and respectable man who arrived from Bilbao and joined our team as an electrician. I got on really well with him; I had fond memories of the Basque region and its people. I was looking out of the office window one day watching Leonardo who was up a ladder, as I had asked him to fix the main generator at the end of the yard.

Bertie walked in, "I see that Zorro is up a ladder with a crowbar tapping a forty-thousand volt power supply."

"Yes, well he's the professional, so he should know what he is doing." I replied.

"Well, I asked him the other day how he got that bald line across the top of his hairy head." Bertie said.

"Yes, I was wondering about that."

"He told me about his first commission when he worked on an oil tanker. When the two main generators went dead, he tried to jump-start them with a car battery. Hence the white electric arch of a hundred-thousand volts that shot across the top of his little Spanish nut. It's similar to what he doing now. " I jumped out of my chair, but it was too late.

When Leonardo returned to consciousness a few days later, he accepted an offer of another job. As a good family man, he sent half of his pay home to his family so he couldn't turn down the offer of a better paid job. I wished him good luck, shook his hand and told him not to worry about it, but he still kept popping in once a week. Leonardo really liked Bertie, the Colonel, Ernie, Scouse Bob and he greatly respected Frank for being a hard worker like himself. He also told me that he loved 'the English banter'; that was the reason why we had so many volunteers in the place; it was fun and more of a theatre of comedy than a place of work. On one of his visits, our Spanish friend did not appear very happy.

"What's up, Zorro?" asked Bertie.

"Where I work, see, they are all thieving bastardos," said Leonardo. "Today see, I lost another hammer and two chisels, last week an electric drill and my toolbox. You all good people, but too many thieving bastardos in this country."

The Colonel interjected, as he was always wise and would try to bring a sense of proportion to any argument, "Steady on Zorro, this isn't Liverpool you know?

"I'm fucking here, I can fucking hear you," answered an affronted Scouse Bob, one of our lorry drivers.

"Where do you work, by the way, Leonardo?" I asked, having realised that I had never asked him before

"Err, how you say? ... Parkhurst prison," he replied to an audience that was not easily stunned, well not until then.

Having established a reputation for recruiting some of life's top oddballs, it was with no surprise that I received a call from central office to say that they were transferring a guy called Mickey to us.

"Why?" I asked.

"He is doing our fucking heads in," was the answer.

"Sounds to be of the required standard, excellent we'll have him, just tell him to howl when he gets to the main gate," I replied.

Mickey was a lovely man and a valued addition to our English production of *The Addam's Family*. We did not dress him up in cotton wool, he was one of the team and he loved it. But life had not dealt him a good hand, in fact the cards were stacked against him; his story was a tragic one. By the age of thirty, Mickey had had a kidney removed along with most of his intestines, added to that, the fact that he was as deaf as a post with an impenetrable Irish brogue and you have a man who has little to be grateful for.

Mickey's job description stated that he was a data manager's assistant; as I had no regard for job descriptions, I made Mickey our new Head of Enquires. I could clearly see he had the qualifications for the job of answering the phone and dealing with complaints. Whenever he answered the phone, he put the volume control on full and attached it to our public tannoy in the yard. He always forgot to turn it down after taking a call, so if any of us answered the telephone, you might as well have had a hunting horn strapped to your eardrum while a shotgun was fired next to it – you would be struck stone deaf for a day or two.

The results fully justified my decision. Our recorded complaints dropped overnight from an average of one-hundred a week to zero. Mickey couldn't hear anyone screaming down the phone line. I remember Bertie acknowledging my managerial expertise when he commented on Mickey's appointment.

"Well bugger me with a butcher's hook and call me Roger." was his particular turn of phrase.

Later that week, Beaker telephoned from the Press Office, "We believe your centre has numerous visitors, including those whom we may call the elite of the entertainment business."

"WHAT?" replied Mickey, holding the phone to his ear and turning the volume up.

Beaker repeated herself, but Mickey's response was the same, "WHAT?" in his distinctive Northern Ireland brogue.

Beaker persisted, "Let me rephrase, I am looking for a celebrity. Do you have one?"

"WHAT?" replied Mickey who was now banging the phone on the table, as he could not hear a word and thought it was broken.

She then shouted at the top of her voice, "I NEED A CELEBRITY!"

Mickey had the tannoy on and the windows nearly blew out as her voice hit a level that would prick the ears of every sheepdog in the country. He responded in his loud Irish brogue, now completely bewildered at such a bizarre request, "BUT, WE DON'T HAVE A CELEBRITY!"

Now Beaker was near to bursting, "CAN YOU GET RIGHTEN? I NEED A CELEBRITY!"

Mickey now completed bewildered replied, "BUT HE'S NOT A CELEBRITY!"

An explosion took place at the other end in the kind of language you would not hear from a docker. Mickey could not hear her eruption and, thinking the line had gone dead, he put the phone down. Poor Beaker then burst into tears and later we heard she had had to change her underwear.

We did indeed have many celebrities using our centre and on one occasion that resulted in me receiving national coverage in the media. That occurred when a famous author was due to visit our centre. It seems that the BBC's Radio Four's *Ladies Time* programme had asked celebrities to do a story about their favourite place in Britain. You could imagine the sound engineer muttering "Bollocks," when one particular celebrity chose a recycling depot in central London, rather than the Isle of Skye or the beautiful town of St Ives.

Nobbie gathered all the staff together, including Scouse Bob, whose every third word was a swear word and that was on a good day.

"Let's all be professional when the BBC crew arrive; you will be live on air, so no sexual comments and no swearing."

"I hear she has great tits," added Bertie.

"Bollocks," added Scouse Bob.

Bollocks to what was a mystery. I had been a big champion of Scouse Bob's job application and when I interviewed him I had wanted to give him a chance, as his form stated he had Tourette's syndrome. I learnt later, that he didn't have the condition at all and he was just a typical 'mickey-mouser', rhyming slang for Scouser, or a resident of Liverpool, who simply swore all the time.

Nobbie continued, "Her books are pulp for the masses, so I am having nothing to do with it," and he went off to the pub.

Bertie turned to me, "Oh dear, it looks like you're driving the celebrity tour bus, dear boy."

"We're all going to get fucking sacked," added Scouse Bob.

The celebrity author turned up with her sound crew. She was a lovely woman, but she was away with the fairies when it came to the real world and so she could easily have joined our team.

"Fucking mad as a bicycle," whispered Scouse Bob to me.

When the sound team had set up their equipment, I set off with the author to tour "her favourite place in the whole world," as she had described it. The first thirty minutes of the broadcast was hardly riveting, until she noticed a visitor to the centre.

"Oh look! An elderly gentleman trying to recycle a fire extinguisher," she said excitedly.

In the background, I could hear Scouse Bob shouting across to him, "Put that back you thieving bastard."

The fear that the others had, that my mischievous nature would prevail, was realised when we then walked over to the glass recycling bays.

"There must be millions of glass bottles here?" she asked

"Yes, residents for miles around are perpetually off their faces, just to empty and recycle the bottles to save our planet," I added.

"Oh, such sacrifice," she said in a now increasingly incredulous tone.

"Oh look! A textile bank," she then added, "What is the strangest thing your team has ever found in there?"

"Oh Christ, someone fucking stop him," I could hear Scouse Bob shout across the yard.

"Well one member of our staff, Bertie, found a one-piece rubber outfit for two, with interconnecting hoses." I said.

"Really, how did they get out of it and why recycle it?" she enquired.

"Well, I think the answer to your first question was that the arse was completely torn out of it, so whatever one of them did, the other wasn't hanging around. As for your second question, a one-piece completely sealed twin rubber suit, now with no arse to it is no longer fit for purpose, hence the recycling bin."

"Is there a moral that we should take away from all this?" she asked trying to hold in her laughter.

"Yes, that even when we are indulging in acts of complete sexual depravity, we should never forget that all actions have an impact on the environment, even when you have a hosepipe attached to your rectum." I added.

"Well, I will certainly takes some messages home with me today," but she could hold her laughter in no longer and desperately indicated to the soundman to cut the recording immediately, before she had an accident and we had to recycle her knickers.

"As a fucking bicycle and that prick is no better!" shouted Scouse Bob.

It's not surprising that the press office went ballistic after the broadcast and I had the head of the company screaming down the phone at me, as poor Beaker took a month off sick. I didn't think much of it, thinking that probably no one had listened to it.

My friend Phil then called to say he had just met his fiancée's parents for the first time that afternoon. His future mother-in-law had decided to tune in to *Ladies Time*, her favourite programme, while they were all having lunch in the garden.

"Well when I heard your voice, I couldn't believe my luck and thought this would put me in the good books with my future in-laws. I told them that you were a friend of mine," Phil said. "Anyway, thanks, mate, my mother-in-law is now in a state of catatonic shock, after listening to you talk about sexual encounters involving all in one body suits with connecting hoses. Especially as the details were narrated by someone I had informed them was a good friend of mine."

"How many times have I told you to be careful who you make friends with?"

I first met Ernie when I had to interview him for a driver's job, which was actually as a replacement for me, as I became the governor and started developing my own little Arkham after Nobbie had told the boss to stick his job where the sun don't shine. In honour of my old boss, Blakey, I was determined to recruit only dysfunctional lunatics, which seemed only right, as it was Blakey who gave me my start in life. Ernie's interview went well until I asked him where he came from, and he replied Knotty Ash, which I knew as the name for a fictitious village invented by the Scouse comedian Ken Dodd, who probably used it as a tax haven.

"Well, unfortunately even I can't accept a reference from one of the Diddymen," (Dodd's fictional residents of the aforementioned town), "but I might be able to get away with one from a garden gnome if you can get one to vouch for you?" I said.

"No wait! Knotty Ash really does exist," he insisted.

I liked this crafty giant from Liverpool, so I decided to check the internet and found to my surprise that Knotty Ash is a small town on the outskirts of Liverpool. The thought occurred to me that by the end of the week I would have the *Man from Atlantis* and *Postman Pat* in the team. I always did like the mickey-mousers, so it's no surprise that along with the sweaty-socks they were always a big part of the official and unofficial teams I recruited over the years.

One young kid joined the team, whom Ernie and I tried to keep on the straight and narrow. Keith had gone down the wrong path with only himself to blame, but he had served his time. His offence was attempted armed robbery. He had entered a bank with a sawn-off shotgun and announced his arrival by firing his gun into the ceiling. His mistake, well apart from trying to turn over a bank, was to shoot directly above his head. He had dislodged a ceiling tile, which then landed on his head rendering him unconscious. He had not secured any work since being released from prison. He was honest with me about his past. He even later asked for his other offences to be taken in account, the most damning of all was that he was a Tottenham fan, but I gave him a second chance.

Ernie's nature was to bring disorder where calm existed but he had a huge heart. I noticed that he would try to keep some of the rough lads in our team on the straight and narrow, so I put Keith under his wing. Ernie was a rogue, but a benevolent one, and I knew he would try and help the kid.

All went well, Keith was a hard worker, though occasionally he would go off the rails on drink, I put that down to the performance of his football team. I would cover his bail sometimes and together with Ernie, gradually get him back on his feet; incidents with the police became less and less frequent.

We all deal with knocks that life throws at us; some of us recover quickly, while others aren't fully back on their feet before the next blow floors them again. Keith was one of those. Keith's wife left him for another man, just when things appeared to be going well for him. I don't know the story behind it, but life could not have been easy for her and the kids over the years. He went to pieces and shortly afterwards he received a restraining order banning him from the family house, after he had turned up drunk and started banging on his wife's door at four in the morning. Keith was one of those characters; who was in permanent conflict with himself.

Following that episode, Keith went on a drinking binge for a month. Ernie and I sought him out one night after work and, when we found him, all three of us went for a few pints. OK, going to the pub was probably not the standard therapist's approach, but you work with what you know. It was actually a great night. Keith started to enjoy himself and said it was the first time he had laughed in months. In fact, he was having so much fun, that he made an impression on a woman at the bar. He hadn't spent a night with a woman since his wife threw him out, so with the drink flowing we encouraged him to forget his troubles and have some fun. That was the last time we saw him alive.

That night back at that woman's place, she covered him in paraffin when he was asleep and set fire to him. She was schizophrenic and from the resultant court case, she said he had done nothing wrong; she just killed him. Ernie and I had done our best and I sometimes wonder whether we were the right mentors for that poor unfortunate kid.

KINDLY ROGUES

"What was it he (Father Jack) used to call the needy? He had a term for them?"

"A shower of bastards"

Ted and Dougal.

Father Ted

Chapter 25 – *That's a Relief*

Benevolent rogues do not grow on trees, thankfully, or half of them would be found with broken necks, having landed head first on the cold ground in the autumn. They are rare creatures. I never met one at school, I only came across H when I worked all those years in the park, but that was a fertile time in my life for such encounters. The Colonel and Bertie were undoubtedly rogues, always up to mischief, always looking for an angle. But they were benevolent, gathering the money and supplies for my first medical deliveries to Romanian orphanages. They, H, Sam and others helped in delivering humanitarian aid to countries around the world.

You're probably wondering where this parallel universe suddenly appeared from, as up till now you have been reading the ramblings of a shallow hedonist. I have to acknowledge that most of the major decisions in life have been influenced by women. My introduction to relief work was no different. In this case Amelia put me on the path of right and virtue. Well, she tried. Amelia and I were together for about five years and, though I can generally be considered as a monument to failed relationships, I have been in two relationships that lasted as long. In my defence, that is longer than a fair number of marriages. Amelia and I shared some great adventures. Each one began when she wanted 'a little favour' – that phrase has, over the years led me to various different continents, into the occasional beating and even made my sharp features a delightful target for snipers.

As I was then working in the recycling centre and Amelia was collecting some clothes for a children's AIDS hospice which was being built by a small English charity in Romania, she asked for some help.

"No problem" I said and called on my motley crew of rogues. Bertie, the Colonel and Francis put the word around and people were happy to donate children's clothes until we had finally gathered enough to fill a truck. Francis led on the fundraising, or extortion, as Scouse Bob called it when Francis had him in a headlock.

"Amelia, here you go. Now when will the truck get here?" I asked.

Amelia replied with "Well!" which meant there was trouble. She also pressed her hands to her hips; this was Amelia's way of announcing that we were well and truly up shit creek. On this occasion "Well!" meant that the truck was no longer available. We now had a mountain of good clean clothes and children on the other side of Europe desperately waiting for them. Amelia's "Well!" was normally met with the standard reply from me, "OK, so what do I have to do?"

Romania is now a democracy, but in those days it was under the dictatorship of Nikolai Ceausescu and Soviet dominance. What money was produced or injected into the system was siphoned off by corrupt officials and squirreled away to various hidden foreign bank accounts. What made the situation worse was that Ceausescu followed Chairman Mao's plan, the Great Leap Forward, which was based on forcing people to have large families to provide labour. The aim was that through an abundance of workers, cheap labour would compensate for a lack of investment in modern technology. The state imposed severe penalties on small families but then provided no support for families when they expanded. When the economy failed to grow and the promised jobs failed to materialise, children were often abandoned on the streets by families who couldn't afford to feed them.

I managed to secure a truck with the help of Francis, who, despite his hard man image, having seen on television the appalling conditions the children had to live in said to me, "I don't care what fucking country they are from, no kid should have to live in shit."

Over the years, whether it was Romania, Bosnia or South America where I drove trucks full of medical aid, benevolent rogues like Francis helped me get my trucks to where they were needed. I was regularly short of funding, but one day a famous comedy actor and international traveller came up to me and handed me a cheque to cover the fuel costs. He didn't, however want any publicity. I will always be indebted to this kind and decent man.

That first trip, it took me a couple of weeks, but with the aid of my benevolent rogues,

I secured a truck, mapped out the best route and got the paperwork in place. All I needed was a co-driver. That was not easy, as I needed someone who like me had a licence to drive trucks, so I put the word out that I needed a volunteer for a few weeks to assist me to drive a truck across Europe. Whilst awaiting a response, I had time to do a little more research on what aid was needed. I discovered that proper medical equipment was scarce and keeping implements sterile was impossible. Injections were given with second-hand needles, so AIDS was widespread amongst the children. I was not alone in wanting to break this cycle and deliver sterile needles and other medicines to children. AIDS was spreading throughout the orphanages and was prevalent in many state institutions across Eastern Europe. I called Mary and her nursing friends to help; by the end of the week, I had received enough to keep one orphanage in fresh needles for three months.

As for finding a co-driver, Amelia had an Australian flatmate, James, who had a driving licence. It took little persuasion to get him on-board, he wanted to help, plus as he delicately put it. "I'm fucked if on this course of sheer lunacy Australia is going to be outdone by a fucken Pom."

That night before the first trip, I knew that all I had to do was to get myself, and the paperwork, to our yard in the morning where the truck was fully loaded; we would then be on our way. I couldn't sleep, but as I watched the sun struggle through the early exhaust fumes of London's dawn traffic I slipped out of bed so as not to disturb Amelia's snoring, mind you she's louder when she's awake, picked up my bag for the trip, my file of customs papers and visas and headed to get the first northern line tube of the day. The journey to the yard was a mere two stops on the tube and a two-minute walk, but to my amazement in the sparsely populated carriage, I suddenly saw Sam who I had not seen since university. I rushed over, picked her up from her seat and with a big hug lifted her into the air. As we were now pulling into my stop, I had to rush so we promised to meet up as soon as I got back. I got off at the next stop and thought, what a great way to start the adventure, meeting Sam was a good omen.

As the tube pulled away, an icy cold feeling hit me; I realised that when I'd lifted Sam into the air I had left the small bag of paperwork on the floor, where it remained. I watched as the tube disappeared into the tunnel. I had to do something fast; I contacted the station manager and had the tube checked at one of the stations down the line, but nothing was found. I even tracked the tube down an hour later and checked it myself, but there was still no sign of the bag or the documents. I telephoned the yard to break the bad news to James that his co-driver was brain-dead. He replied with that Australian phrase that I would hear a lot over the coming weeks.

"She'll be right."

I turned up at the yard feeling lower than I had ever felt in my life. I had let James down, all those people who'd donated clothes and money and above all I'd disappointed the children who were waiting for the aid. To make matters worse, the surplus medicines from neighbouring hospitals, which would help ward off disease and potentially save lives had a limited lifespan of only a few weeks. I sat at my desk in the office, thinking why had so many people had placed their faith in the world's biggest arsehole. Even Bertie couldn't add a sexual connotation to my plight; though Scouse Bob was able to express what everyone else was thinking: "Prick!"

The telephone rang. It was Sam, "Hello Angel-face," that was her term for me and another to add to my dropdown menu of nicknames. "Did you leave a bag of papers on the tube?"

"You have them?" I asked.

"Yes darling, you silly boy I thought it was yours when I saw the bag, so I picked it up."

"Oh, I fucking love you!" I shouted which was rare utterance for me and not even declared in moments of extreme passion, "I will be straight over, where are you?"

Within an hour I had the bag of papers, James and I said our goodbyes to Amelia,

Bertie, Ernie, Francis and the Colonel. We loaded up the final supplies of ham sandwiches and a bucket of vegemite on to the truck and we set off on our epic journey. Once again, I thought the omens were good, which was unusual, as unlike James I wasn't spiritual and certainly didn't believe as he put it in 'one's karma'. Thirty minutes later, the truck broke down in the middle of the Shepherd's Bush roundabout. We sat in the vehicle waiting for the breakdown services and James said, "She'll be right."

"Bollocks," I replied with the English equivalent.

After we made a very dodgy repair to the engine, the dodgy brothers, as James now called us, were back on the road and we made it to Dover and caught the ferry to the continent with four minutes to spare. It was only midday of day one and I was fit to collapse with exhaustion, with any thoughts of omens well and truly eradicated from my thinking forever.

By the time we reached Germany, we decided that we had better put what money we had into getting the vehicle fully serviced before we headed into Eastern Europe. In doing so, we witnessed two sides of the German persona. Firstly, when we asked the German mechanics to do a full service, it was an almost-surgical operation carried out with the utmost efficiency and professionalism. We witnessed another aspect of their makeup when James sat down cross-legged on the garage forecourt and began to fry sausages on our camp fire gas burner. The Germans were not impressed and scolded James, but to no effect. So, true to type, our Germanic saviours were both efficient and humourless.

When we reached the Hungarian border, I formed a similar stereotypical view of Hungarians, though minus the efficiency. There was a backlog of vehicles, so I got out of our truck to see what was holding us up. Hungarian vehicles were getting through, but a German convoy of six articulated trucks in front of us was the focus of the Hungarian customs officers. The German trucks were then ordered off the road and told to park up in a siding. I decided to drive our truck up into the space, but a stand-off began as we were next to be challenged by the border officials. They wanted all the aid trucks to turn around and head back to their country of origin to get an additional stamp or they would not be allowed through the border.

"Bollocks to that James, we were lucky to get this far," I said to my dodgy brother.

"Fuck em!" he replied with another of those quaint Aussie expressions.

I positioned our truck across both lanes on the border and refused to move. It's not just the French who have the monopoly on being awkward. I got involved in a very loud screaming match with the border guards as well as the Hungarian drivers who were also blocked by our truck. As this was going on, I recognised a familiar smell and I looked over at James. He was sitting cross-legged on the border entrance with the gas stove fired up and sausages on the go again. The guards seemed well prepared to rant and rave at me, but that completely threw them. James was as cool as ever and refused to acknowledge one guard when he stormed over and starting screaming at him.

The German convoy drivers started their truck engines and began to turn around on the crowded border road to head back to Berlin. I challenged the German convoy leader and pleaded with him not to turn around, but he said that if new stamps were needed then they had no choice but to return to Berlin. Many of the vehicles contained medicines, which like ours had a short shelf-life of only a few weeks. We all knew that those vital supplies, along with the lives they were intended to save, were lost. My respect for the Germans was being re-evaluated; they had more clout than us, bigger trucks and more men and I was sure they could have got through, but in the end they were only obeying orders. Perhaps, they didn't want to cause a major incident and by sacrificing all that aid, they felt they would not endanger future convoys. All I knew was that there were no victors that day. I turned to the one of border guards, who was still screaming at me.

"Ok, the Germans are gone," I said, "but you're stuck with me and I'm a fucking nightmare, and as for my dodgy Aussie brother James over there, he will never part from his fry-up. We are not going anywhere, so you had better get ready to deal with the three-hundred angry Hungarian truck drivers behind us. They will not discriminate and will easily lynch you idiots, along with us." The border guards didn't understand English, but it was clear from our actions that they would have to shoot us.

Just then, there was a loud noise and the guard turned to look at our vehicle, which I had to admit it was more akin to that of a clown's car, as the back door fell off and landed on the ground. Eventually, after James handed over a *The Sun* 'topless page three girl' calendar, a donation from Bertie, we were allowed across the border. I still remember so clearly those German trucks as they turned around. It was one of the saddest sights I have ever seen.

One principle that James and I agreed on, if you ignore the loss of April's topless Debbie and her eleven companions, was that we would not pay any bribes to get our aid through. This is a principle I have adhered to ever since. I have seen too much abuse over the years and I know full well that, particularly in war zones, any bribe, whether it's a portion of the aid you are carrying or a financial incentive will end up perpetuating the conflict. I have turned trucks around rather than pay and then found other routes to get supplies through. I make no apology for my purist view.

Others will say what harm does it do when sometimes the bribe is inconsequential, compared to the objective? I have argued on this point as I believe that it taints all that you are trying to do in using resources you have been given in good faith to then grease a corrupt palm. It also means that every aid truck that follows has to pay as well and probably at a higher price. On a larger scale, paying bribes perpetuates a culture of corruption and as a result it may be to the financial advantage of the belligerents to keep a conflict going. More aid is then required so that more lucrative bribes will follow. I am not just talking about independent aid supplies, I have seen international governments and their agencies continue to put silver into horny palms when they could instead exercise their considerable muscle and challenge corrupt regimes and their stooges.

It was a tough journey, but it was helped as we did share a sense of humour, though not everyone saw the joke. At the Romanian border, a very large gruff female customs official approached our vehicle. As she did so, James awoke from his sleep in the bed behind the driver's seat; he sat up to lean forward and look at her.

"I bet she has one on her the size of Tasmania," he commented, no doubt referring to the region of her pubic hair. At which point I exploded with laughter, unfortunately directly into her face, which led to James and I being carted off to the border office for questioning by a now very angry and no doubt very hirsute border guard.

My dodgy brother and I encountered a number of obstacles during the journey; including border guards who did everything they could to make life difficult and relieve us of much or all of our cargo. After an eventful two weeks, we finally made it to Bucharest and met Malcolm, the head of the project. He and his team thanked us for getting the desperately needed supplies through to the orphanage. He then treated us to a drink in an underground club beneath Bucharest's biggest hotel, The International. As we drank our welcome beers, he told us how he and some friends in the Criminal Investigation Department (CID) back in England had got involved and how, with their help, he had got the operation up and running.

"Ah well, they're not known as 'Cunts in Disguise' out of affection. I would steer clear of them," I informed Malcolm.

"I'm head of the London Division," he said, leaning towards me.

"We have probably met in a professional capacity," I added and thought to myself that I would have a few words with Amelia when I got back, for not providing me with that little snippet of information. Loftus would have laughed if he heard I was working alongside the 'Choirboys', as he called the police after Joseph Wambaugh's novel about New York's finest. The only book I ever saw him read. The three of us enjoyed a good evening swapping stories, but mainly hearing Malcolm's harrowing tales of the plight of the kids in the state institutions.

At the end of the evening, we all shook hands and I said to Malcolm, "Ah well, it could be worse, you could have been God-squad."

"I am an ordained priest," replied Malcolm, as James and I looked at each other, but we could see that he wasn't joking.

I thought to myself, 'Amelia, you wait until I get home,' then I heard a loud laugh from somewhere at the back of the club, which reminded me of my old friend Loftus again. He would have loved it all.

For the next week, the dodgy brothers and Malcolm's team of volunteers got stuck into the renovation of an old block of flats to turn it into an orphanage and provide a clean, healthy and safe environment for at least some of the children. The job itself would take years, but though James and I were merely passing through, others were there for the duration, making incredible personal sacrifices along the way. They had integrated well with the locals, which was essential for the survival of the orphanage. If the locals did not accept that the home was needed, it would be stripped of everything as soon as we left, for they had nothing except the canvassed tents they lived under. The orphanage was not supported by government or its officials, as it was located in a gipsy area on the outskirts of Bucharest. This added to its isolation, as it was officially recorded as a 'no-go area', and gypsies were despised and persecuted by the authorities. It really was a meeting of the cultures, with western volunteers working day and night alongside gypsy families, to deliver this new children's orphanage in one of the bleakest and most deprived areas on the planet.

James and I wanted to get to know the locals a bit better, so using the universal language of football, we organised games followed by a few beers each evening. We were thinking that a combination of football and alcohol could not go far wrong, well that was until my dodgy brother lit a fart through his jeans for a laugh and cleared the room faster than a bearded passenger on an airplane asking for a match to light the fuse in his shoe. The locals did not return for a day or so, in terror of the western monster who could breathe fire through his backside.

A few months later, James went back to Australia and he went on to work with disabled children. We keep in touch, but he remains the same old James. For example, the pay for social work is the same as in this country, so he has to do a number of other jobs to subsidise his wage. One of these involved him doing a life class for some female art students. He described to me in his own inimitable manner how, whenever he caught the eye of a good-looking artist, he could then hear the sound of her eraser rubbing out to reflect his change in 'pose'.

I have met James only once since he went back home, and that was when I went out to Australia ten years after our aid trip. We met in Melbourne, spent the night propping up a bar, as we had done in Romania, and sharing our stories of the highs rather than the lows. He asked how Amelia and I were doing.

"We ended it a few years ago," I said, "I think her therapist gave her a drug that gave her sanity back. Bastard." I returned the question, "What of you James, you were never short of a woman somewhere?"

"Well, I'm a bit like you in that respect, not big on commitment, not yet, maybe one day," he answered.

Ten pints later we returned to the subject of women, "Dodgy Brother, what do you look for in a woman?" he asked.

"Well my good friend Jockey'" but again more of this bundle of testosterone lately, "always answers that question with 'My cock', but for me it's the same as most men, kindness, big heart, warm smile and goes like a train, what about you?"

He pondered as he blew smoke across the bar and as cool as ever replied, "Low self-esteem."

I knew any long-term partner of James would be smarter than him - they would have to be – but in my experience no man had greater respect for the opposite sex, but he was a rogue for he could make fun of himself and his ready wit kept you on your toes.

My dodgy brother James remained calm and philosophical throughout our journey and his approach got me through the incident on the border and many more situations that followed on our one and only convoy together. We are as different as two people can be. He was a spiritual longhaired pacifist, while I was a clean-cut, volatile, 'fuck em, I want results now' kind of guy. We were a perfect combination and I learnt a lot from James that held me in good stead for further adventures to come. Thank you my dodgy brother.

I made many more trips to Romania over the years and I always succeeded in getting all of the medical supplies to their destination. This was thanks to my benevolent rogues, plus thorough preparation; an abundance of stamina and an incredible amount of luck of which I have had more than my fair share.

As an example, on my next trip which I had to do alone, I was challenged by Hungarian locals to leap from some high rocks into a local lake. I agreed, but it was more about saving the cargo than about my bravado. I needed the help of some locals to navigate our truck through an unofficial mountain pass, which was now closed due to the thick fall of winter snow. They had agreed to help, but first I had to prove to them that I had balls. I had to earn their respect if I was to secure their help in keeping my truck on the road as it crossed the mountain. The challenge was to jump from the rock face, clear the boulders below and enter the lake's water while avoiding the various obstacles which had been dumped in it over time. I dived and eventually emerged unscathed, although covered in mud and shit; maybe that is why some say landing in shit is lucky.

The next diver also missed the rocks at the base, but then impaled himself on a railing hidden just below the water. I dived back in and lifted his leg off the spike before he nearly drowned. That day it was sheer luck rather than guile that saved me. After giving Yulcan, the injured diver, a tetanus injection, we navigated the truck laden with aid over the mountain pass. Even Yulcan, along with his father and uncles, was pushing our truck away from the edge, as I tried to steer it on the ice. As with anything, apart from grief, luck is not inexhaustible and years later in South America, mine ran out.

On my trips, I worked alongside volunteers who understood that you did not do humanitarian relief work and expect gratitude. In general, you just got on with it. The people you tried to help owed you and the world nothing. Yet, I was staggered by the attitude of some volunteers who expected that those they intended to help would immediately adopt their values, just because they had painted a wall or plumbed in a toilet. As an example, things would often go missing and someone would rage about how much they had given 'these people' and that 'they then repay you by stealing'. The most sensible thing was never to bring anything of value with you and, if you did and it was stolen, then tough, live with it. Often we would we would try to construct an orphanage or a hospital in the middle of complete destitution, surrounded by people who could barely afford to eat. If a wristwatch or a pot of paint presented itself, your brothers and sisters were starving and you could sell it for food, what would you do?

With the fall of Ceausescu and the communist regime, Romania had a chance to break free of its misery. It took a lot longer than was necessary. Those at the highest level refused to acknowledge that there were thousands of children, many dying of AIDS, in Romania's orphanages and run down hospitals. As one official told me, "We know what is best, for such stories would frighten away western investors". Similar men in official positions around the world 'know what is best' and have set up government programmes to hide children away in institutions on the outskirts of cities. Children are often forced to hide in underground sewers for fear that they frighten investors and tourists. Some officials have gone further and set up 'clear-up taskforces' or, as they are more commonly known, 'death squads', in some South American countries.

There is a common view, expressed by international agencies that individuals and unofficial groups of people who just turn up in countries with aid, do more harm than good. But what happens when the rest of the world sits back, usually due to politics rather than resources and refuses to do anything to prevent the blatant abuse of the vulnerable? I and others couldn't stand back, but looking back, I have to agree that on too many an occasion I have encountered aid workers who need as much therapy as those they are trying to help. Also, much of the aid collected was not needed and in fact, used up resources to get it to where it wasn't required. There is a story that after an outbreak of drought in a region in Africa, one crate of aid from California was found to contain an exercise bike. I have also seen aid workers pushing through their own agenda: some pushing a political ideology or religious zealots strapping bibles to aid packages.

I found myself working alongside one particular character. Our strange alliance was formed when I had to get an emergency supply of medical supplies to a children's AIDS hospice just on the outskirts of Bucharest. I had no co-driver. Even more importantly, I needed a large refrigeration truck, as the effectiveness of some of the medicines would be greatly diminished if not kept at near-zero temperatures. Enter Mr Simeon Higginbottom. I heard of a man in Manchester who had such a truck, he ran relief convoys and all he required was that the costs of the vehicle were covered. I checked him out through my police contacts and got a call from Mike, who did the check.

"He's clean, no criminal record, no reported incidents with children and, as per your query, no wife or dependent children." Mike said.

"Excellent."

"Not really. He's God-squad and a total nut job."

"Thanks for the warning Mike, but I have no choice." I said.

I telephoned Simeon Higginbottom only to learn that he did have a 'payment' as he put it, and that was to distribute bibles when we got there. Knowing the families at the other end, I was pretty certain his offer of bibles would receive a warm response and I was right.

Three days later, though I was pleased to see his truck enter the recycling yard, all I could say, along with Scouse Bob, was "Bollocks." There in four-foot-high letters on the side of the truck were the words 'JESUS WILL SAVE YOU', with a huge red crucifix positioned underneath on both sides of the vehicle.

Bertie said, "Well, fuck my old boots!" and could not contain his laughter. He then turned to me and said, "It should read 'JESUS FUCKING HATES YOU'," making reference to the hell I was going to suffer over the next few weeks travelling across Europe.

The Colonel was also laughing, but he saw this recent development from a different perspective, for as he said, "I feel sorry for your colleague, he will be incinerated when the lightning bolt from his god hits the vehicle when you get into it." He turned to the others, "Back away from the mobile-church, chaps, we don't want any casualties from the back-draft when it bursts into flames."

Bertie, the Colonel, Ernie, Francis, Kathleen and many more helped load the truck, but Simeon kept telling Scouse Bob to mind his blaspheming, which resulted each time in Scouse Bob apologising.

"Fuck yes, ah fucking sorry about that, I didn't even fucking realise I saying it, what a cunt."

I looked out of the window of the vehicle as we set off. Bertie later told me that I had a face on me like a little kid on his first day at school, knowing that each day from then I was going to get battered and robbed of my pocket money.

That was a long trip and I felt that Bertie was right and some new form of torture had been devised for me. Simeon played Billy Graham recordings on the cassette player all day, every day. Although as part of the contract, I had agreed he could play anything he liked, once a cassette was played I subtly throw it out the window, as I believed my contract was fulfilled. However, despite liberally distributing cassettes in ditches across Europe, the torture continued, as his supply was seemingly endless.

On day four, after being trapped at the Romanian border for thirty-six hours, Simeon looked at my tortured face and said he had a treat for me. As I looked out of the window, I prayed that it was something less painful, perhaps a nice bit of water-boarding. My prayers were once again unanswered, as Simeon pushed a Cliff Richard tape into the cassette player on the dashboard positioned just beneath the open-armed statue of Jesus.

"I have every song he ever made here with me you know," he said with a huge beaming smile. I am not a violent man, but the thought of throwing him out of the vehicle under a passing truck kept me going throughout the rest of the day.

As with a number of members of God-squad, there is a hidden vice or secret guilt that leads to a person praising the Lord with public zeal. It's like a cloak of righteousness covering the stock whips and leather gumboots underneath. Simeon was no different and in his case his vice was women; he made a pass at every woman we met as we made our way across Europe. Fortunately, the checks Mike made were thorough, as he never made an advance on any underage girls or he would have joined his Cliff Richard tapes in a ditch. That appeared to be the only line he would not step over as no grown woman was spared his advances. Every female border guard, the traffic cop that stopped me for speeding, the old woman rummaging in a dustbin by the motorway outside Bucharest, all were propositioned by 'God's messenger', as he called himself.

Just before we reached the Romanian border, we pulled into a petrol station. It was operated by two large women in their sixties; they were manning the petrol-pumps. They were wearing the standard dress for that part of the world, the obligatory headscarf, crumbling cheap plastic shoe covers and unwashed dungarees held together with petrol-saturated aprons.

"I'll fill you up, if you fill me up," was Simeon's welcome to our pump-attendants.

He disappeared with both of them for twenty minutes; I thought, as they were grown women, that whatever deal he did with them was their business. I was just happy to sit in the truck in perfect silence, having now frisbeed the last copy of Cliff's *Summer Holiday* across the forecourt into a field. Bear was in tears months later, when I told him how many of his all-time favourite classics had been recycled as winter shoes for goats in Eastern Europe.

When Simeon got back into the truck and we drove out, he didn't wave to his pump-attendants and they didn't give him a second look. He produced another cassette and we continued as if nothing had happened. Everything was as it had been before and the incident was secretly filed away, while he continued his good works. I later drew in large script underneath the 'JESUS WILL SAVE YOU' sign (Aka THE VOYAGE OF REDEMPTION).

Once again, every single piece of aid we loaded onto the truck in England reached the orphanage and I had to admit that Simeon knew his business. On the journey, we encountered robbers on numerous occasions trying to break into the back of the vehicle or siphon off diesel when we were parked. It meant that one of us always had to be awake when we had to park to get provisions or have a toilet stop, but most of the time we kept on the move, which was safer. One night, we found ourselves surrounded by some men who were eager to sell us illicit fuel, which we refused. They were not used to receiving a 'no thank you', so when one of them produced a knife, my boot and a follow up right hook proved to be an effective method of delivering our message. Simeon had worked his way behind another man who was attempting to move towards me with another knife; he put him in a headlock until he dropped his homemade knife on to the tarmac of the hard shoulder. As the other assailants ran off, the commotion gave me an opportunity to jettison more cassettes out of the window. They landed in front of grazing cattle in the nearby field; they looked in serious need of the word of God. By then, I was more in fear for my sanity than I was for my life.

Simeon knew some tricks, which when I look back on them now, could have got us killed or at the very least jailed. We were desperate to get through one border, as we heard on his Citizens Band radio that if we stayed in the truck queue we were in, it would take us about five days to get over the border. Our supplies, mainly insulin, were urgently needed, as the hospital only had a couple of days worth left in their reserves. Simeon went to a secret compartment inside the cooling generator of his truck and took out two white vinyl stickers, at which point a load of cassette tapes fell out. The crafty bastard, I thought, the next herd of cows will have some new grazing material tonight. The stickers were the same colour as the truck and he carefully attached each vinyl sticker along the bottom of each red crucifix. He then got back in the truck and drove our vehicle off the road; he put his foot down as we drove at a forty-degree angle at full-pelt along the ditch on the side of the road.

All the truckers were looking at us and waving their fists in anger as we raced past them in the queue. If any of them had been a little quicker, they could have rammed us with their vehicle and knocked us fully on one side, easily killing us both. Simeon's ploy provided us with enough respite, so that after driving for a full half-hour in the ditch, he drove our vehicle back up on to the road just in front of the border. He then leapt out and pulled the two pieces of white tape off each red cross before any other driver grabbed us. We were the next truck to be waved through.

As for the bibles that Simeon delivered to the orphanage and the local families, they received a very warm welcome, as they kept the fires going throughout a very cold winter.

Again, you will see that my criterion to define someone as a 'benevolent rogue' is not based on defined characteristics. James, as a pacifist, was directly the opposite of H, Larry and Tommy Spitfire, whose philosophy is "Kill em all. Let God sort is out." Maureen would not be seen dead without her make-up and a copy of *Cosmopolitan* in her *Gucci* bag whilst berating high court judges, while you would never find Sam spending three hours picking a dress in Harvey Nicks, when a bottle of *Bollinger* was crying out for her attention in the restaurant. It may also explain why so few of my benevolent rogues get on with each other.

Chapter 26 – *Know Your Left From Your Right*

The death of Marshal Tito, leader of Yugoslavia, meant that his iron hand was no longer holding the lid on the pressure cooker of ethnic tensions; the region erupted. Demands for independence stretched from Croatia to Bosnia-Herzegovina and led to demonstrations on the street; all were trying to break free from the artificially created Yugoslavian state formed after World War 1. These incidents were seized on by the incumbent Serbian politicians, Slobodan Milosevic and Radovan Karadzic, who saw an opportunity to create a Greater Serbian state. Tito had never allowed any ethnic minority within Yugoslavia the upper hand, but now these Serbian leaders saw the chance to take overall control of their neighbours. This led to civil war and the attempted genocide of Bosnians.

I was asked by a charity to use my recently developed skills in Romania to get medical supplies into Bosnia, particularly to take insulin for children to the hospitals now that their supplies were cut off. Time was short; Serbian troops had closed all of the borders leading into Bosnia. My involvement in delivering aid led to the loss of many friends. Some lost their lives, while I lost the friendship of others.

The 'Left' in Britain was rightly proud of the actions taken in the thirties by volunteers such as George Orwell, who had joined the Spanish forces of the communists and socialists and fought General Franco's fascists. However, many on the younger wing of the 'Left' in Britain in the nineties saw the genocide of Bosnian Muslims differently. To them, the enemy was the United States. A number of left-wing parties in the post-war world had become apologists for the dictatorships of Joseph Stalin and Chairman Mao. Some did so for Milosevic and Karadzic, interpreting their 'socialist' and 'communist' labels at face value.

This belief was later seen as fully vindicated when President Bill Clinton condemned the genocidal acts by the Serbian dictators. Their support for Milosevic and Karadzic was exonerated, as it was anti-American. Even today you will find many left-wingers have a greater hatred for the previous incumbent of the White House, President Bush Junior, and ex-

British Prime Minister, Tony Blair, than for Osama Bin Laden. Personally, I know which one I would view as the most dangerous of the three.

Then there were those on the 'Right'. Now I must be honest, I never had any friends in far-right loony fringes, such as the National Front, Millwall or the Young Conservatives. However, I made enemies of a number of rightists who saw the war in the Balkans as a new Crusade and the Christian Serbs as crusaders on the forefront of a new religious war against Muslims. One right-wing fanatic called me "a fucking raghead lover."

In terms of those who expressed the view that I was wrong to deliver medical aid, the most common objection came from those who did not take a political stance. In this case, people who knew little about the conflict would say "You should help your own."

That argument actually had more substance that any argument from any politico. But my view was that what was happening in Romania, Bosnia and later in other countries and in Britain was about trying to help the vulnerable, those who were caught in something that was not of their making. Francis, who once told me proudly that he had never read a book in his life, once said it far better than I ever could when we were drinking at a bar, "You wouldn't let someone treat an animal like that, so when families are going to be wiped out, fuck politics." When getting together vehicles and supplies for Romania and Bosnia, no one helped me more than this feisty, totally illiterate and sometimes dangerous individual.

As I said, I do believe that the efforts made by individuals such as myself did make a difference in the early days of the Romanian and Bosnian aid efforts, but, at the same time these efforts, even with the best will in world, were pretty chaotic. After years of getting involved in such work, I became more professional; much later on I helped to set up a high risk family relocation programme in the United States. But in the early days of the Bosnian war, if a film had been made about the various convoys set up, it would have been entitled, 'Carry on up the Balkans.'

Many others from all sectors of society saw what was happening in the Bosnian conflict for what it was, genocide, and they decided to do something. If the composition of those critics who attacked me ranged from the left to the right, that was nothing compared to the variety of those who banded together to help. If you remember the Hanna-Barbera cartoon *Wacky Races*, then you have a good idea of what those first convoys were like. There were splinter groups of old fighters on the 'left' who had always opposed fascism; there were Bosnian Muslims who wanted to stop the genocide of their people and there were liberals and conservatives who saw the need to protect the democratic ideals of Europe. Finally, there were young people like myself with no political ideology to champion, who watched what was happening on the TV news reports and saw that it was the weak, the old, the infirm and the children who were suffering most. This most diverse collection of individuals I have ever encountered all shared one objective, to help anyway they could.

After completing my first independent aid delivery into Bosnia, I was asked to be part of a larger convoy to deliver more aid to Bosnia. Unlike Romania, this was a war, and I knew that I was lucky to have succeeded in my first foray, so I agreed. The convoy comprised of council trade unionists and Bosnian exiles who formed an alliance with a group of anarchists; it was a crazy mix united by one shared objective, to somehow help. The anarchists travelled in a huge black truck marked in big bright letters, which read 'Goths of War'. Whoever signed the agreement for the anarchists to agree to work alongside the rest of us was probably immediately expelled for such non-anarchistic behaviour. Because I was known for my relief work in Romania, I was regarded in various circles as someone with operational experience. I later learnt that the anarchists had issues with anyone who had any previous involvement in a convoy, believing that it might make them a leader. I was an exception. Somehow, they had checked me out and were very impressed to discover that on a number of my previous relief operations I did not 'officially exist.'

Looking back over the years, I've formed some similarly strange alliances. These ranged from friends in military intelligence and sympathetic police officers who carried out criminal checks for me on prospective co-drivers joining me on humanitarian convoys; to forgers in Pentonville prison who assisted me in drawing up papers to get medical supplies across hostile borders. Just before one trip, I was abducted or as one of the henchmen called it 'forcibly introduced' to one notorious gangster who had heard what I was doing. To my surprise, rather than attempting to secure the medical drugs from my cargo, which were worth a small fortune on the black market, he offered me a roadworthy truck, money and had his henchmen to help me load it. I found out later, that his only daughter had died from diabetes the year before and his wife told him they must help. His only request was that I kept his identity quiet in case his rivals heard of his humanitarian gesture and took it a weakness. As one of his henchman warned me, "If my boss' name ever gets out, you will pray for me to tear your throat out with my hands by the time I have finished with you."

"No postcard then?" I asked, but he didn't smile; most men in such professions are humourless fellows.

When I was getting my truck together for that first convoy to Bosnia, I was looking for someone of James' caliber. I put the word around and Wayne, a New Zealander, stepped forward or rather staggered forward. It sounded strangely similar to my Dodgy Brother James' original offer to help.

"I heard an Aussie helped you in past, so a Kiwi can do it in his fucken stride." Wayne said by way of introduction.

Wayne was a mechanic, and after my experience on the first trip where we lost parts quicker than Barnaby, I welcomed his help. I did expect one or two problems, as Wayne had a reputation for being a heavy boozer, but I was in no position to be choosy and without him I would have failed. I owe a lot to our antipodean brethren; whether they are Kiwi or Aussie, these benevolent rogues slapped me on the back when I was low and said, "She'll be right."

The convoy set off and made its way through Europe with the usual unforeseen stopovers and mechanical breakdowns, but things got tougher once we entered Bosnia. After having negotiated Croatia, driven alongside the Adriatic and back up towards Tulsa, we then had to negotiate our trucks through Snipers' Alley. This was the most perilous part of our journey so far, as we knew that Serbian marksman were atop the mountains with their rifles bearing down on the only road through the pass. We waited until darkness had fallen and we were fortunate that it was cloudy and the moon hardly provided any light as we moved silently through the mountain's meandering road.

Just at that moment, Dr Bill, another volunteer, who was sitting between Wayne and me, decided to push in the cassette into the music player on the dashboard. That was not a good idea as any sound might alert a sniper. My music selection didn't help. Big Audio Dynamite's classic nineties track *Medicine Show* burst forth on the cassette player at full volume. If you are not familiar with the tune, it contains the sound effects of explosions and gunshots from the soundtrack of Sergio Leone's epic spaghetti western *The Good, the Bad and the Ugly*. This resulted in complete pandemonium; Dr Bill was now of the belief that all hell had broken loose and we were under attack. He started screaming and tried to get out of the vehicle, but I pulled him back and muffled his cries by restraining him by the neck with my arm. The poor doctor was already in the throes of a mental breakdown from the stress of the journey so far, but my music selection sent him totally over the edge. My good friend Terry would not be the least bit surprised.

We got through Snipers' Alley unscathed, but trying to get twenty trucks through such a death trap sent all the normal people to their beds later that night with nervous exhaustion. Wayne and I however managed to find a little bar behind a border post, which was our way of dealing with trauma.

The next day, Dr Bill's mental breakdown was now clearly evident when members of a local Bosnian militia, threw open the door of our truck only to be greeted by Dr Bill.

"Are you the dancers?" he asked them.

Whatever world Dr Bill was in, at least he was not afraid to explore the light entertainment. Fortunately, Wayne had nothing to add as he was in a drunken coma in the back of the truck. Meanwhile, I had to silence the anarchists, who had refused to recognise the authority of the leader of the militia; I just managed to talk our way out of being shot.

Our convoy was then held back on the outskirts of Mostar, as the area was under constant artillery bombardment. It was so surreal; I found it impossible to grasp that people were still inside the buildings collapsing and burning in front of me. It was like watching children lobbing missiles at a Lego set, only people were trapped inside.

We finally got through the border and a few days later we reached Tulsa. If I had thought that dodging snipers, avoiding landmines and bridges blowing up as we drove over them was the worst of it, I could not have been more wrong.

As I unloaded one of the last boxes of insulin from out truck into the town's main warehouse, one of the local children, Rumen, ran up to me.

"Your friend is drunk and in danger." he cried.

"OK, tell the manager I'll pay his bar bill," I sighed.

"No, he has lost face."

"Too late to worry about my drunken friend's reputation," I said.

"No, he fell off cliff and his face gone, but I have to go or I will miss concert."

"Face gone and … CONCERT, WHAT CONCERT?" I asked.

Quickly, Rumen explained that Wayne had fallen off a cliff. I could easily understand, as for Wayne getting drunk and falling off a cliff was as easy as falling …well you understand. But, as it was dark and a curfew was in place, I was more worried about a concert that was taking place.

"The concert, Wayne, which direction?" I asked.

"I will show you, I want to see the lights of the concert in the sky."

"Lights in the sky? Christ!" I said. "No you won't, you will stay here and get everyone to take shelter, now just point me in the direction of the concert."

"Just follow the beams in the sky," said the enthralled kid.

I looked up and saw lights darting across the pitch-black night, just then the Serbian artillery opened fire from the hills above.

The Goths of War, unbeknown to any of us, had packed concert amplifiers along with their aid and searchlights to light up the sky above to attract a large audience. Their approach was laudable in trying to bring music to those traumatized by the conflict, and normally I would applaud such an initiative as music is a powerful therapeutic tool – but not to a town under siege. Enemy artillery was positioned above the town waiting for the slightest sound or light to aim at; I knew the town would be flattened by dawn. We were firmly in the spotlight, a spotlight we had provided.

I drove my truck at full speed, which was dangerous enough, but I was driving in pitch-black darkness; I had to leave the headlights off for fear of attracting fire. I shot off the road twice and had to wait for the next explosion in order to find the road and get back on it again. When I reached the concert, all hell was breaking loose. The Goths of War had started their music and brought all the towns' young people out into the open. Rockets were landing everywhere and the kids were running in all directions. I shouted to them to get into my empty truck. As they piled into the truck, I ran towards the beach to find what was left of Wayne. The Goths of War quickly realised that they had exposed the children to enemy fire and were ferrying kids into their truck and out of harm's way.

I found what was left of Wayne, his face was covered in blood and his trousers were in tatters; what was on show did no credit to the men of his nation. I threw my exposed and drunken co-driver over my shoulder and ran back to the truck along with the local kids who had been trying to patch up his face with seaweed.

When I got back to the truck, it was now packed full of children and teenagers screaming and crying; they were terrified at the continuing explosions landing around us.

"Sorry kids, if you thought it couldn't get worse, look at the face on this drunken idiot," I said as I threw Wayne on the floor of the truck. For a brief moment the sorry state of his undergarments appeared to distract them from a new barrage of mortar attacks.

After an hour of driving in total darkness, the Goths of War and I got our trucks back into town. Incredibly, though a few kids had injuries, no one had been killed or severely hurt. When I stormed over to the Goths of War to set fire to their Mohicans, I realised that there was no point. It was obvious from their faces that they were in shock at having nearly killed us all. I too learnt a valuable lesson, in that I would never again be part of any humanitarian convoy until I had had everyone vetted by Mike and his police colleagues, not just my co-drivers. Despite my earlier encounters, I had secured some good friends amongst the 'choirboys'.

At this point, I had better clarify the cause of the injuries to Wayne. You would expect someone who was right in the thick of the action to have suffered his injuries as a result of enemy fire. However, though sniper fire and exploding artillery shells were a harmful option for anyone darting for cover on the beach, my drunken friend had been on the edge of a cliff trying to dive off it. Of course, he had to go one stage further than being blown up along with everyone else and enhanced the risk to his life whenever possible. That would explain why bungee jumping was invented in New Zealand; any tourist could simply jump off a bridge into the water, whereas your average Kiwi needed to create a further element of danger by tying an elastic band around his ankle in an attempt to shoot his bollocks through his eye sockets. While rockets were landing all around him, my friend decided to jump off the cliff in the belief that he could clear the rocks and dive headfirst into the waters below. My experience of the railing and the spike in Hungary only served to whet his appetite.

On viewing the cliff a few days later, I realised that he might have cleared the cliff-face, provided that one of the local Serbs had been kind enough to fire him out of his cannon. But being smashed out of his skull and unable to walk in a straight line, let alone take a running jump, his condition explained why he had hit the cliff; face first, a yard into his dive.

Not only did he hit the cliff, he bounced down using his face to soften the repeated impact. Strangely, it was the drink that saved him; if he had been sober his body would have stiffened with shock and broken like a twig. Of course if he hadn't been drunk he wouldn't have tried to dive off the cliff in the first place.

When Wayne decided to join me on my second convoy to Bosnia, his girlfriend, Caroline, asked me to make sure he did not injure himself again. Mind you, I think Wayne thought this was to be his first trip: he could remember very little of the last one. So we headed once again to Bosnia with a truck-full of insulin. We encountered the first of many problems when our papers did not arrive in Austria; we could not go over the Hungarian border until they were faxed through the following day. We stayed the night in a quiet hamlet where there was no indication of the mayhem that was to unfold later that night, unless you take into account that Wayne and I were left to our own devices.

We decided to have a few drinks and a good meal in order to be ready to take on the next day's challenges. Naturally, by ten o'clock, we were absolutely bollocksed and drinking with some locals in the only club in town. After we had left the club, Wayne decided to show off his athletic prowess to passersby and he jumped on a bike left unchained by a nearby lamppost. He started very well, until he picked up speed, got the front tire jammed in a tramline and landed face-down on the gravel, taking the other half of his face off. His audition for 'Dawn of the Zombie's' was a great success, as the people passing by screamed and ran off. Back in the hotel I picked lumps of gravel out of his face with a pair of tweezers from our medical kit, while cooking spaghetti bolognaise on our little portable gas cooker in the bedroom.

In the morning, we found a quiet little café to have breakfast in while we waited for our documents to come through. Wayne's torn face had healed into a crust during the night, but unfortunately any facial contortion was painful. He couldn't crack a fart without the newly formed crust over half his face cracking open. I decided to place our orders with the waitress who was trying her best not to look at Wayne.

"May we have two cappuccinos, one with a straw please?" I asked.

Wayne mumbled out of the fifty percent of his mouth that he could move, "Baaa...stard."

The locals, though very polite, could not help but occasionally glance over at the two strangers. I played the role of the polite Englishman and exchanged pleasantries with the waitress, while my companion spoke some strange dialect where every third word was "fuck", "boll...ocks" or "pom...mie-fuck...witt!" I just wished Scouse Bob had been there to converse with him. I picked up a local paper to see what exciting events had taken place and proceeded to relate the headlines to my crusty faced companion.

"'LOCAL MAN PLANTS SEED IN POT PLANT'. 'MAYOR DECLARES A NATIONAL HOLIDAY TO CELEBRATE A NEW APPLE STRUDEL SHAPE'" then my eyes opened wide with excitement.

"What ... is ... it?" my crusty-faced companion mumbled.

"'MAJOR CRIMEWAVE IN TOWN LAST NIGHT'", my battered-faced friend leant forward, sticking his straw through his crust.

He produced a muffled, yet still loud, "Ffffuck!"

"The headline continues," and I looked directly at my friend and continued to read out the headline, "ELEPHANT MAN STEALS BICYCLE!"

"Bas...tard..," he replied, collapsing into hysterics, which painfully broke his face into four sections. That resulted in such obscenities that even Scouse Bob would have covered his ears. Wayne would get his revenge later, although he did not realise it at the time.

On the journey back, after delivering our aid safely to the hospital in Mostar, it was my turn to sleep. We took turns to sleep on a raised platform in the back of the truck wrapped up in sleeping bags. Whilst I slept, Wayne, who is an excellent driver, had to break suddenly to avoid another truck as it veered off its side of the road and drove directly at us. His manoeuvre projected me off the raised platform and I dropped six feet on to the truck's bag of tools below. As I was still asleep when Wayne slammed the breaks on, I was unable to prepare myself and I landed face down on the tool-bag. I was unable to untangle myself from my sleeping bag and I was in so much pain that I passed out. Wayne opened up the back shutter six hours later, to find me still encased in my sleeping bag and lying on top of the green canvas tool-bag like a ruptured caterpillar bent over a large leaf.

My following trips carried on in much the same vein, except that I learnt a little bit more each time. Over the years my initial cack-handed attempts evolved into a more considered approach. Every truck I drove delivered its cargo intact and for that I have to thank my Dodgy Brother James, Wayne and numerous other benevolent rogues. I saw many horrendous things and met some dangerous characters during those years as I travelled back and forth with medical aid during war, but I believe I only met one man who was evil in its purest sense.

I have encountered many other cruel and malicious men in my life and on one aid trip to Romania, I came across some children who were new to the orphanage I was working with; the medical checks quickly uncovered that they had all been severely sexually abused in their previous institute. I decided to pay a visit and meet the director of that institute; a man called Convic. He was aggressive from the start and, when I demanded to see the children of his orphanage, he tried to throw me out along with Romana, the Romanian nurse who had come with me. A scuffle then took place and he fell through the window. He survived and, from his hospital bed, he demanded that I be arrested.

To my good fortune, Karl, the investigating detective involved in my case was a good and brave man. Karl later told me that he had found that many of the children under Convic's care had been regularly abused, and amongst their abusers were some powerful government officials. Karl went to his superiors and reported the abuse. Convic soon dropped his charges against me and, though no charges were levied at him,

Karl informed me that he would never be allowed to work with children again. That point was irrelevant anyway, as Convic was paralysed from the neck down due to his fall. The detective officer warned me to be on my guard, as some associates of Convic's were of the view that it was not an accident. Convic was a criminal, he was motivated by avarice and he was willing to permit the vilest obscenities to happen to innocent children for money, but the most dangerous man I encountered during the war in the Balkans was motivated by the pure pleasure of killing.

WARRING ROGUES

"They challenged the world order in that our government knew best; indeed they challenged everything I had been taught to believe in. Criminals to some, mavericks to others; mistrusted, even hated by our superiors, but when these characters were mentioned a few of us would turn our heads away and smile."

Private Claire Manning's diary,
US Army
Bosnia 1991

Chapter 27 – *"Englander!"*

I had met men who loved a fight, a few who tried to maim me and some who had even attempted to kill me, but I had never met anyone whose sole purpose in life was to kill. I met this person, as I was returning from delivering medical aid to a recently bombed hospital in Sarajevo. José, my French co-driver, and I had just about avoided the Serbian patrols and we believed that we were over the worst; we headed back south and made our way up the Adriatic coast to Split. On arrival, I left José to get some well-earned sleep, while I headed to a bar in town, which, I knew, prepared a blood-rich fulsome steak.

During the war, the population of the Croatian town of Split was a strange mixture of regular troops, committed freedom fighters and mercenaries. Many of the mercenaries were hired by Serbian or Croatian forces; they were plentiful and cheap to hire. Some were remnants of old conflicts, some were ex-paramilitaries from Northern Ireland, others were ex-policeman from South Africa's apartheid police force. There was a new influx of armed killers who had been driven out of their homelands by the Russian army in Chechnya or by the Israelis in Palestine. As individuals, they ranged from psychopaths willing to murder for pleasure to those who had adopted this career. There were some who couldn't handle a life without war and a majority who were simply there for the money. One London road-sweeper was discovered working as a mercenary in Croatia at the height of the war. He was dismissed by his council employers, not for his crimes, but for drawing sick-pay while he was seen on the evening news and clearly fit to work.

No outsider can contemplate the effects of war can have on those fighting one. At all times you are in the proximity of death; the possibility of killing or being killed is one's only concern. I never fought in a war, thank God. I'm not sure my sanity, such as it is, would survive, but I witnessed how the vilest elements of human nature gorge themselves in the carnage. Thankfully, my faith in mankind remains because of people like Rory and the other benevolent rogues I met during the war. They lived by their moral code and were never corrupted by the horrors they witnessed.

As I entered the main bar in the centre of town, I saw a man in full Highland regalia and an AK47 propped against the bar beside him. His name was Lazarus and he stood at about six-foot-three-inches tall and was built like a shithouse in the Gorbals area of Glasgow. With no insult intended to the great Scot, but with his build, his expertly maintained grey beard, his gait and the frequent jutting movements, he looked like Sean Connery on acid. I had seen him once before in Split when another driver had warned me to stay clear of him. I knew the man to be a mercenary and that he was on the run after breaking his wife's jaw back in Glasgow.

He introduced himself to me, as he seemed to have heard of me and addressed me by a new name, which in future was the signal that he was about to kill me.

"Englander!" he exclaimed as he extended his hand. I shook it, so as not to aggravate the situation; my intention was to drink my beer and leave as quickly as possible.

His grasp was like having your hand plunged into rapidly-drying cement as it slowly crushed the life from my fingers. He placed a bottle of beer in front of me. It quickly occurred to me that he thought I was a mercenary, but I was bemused as to why he would think that; I did not have a military bearing, I had no weapon and I lacked a head that was the shape of a cornflake packet. He held centre-stage; he was obviously known to everyone in the bar, except me. His infamy was based on his reputation as 'a full-timer', for when he wasn't fighting for the Croats, he would join the Serbs during any lull in Croat engagements.

At that point Rory, another Scotsman, but a convoy-driver, appeared from the group behind Lazarus and grabbed my free hand like an old friend. I was relieved to see him. I had met Rory briefly in the few towns that drivers used as stopovers, before driving trucks over the border. Rory was a legend to those in the area of international aid work. Originally, he had been in the British marines in the first Bush war in Iraq. President Bush Senior halted the allied march on Baghdad when it became an embarrassing turkey shoot and the western armies withdrew. When orders were given for the total withdrawal of allied troops, Rory refused to leave the Kurds to be massacred by a vengeful Saddam Hussein. He voiced what western politicians and his military superiors knew would happen, but dare not say publicly.

With his regiment gone, Rory had fought alongside the Kurds even when they were the victims of biological warfare. Eventually he had to flee when the allies came back into Iraq. Senior generals from both sides wanted him badly; he was officially branded as AWOL. Many of his fellow soldiers admired him for what he had done and for his bravery in fighting with the Kurds against Saddam's superior forces; they helped him escape. Hence my bemusement as to what was he doing with this lunatic.

"Lazarus, this is the guy I was telling you about, he's one of us," said Rory.

Now I turned on Rory, who was a good six-foot-one and similarly built like a shithouse, though more of the Edinburgh variety – more refined and lighten in tone, but still a shithouse that was able to withstand a nuclear attack. The Edinburgh accent always sound to me as if it comes from afar, almost muffled, as if they are talking to you from the bottom of a deep well, while Glaswegians sound like they are standing right next to you, shouting angrily down at someone who has stupidly fallen into a hole. I had no idea what these two giants were talking about, but then Rory leant back out of sight of Lazarus and gave me a subtle wink. With that gesture, I knew that Rory was spinning a tale and had not gone mad.

Then Lazarus unveiled a photo album and placed it in front of me. At first, I could not make out what the subjects were.

"This was Sarajevo," Lazarus said, "Fucking must have killed fifty of the dirty Muslim cunts that day. I never understood all this fucking about, torturing and raping; it wastes time when you could be moving on to the next fucker."

The pictures in front of me, in a Woolworths' family photo album, were of the corpses of men, women and children. No one was spared when Lazarus was at work.

"But there are times when I really take a dislike to someone, I put my gun down, put my knife back into its sheath and then I take my time as I ring their fucking neck with these,"

Lazarus said as he lifted his cement-shovel-sized hands to my face and added, "No one has ever escaped from these." I had no doubt that many had been strangled by those hands and the last thing they would have seen were Lazarus's eyes inches from theirs and his smile of pleasure at killing them.

I was getting the feeling back in my hand and I slowly raised it to grab the bottle of beer from the bar and jam it into his concrete column of a throat. I have been involved in a few nasty incidents, but that was the only time I ever wished to take another man's life. If you believe that all people are basically good and can be redeemed, then I would agree with you, with the one exception of Lazarus.

More of Lazarus's militia had turned up in the bar and Rory was indicating by a shake of his head that it was no time for heroics. However, I was losing it and my rational thought process disappeared as the red mist descended. I just wanted to do something to damage the lunatic and delay his next killing spree, if only for a short time. Rory read the signs, unlike Lazarus who was laughing and at ease as he continued leafing through his photo album. Rory saw what I was going to do and he put his arm around my neck, in a playful wrestling manner, shouting loudly that he wanted to talk to me about when I moved in on his girlfriend in Dubrovnik. He dragged me out the door. Out of view he then tightened his arm-lock on my neck until I was about to black out, forcing me to release the bottle, which crashed on to the cobbled street.

"There's too much to lose, Johnny boy, just to satisfy your brief moment of heroics." Rory said. "You attack him, and every volunteer will be fucked for your moment of glory, you stupid cunt."

I had nearly lost it for the first time in years, but his grip was like a mechanical grab and I slowly began to calm down, seeing the sense in his reasoning. With the bottle gone, he released me and I dropped to the ground. I kicked out as I did so, but barely made any impact as I struck his leg.

"So we just fucking leave him to carry on murdering people, while we try to keep people alive so they can be used for target practice?" I asked as I gasped for air.

I had not cooled down too much; my hand was around Rory's knife, which I had liberated from his boot as I dropped to the ground.

"He will be stopped. I swear he will, but not here," said Rory and I knew from his calm and unhesitating words that he meant it. I no longer argued. I knew he was right and that it wasn't the time. I handed Rory his knife back.

"Well fuck me, not bad, Johnny boy," was all he said.

I met Lazarus two more times in as many years and, after our brief encounter in Split, he knew who I was and I became someone he took a 'dislike to.' I just about got out alive following each encounter, but the second time was close and I only escaped with the aid of a can of petrol. Rory said that the legend was that I was the only person to ever free themselves from Lazarus's grip, but he smiled as he added "The fact that he was on fire helped."

Rory and I occasionally completed humanitarian aid trips together. It once again involved applying to friends in low places to secure the necessary paperwork. However, it was even more difficult as Rory, like many other foreigners in the Balkans, was on the run. I grew to have the utmost respect for that man; he did what he thought was right, losing contact with his comrades in the marines because of it, and he was publicly declared a deserter by his own country. The most difficult aspect of our relief work was that Rory was well known. Sometimes he would have to leave our truck when his notoriety brought too much attention and risked our load being impounded by border guards. Years later, my reputation became more of a hindrance than a benefit and, like Rory, I had to take the tough decision that it was best if I was no longer involved. It happened too late for Rory.

On one of our convoys together, after delivering our medical cargo, we were asked to take some children out of a nearby hospital that was under daily rocket bombardment. To this day I can't work out if the Serbs or the Croats were bombing the hospital, but it was of little interest to the attackers as the children inside were a mixture of Serbians, Croats and Bosnians from a nearby town where their families had once lived, on the whole, happily together.

Rory and I prepared the vehicle for the journey ahead, firstly by riveting some metal plates to the hospital bus, which we would take with us alongside our truck. We did the same for our truck with what was left, which in itself would provide cover for us. Where the road was wide enough, we would drive alongside the hospital bus, providing an additional shield from enemy fire. We took the metal plates from some bombed-out trucks and the remnants of a nearby burnt-out tank. The tank must have been a cut-price self-assembly flat-pack model, as its armour when hit by mortar-fire hadn't put up much resistance. I was worried that the plates offered little protection for the children in the bus, so I found some tin steel plates which had been used to cover the hospital windows. I then proceeded to cut those into rectangular shapes and welded them together to make little oblong boxes. My purpose was to add additional safety shields as well as to secure safely the babies we had on the hospital bus. I called these metal boxes, Little Nannies, as being a Bond fan, in homage to the 'Little Nellie Gyrocopter' from You Only Live Twice, to protect our little passengers from the attacks of larger forces.

Strangely, one of the doctors who came with us told us that our presence made everyone feel safe for the first time in weeks. Clearly they didn't know us. That night, the older children and babies slept soundly in various parts of the cellar, the one part of the hospital that was relatively undamaged. The doctors and nurses slept on the floor next to the younger children, as Rory and I had taken control. As I was working away in the part of the cellar furthest away from everyone, Rory walked in, after given a quiet tap on the door.

"Excellent," he said, "I'm in the middle of a war-zone with no back-up and I have to get twenty babies, plus some doctors and nurses out in one piece. So what does God do? He gives me the help of a world renowned lunatic, who picks this moment to have a complete mental breakdown," he laughed, as he picked up one of the Little Nannies I had just made. He turned to me and whispered, "Armoured-plated baby cribs, Mothercare will never go for it," he laughed peering down the assembly line of metal cots.

"You've never been in a crèche in Liverpool," I replied as I continued my task.

With the sun down and the moon hidden behind some large clouds, we woke everyone up, loaded the last essential items on to the truck and the bus and slowly opened the doors to take our vehicles and passengers into the unknown. Rory had had less sleep than I, as he had caught the fancy of one of the nurses, but he was fully focused on making sure that we had everything and that no one was left behind. The truck was heavy and Rory's humour well and truly evaporated. He worried that the weight of the hospital bus would sink us into the muddy roads ahead or, worse, lead to bridges collapsing under us. He cursed me for the additional weight of the metal plates.

For eight hours, which felt like a month, we drove in pitch-darkness, slowly along the road leading from the hospital, until dawn started to break. I looked over my shoulder from the driver's seat and in the morning sunlight I saw the strain on the faces of the doctors, the nurses and some of the older children, they were wondering if they would survive the day. The comfort that Rory and I had brought had evaporated; one of the nurses started to cry as the sun's rays pierced through the bus. Others reflected dangerously off our windows. We were now out in the open and vulnerable.

One doctor took over the driving of the hospital bus, while Rory drove our relief truck and I mounted a motorbike with a sidecar attached, to run ahead and scout for the safest route. I had adapted the motorbike and armoured its sidecar, which I christened Little Nellie after the original and which also contained a spare Little Nanny, on the off chance that it might come in handy on our journey. It was lucky that I had.

The babies seemed content enough in their Little Nannies; they were firmly wrapped up in their quilts, which I had also cut to size from hospital blankets. I had also made sure that the base of the boxes was shorter in length than the sides to leave a gap through which we could wash the crap out of the boxes using a tube from our water tanks. The doctors and nurses held babies as we hung the Little Nannies out of the bus windows to clean them on the move. Rory didn't complain about this little adaptation of mine, as he knew it was essential to keep the bus moving – any stops made us an easier target for snipers. It also helped keep the bus as hygienic as possible, as some of the children and babies had severe infections and the spread of further disease was the last thing we needed.

We had many a close escape over the next few days, as we manoeuvered through the mountains, but one episode truly captures why Rory was a benevolent rogue. Rory took a turn scouting ahead on Little Nellie. As he checked the condition of the next bridge on our route, he heard the cries of a baby coming from a nearby derelict building. On entering he found a baby girl amongst the debris and the lifeless bodies of what were probably her family. With his new passenger wrapped up tightly in its Little Nanny, he set off at full pelt, dodging snipers and negotiating minefields as he made his way back to us. He came across a truckload of US troops on a mountain road, while up on the mountains were more enemy snipers training their guns on the only passage through the mountain pass. Rory then drove at the US truck and shouted abuse at the Americans in Serbo-Croat; an attempt to get them to open fire on him. This he believed would convince the enemy snipers not to shoot, thinking that he was one of them now that the allied troops were shooting at him. It worked. Rory drove through the pass with bullets whizzing around him, but he got the little girl back safely to our truck and a few days later to the safety of a hospital in Split.

When he told me his strategy as I was digging bullets out of Little Nelly, I said he took a hell of chance that the Americans wouldn't have killed them.

"They couldn't hit a cow's arse with a banjo,"

Rory had a greater respect for the experience of the older enemy troops firmly positioned above him in the mountains, than American teenagers trying to aim their rifles at him in a moving truck. Rory had courage, daring and that element of madness, in putting the little girl and himself directly in the line of fire. It saved them both.

Days later we made it safely to a secure hospital. The truck had taken a few rounds in its sides and we'd lost a few windows, along with the windscreen but everyone made it. Neither Rory nor I had had any sleep over those days, but our relief at getting our passengers to a place of safety is impossible to describe. We burst into hysterical laughter at the lunacy of what we had done, but everyone was safe and it was just the release of the tension that had been building over recent days. We had emptied the bus and were stripping it of its metal plates when one of the doctors came over to me and gave me one of the Little Nannies. He indicated to Rory and me, the outline of a couple of deep indentations on the metal plates covering two Little Nannies. The doctor then gave each of us a bear hug, while we just kept looking at each other and then back to the Little Nannies in our hands. Bond may have taken out four helicopters with his Little Nellie, but our Little Nannies had hurt no one and Rory had perhaps saved a least one little lady who was now asleep in a warm and more comfortable cot in the hospital opposite.

"One day I will give you this as present for your first kid," I said to Rory, "who knows maybe you and your nurse? She seems to like you, the poor deluded lunatic."

Rory started laughing again and he said, "I think you're right. I think she likes me, maybe it's time to settle down. I like the people and I like it here."

I then made an observation, "Before you do, just one note of caution," I pointed at Rory's new love; she was kissing a man who was probably her husband, while mobbed by three kids.

"Bollocks, ah well it's a war and people do desperate things for comfort when they think they think their life is going to end," said Rory.

"How true, it shows how desperate that poor woman's situation had become, that she climbed on top of you."

"Fuck off Johnny boy," he said. Rory shrugged his shoulders and then his face lit up.

"Fancy a drink? I know the barmaid in that café over there," he said, pointing in the direction of a little bistro.

As we set off to what was left of the café, (the roof had been blown off) I noticed for the first time a sign that Rory had painted along the side of the truck.

'LITTLE NANNIES FOR THE LITTLE GENERALS IN YOUR LIFE'- LIVERPOOL MOTHERCARE 1989.'

"Well that will keep the shop's metal detectors busy," I added as we passed.

"It's Liverpool, they will be anyway," added Rory.

Rory was a true benevolent rogue. Though I concentrate more on the roguish nature of the characters I have encountered, their moral courage to do what is right, whatever the risk, defines them. It is why I wrote this book, so their stories do not die along with them and their children can know how special their parents were.

I thought at the time how lucky I was to know benevolent rogues like Rory. However, later that night we got into a massive argument about which one of us the barmaid fancied.

Chapter 28 – _**If Only I Had Learnt Latin at School**_

Rory and I kept in contact and met up where we could, but it was difficult as he was on the run and hunted by so many governments, not just our own. Three years after the end of the Bosnian war, I was attending his wedding. It was good to meet his family and Luisa, his Argentinean bride, who asked me about the strange wedding present I'd brought. I could see that she wasn't greatly impressed with a five-sided battered box with a big hole in the base and rather jagged strips of metal sticking out from it. Rory, like so many other good men, never talked of his adventures.

After the ceremony, Rory came over and had a drink with me at the bar, "When are you getting married, Johnny Boy?" he asked.

"Who would have me?" I replied.

"A good point well put," he said and we carried on the evening's celebrations.

A few years later I received a call late one night from Rory.

"Christ Rory, it's three in the morning!" I said.

"Not where I am." There was no point in outlining the transatlantic time differences.

"I give up, which dictator have you upset now?" I asked.

"This is different, there's this Texan millionaire who wants to get all the lead characters, who delivered humanitarian aid into war zones, together on his ranch in Texas."

"Is there no end to Bush Junior's plans for world domination," I sighed.

"No, this is a different megalomaniac." Rory replied.

"OK, but you're not seriously considering going? You'll be arrested and extradited back to England as soon as your foot touches US soil."

"Yes, but it's worth the risk as there is a real chance he will fund future relief convoys on a major scale."

"Good luck," I said.

"No, he wants us both there." Rory persisted.

"Look I love Arnold Schwarzenegger films as much as the next man, but I have no time to indulge some millionaire who wants to be surrounded by a kind of League of Extraordinary Lunatics."

"He will pay for the flights, cover our expenses, it will be a break and a good laugh at the very least." I knew that this meant nothing to Rory, and he was trying all his persuasive skills and concentrating on my weaknesses as a hedonist to make me want to join him.

"What is the true reason?" I asked, "These characters with more money than sense are always popping up," I replied.

He paused, "I'm broke and I am in the middle of a big South American convoy. He will pay me an agent's fee, which will cover my fuel and provisions and get my convoy going, providing I get ten or twelve of the main people there and his list includes you."

"Fuck it Rory, I'll give you the money." I said.

"It's bigger than that, he's a millionaire. If this works he is willing to fund all of my humanitarian aid convoys, – he's a kind of philanthropist who has made his money and now wants to plough some of it back into helping others."

"Bollocks, this is *A-Team* stuff." I argued.

"Look, he mentioned you by name," Rory continued, "also if you don't go, fuck it I'm not going either, it's too dangerous."

That was a different line for Rory; he had never feared anything in his life. Maybe he was playing to my other weakness, to make me take on a challenge, or maybe now that he had a family he had changed and was in fear, not for himself but for them.

"Dangerous eh? Worried about too many calories in your free surf and turf," I goaded him trying to get to the other reason for going, which I knew he wasn't telling me.

As I refused to budge he told me, "Lazarus is rumored to be heading there and he will kill."

"But, the last I heard of him, he was offering his skills to both sides in the civil war in

Angola. Anyway, I thought you said the Texan wanted only humanitarian aid drivers?" I replied.

"They kicked Lazarus out for killing one of his sponsors. As for the Texan, he's a little confused about where to draw the line between humanitarian aid workers and those who have experience of war, including mercenaries. That's the problem," said Rory.

"Let me get this right," I said, "You want us to go to America and meet up with some other lunatics for the benefit of some Texan oilman going through a mid-life crisis? Plus, at the same time, we have the option of being strangled by a six-foot-three psychopath in a skirt? As a recruitment agent you're not selling it."

Rory exploded, "It's not a skirt, it's fucken a kilt," he continued, "and he doesn't strangle people anymore; his new method is to push his thumbs through your eye sockets until he punctures your brain."

"Ah well, that makes all the difference, why didn't you say so, I'll put my spurs on. Now where did I put my horse?"

"Fuck off, I'm serious, I know you don't do this stuff anymore, but you could help secure the finance to help the rest of us …" the line went quiet for a while and then he said, "If Lazarus turns up, I need you there. I can't take him alone."

He was right. I had turned my back on it all, because since Bosnia, the world was different. The simplicity of nation states, which I spoke about with the old Indian, Amran, around a campfire many years before had gone. The invasion of Iraq to tackle global terrorism, had failed. Saddam was gone, and later Bin Laden, but terrorist cells move like a virus to wherever security is lacking. The United States, Russia and even China saw that the world was more complicated now. National borders were irrelevant and terrorism was global.

Aid was seen as a vital economic instrument in fighting poverty and hunger, but it had also become a weapon in international conflicts, the premise being that it undermined the support on the ground for terrorists and it would 'win the peace' as well as opening new markets for economic expansion. Between the superpowers, most corners of world were coming under an aid umbrella, and so the services of benevolent rogues are no longer required. It is the level of aid and what strings are attached to it that has become the issue now.

I was not the same man I had been twenty years ago. I did not have the stamina anymore; my last trips had been harder and the recent loss of my good friends José and his girlfriend Maria had finally pierced the defensive system of this emotionally-stunted individual. It was not self-pity, but rather an honest assessment that my zeal and determination to achieve at all costs were now in question. Like a punch-drunk boxer well past his prime, I had to be honest and acknowledge that I was a liability on any relief mission.

But Rory needed me for what was more of a social networking event than an aid convoy. I decided to go and help Rory and the others raise funds for future convoys. There was no long-term commitment on my part. Maybe just a few days in the sun on a Texas ranch with free food and booze would be just what the doctor ordered. Especially as my doctor was dying of cirrhosis, so I had no doubt he would prescribe this trip. But, I was also drawn by another thought, a final encounter with Lazarus.

I had been to Texas before, but when we landed we might as well have been anywhere; we were ushered into a limousine and taken straight out to the millionaire's ranch, so as not to draw attention to ourselves. Rory and I were met by a collection of mercenaries; private security guards and friends from previous convoys. It was, as you would expect, an uneasy gathering. Some of the mercenaries and those members of private militias had grown fat off the conflicts; I had no doubt that some had even killed those we were trying to help. They probably had Rory and me in their sights on that occasion.

The Texan was a little man called Doug, he wore no cowboy boots and no big Stetson hat and he impressed me as a decent man, who appeared sincere in his intention to get humanitarian aid through to where it was needed. But, Doug was at a cross-road. He could use his wealth and highly placed contacts to get humanitarian aid through by using the peaceful efforts proposed by Rory and his fellow relief workers, or he could choose to finance armed security personal militias and similar companies, to get aid through using the superior gun-power offered by our well-armed fellow guests.

Rory and I knew some of the hired guns and they knew us. We all kept a respectful distance that weekend, knowing that if it kicked off, the causalities would be high. Rory and I tried our best to persuade Doug to not go down the hired mercenary route, but we failed. I could see that Doug was becoming enamored by the guns and entranced by the macho stories of the mercenaries. Rory and I were tested to the limits of our endurance, we were far from pacifists, but all of their tales were based on violence. On the last night before we left, Rory had been drinking hard; he was despondent that his mission to get Doug to fund future convoys had failed. Even worse, that those funds were going to resource armed militias, which would lead to more suffering for those caught in the conflicts. That was at a time before what we see now in Iraq and Afghanistan, and some say Pakistan, where private security services like 'Blackwater', thrive. Those private security companies now deliver a large part of the security services as part of the United States' effort to provide security for many government buildings. Their work includes delivering humanitarian aid on occasion. What I experienced at our gathering in Texas was a taste of things to come.

A fight broke out between Rory and an ex-South African policeman, who had become a mercenary in Rwanda. It was short and bloody, but by the time we pulled them apart, Rory had been stabbed in the side and the mercenary was unconscious with a broken jaw. I took Rory out of there early the next morning, along with all of those who thought as we did and we left Doug with his hired guns. Doug ran after us to say that we should all work together for the greater good. Doug knew he had failed to change our minds but he gave Rory, who was holding his wound together in the back of the limousine, a large cheque to finance his next convoy in South America. He turned to me and asked what was so wrong in financing an armed force to get supplies through to those who would die without them.

"If you choose to enlist those lunatics in there, you may succeed in getting supplies through, but you will be just another armed militia force amongst many. Many will die at the hands of your bastards and the anger and resentment you encounter when you go in will have grown worse by the time you leave. Your intentions are good, but you're about to start something that you can't control," I replied. It was the longest speech I had given since I was stuck on the Hungarian border with my Dodgy Brother James many years earlier, but I couldn't change his mind; he was convinced that delivering aid at the point of a gun was right.

A few years after that weekend, Rory died and Doug sent me a message to ask me to attend another gathering, this time in London. I felt that I owed it to Rory to try and persuade Doug to abandon his militia approach, but once again I failed. Even though I only attended the dinner in the secret rendezvous around the corner from the Ritz, I still ended up with a bruised jaw and a few cracked ribs; it all kicked off before even the starters had arrived. Doug told me, just as the fight erupted from a heated argument between two ex-Russian soldiers, that my final words had had a profound effect on him, but he told me that though he had a number of militias in place to go anywhere in the world, he was in control and nothing had changed. I knew otherwise; for a man who protested peace, every encounter erupted into violence.

After the fight was eventually broken up by Doug's personal security guards, I turned and excavated Doug from under the table. He readjusted his clothing and put his hat on.

"Cowboy boots, a Stetson hat, is this a new look?" I asked him.

"Yes, do you like it Righten?" he said.

"No, but it suits you now," I said, as I shook hands with him for the last time and I walked off slowly in the direction of Piccadilly, nursing my injuries.

It is funny that when someone talks of Americans thinking they have a God-given right to alter the world to match their vision; it's really only Texans we mean. Men such as Charlie Wilson in Afghanistan and George Bush Junior were up to change just about everywhere where there wasn't a church. Even Sir Allen Stanford had to muscle in on the game of cricket. I am not too worried about that last one; I never liked cricket anyway. Some see Sarah Palin posing a similar threat, but I don't, as she is not Texan and she couldn't find Iran on a map.

I never did meet Lazarus again after Bosnia. I heard that both the Croat and the Serbian commanders were sick of the violence and could no longer stomach his killing sprees. Later, I heard that Lazarus had been killed in a hotel room in Austria. He had been stabbed to death with a British Army knife and with such force that the blade of the handle had driven up and become embedded in his skull. I remember Rory asking me, shortly after hearing the news, if I had any idea who had killed Lazarus. I asked him to repeat the question; it was hard to make out what he was saying at the time as his jaw had been broken in some recent altercation. Somehow the Austrian police had bungled the crime scene and lost the DNA samples of the suspect, they had closed the case. Rumour has it that some rich Texan had taken a personal interest in the matter. Looking back on Lazarus's demise, my thoughts were that it was untimely: I wished it had been sooner as there would have been fewer photos in his album.

Rory was a direct contrast to hired killers like Lazarus. He lived according to his personal code. Doug immediately arranged a freight carrier to send Rory's coffin back to Scotland. That twenty-two hour flight was the loneliest journey I have ever taken as the only cargo was me, the crate I was sitting on and my good friend's coffin. Two days later, when I saw him laid out in a box in his parents' house in Edinburgh, the mad bastard I remembered was not the lifeless body in front of me. It was not that the body did not look like Rory; it did in a strange, glazed and pumped up fashion. It was just that Rory had had energy in abundance; he was always challenging people and bouncing around powered by his constant drive and passion. He wasn't that sedate figure laid out in his own, larger, Little Nanny. Rory must have lived three lives, so I was not sad about his death in the way that I am whenever I see a coffin bearing a child, knowing they have never had a chance to truly experience life.

I left my wedding present next to him and said, "Goodbye old friend," and I closed the lid, as I was the last of the mourners.

The success of those times was due to those benevolent rogues who found the money, secured the border papers, got the medicine supplies and then put their lives at risk to get the aid through. This involved personal sacrifices, but sacrifice leads to suffering and often not just for the person involved, but also for those who love them. I could never commit to marriage or fathering children, as it was likely that I would not be there when they needed me. For this reason, I had a firm policy of never taking anyone on a convoy if they were married or had a family. A number of good friends tried to help, but I hope they forgave me for refusing their kind offers.

A fair few of those people who did voluntary work abroad were tortured souls and needed as much therapy as the people they tried to help. You could spot some of them straight away. I remember in the early days, one woman had recently lost her child and was so desperate to find an outlet for her frustrated maternal instincts that the children in one Romanian hospice were terrified of her. They had never experienced intimacy, let alone this strange woman who just wanted to hug them and smother them with love and affection. It was heart-breaking to see her approach one little boy who ran off every time she tried to comfort him. When she returned to England, she took her own life, by jumping off a bridge.

Some of those I worked with suffered psychological damage from what they experienced. If you've walked into a children's ward in a hospice where kids are suffering a painful death from AIDS or you see that they have been violently abused, it affects you. I remember a kid called Robert; he lived in a Romanian orphanage and was permanently in a rage at everyone. The things that had been inflicted upon him are too unbearable to put to paper. When help had belatedly arrived and his abusers had been banished, but not punished, he continued to strike out. One of the nurses, Roma, had tried to get close to Robert but he had stabbed a pair of scissors into her hand for her efforts. Roma, in desperation, asked if I would try. She had watched me do a puppet show with a pair of socks while hiding under a kitchen hatch. My audience had been the children who were in the dining room while they were having lunch and Roma thought that Robert, though he hadn't smiled, had been captivated by my show.

I said that he was more likely thinking, how dumb I was in not knowing how to put on a pair of socks. However, she was persuasive and we both went to see Robert. Immediately the little boy of no more than five lashed out with such violence that Roma jumped back. He kept trying to kick my shins, I avoided his feet most of the time but I was determined not to restrain him. After a good ten minutes of lashing out at me, he started to get exhausted. I then knelt down to look him directly in the eyes, so that my height did not intimidate him, but also because his boot had occasionally had connected and my shins were hurting. It may not have been the best strategy: he did land a couple of good punches on me, but I was quick and mostly ducked and dived out of the way of his fists. That went on for some time and he knew how to hurt, as he had learnt from experts. I did not restrain him, but stayed firm, refusing to remove my gaze from his even as I ducked and weaved. Then he started to stagger his punches, holding back after each one to see if I would retain his gaze. His punches grew less frequent as he was tiring.

Then, after one real belter connected to my left eye, he fell towards me completely exhausted. It was as if I was holding myself. I had been that angry little boy who had believed he had been abandoned. But, I had never experienced anything on the scale that poor Robert had suffered. He cried as I lifted him up into Roma's arms. I left them together, but one brief moment doesn't heal that kind of anger; that only happens in the movies. Later that same day, he stabbed Roma in the face as she was feeding him; he just missed her eye with his dinner fork.

I survived roughly intact; I think because my gallows humour saved me, and I'm not a big thinker. I think that self-pity would be an indulgence, considering everything that I have seen others go through. For that attitude, I often seemed to others to be heartless and insensitive in not breaking down when confronted by the horrors of life. However, whatever defensive mechanism I had, it did not mean that I could not see the pain in others. There was couple I worked with over the years: José who co-drove with me on one relief trip to Bosnia and his girlfriend Maria. Together, during their time in Albania, they had discovered some children who had been chained to their beds and repeatedly abused; both José and Maria were naturally emotionally and mentally scarred by it. They continued to help such children, but when I met them afterwards I could see it slowly burning them out. José was not as strong as Maria was, and he would succumb to severe bouts of depression. Maria would nurse him through each episode, but it sapped her strength and dragged her down further each time.

One day I received a call from them to say that they had decided to give up the voluntary work and raise their own family. Maria was living at that time with her family in Paris and she decided to move to Madrid, settle down with José and start a family. She was happy, but José was going through another very severe bout of depression. I was invited to their wedding later that summer and I said that I could not commit immediately, as I would have to check Chelsea's fixtures. Both of them laughed at that, laughter came easily to Maria, but was rare in José. A month before the wedding, José took a fatal overdose.

I went to Madrid to help Maria with the funeral; that task was becoming all too familiar for me. She was crushed. They had been together nearly every day throughout the ten years I had known them, apart from my and José's trip to Bosnia. I went to Madrid as much to ensure she was able to pull through, as to help with the practicalities of the funeral. I then drove Maria and her belongings back to her family in Paris. I tried to raise her spirits and drove slowly, as I knew that she was not very close to her father and the last thing she needed to hear from him was that, 'She was better off without that loser', as he had often told Maria.

On the journey, we shared stories about José, but I feared that her thoughts were to join him. By the time I got back to England, after delivering her to her family, I had a message saying that Maria had followed José by slicing her wrists open in a hotel bath following an argument with her father. I have lost count of the numerous ways in which I have known people to take their own lives. That day, I called my Kids together and got very drunk, without telling them why. I don't understand why someone would take their own life: it's short and my view is that you must live it to the full. But just because you don't understand someone, doesn't mean you don't miss them when they're gone.

Rory's death was the result of his falling asleep at the wheel of his car and shooting off the road into a wall causing his vehicle's petrol tank to explode. There was nothing suspicious about his death; he was simply exhausted after travelling across South America and preparing for his last convoy. Fatigue and mental exhaustion have caused the deaths of more of my friends in the field of humanitarian work, than the gun and or the blade. Luisa's parents flew in for the funeral, but expressed their anger at Rory for continuing to press on with his project even though he had a young family. I understood their rage: now their daughter was alone to raise her two boys. I also understood Rory's reasons for completing his task, although I had tried to persuade him not to do this one last medical convoy. At least Luisa and I had lived the adventure with Rory, but I was sorry for his two children who lost him before they had really experienced the humanity, the strength and the integrity of the man who was their father.

After the funeral I learnt from Luisa, that she was determined to complete Rory's final project, as Doug had confirmed that he would honour his promise to fund it. The problem was that she had two kids to look after alone and she couldn't drive a truck; in fact she didn't have a driving license. After we had a heated debate following the funeral, she reluctantly agreed that I would complete my friend's last wish. The problem was then, for the first time ever, I had a feeling that I might fail.

A former Vice President of the United States, Dan Quayle, supposedly once said on a visit to Latin America that "I wished I had studied Latin harder at school." My knowledge of the region was just a little beyond his. I had to make myself familiar with Rory's logistics plan which was in three stages; one truck would go from Rio to the Paraguayan border and then the medical aid was to be transferred to the Paraguayan truck and delivered to a hospital in the capital, Asunción. After that, the second truck, which contained medical equipment, would leave Santiago, the Chilean capital, for Lima, the Peruvian capital. A few days later, the third and final truck carrying similar medical equipment, but for a maternity unit, would go from the Argentinean capital, Buenos Aires, to Montevideo, capital of Uruguay, and from there to collect specialist medicines to go back to Argentina. I had under a month to complete the preparations, before the first truck was on its way.

From the beginning things started to go wrong. Having checked out both of Rory's co-drivers through my police contacts, I discovered that both had had previous convictions for drugs. Nothing serious, but enough to get us detained at borders if checked and, more importantly, to risk the cargo being confiscated. The aid would be of increased interest as cases of needles and drugs would have a high value on the black market. At the last moment, my contact in Brazil informed me that Rory had had a back-up driver. He was the son of a leading politician and it turned out that he had a bodyguard to accompany him who could also help with the driving. I had them both checked out and, apart from the bodyguard once being charged with bank robbery, both were clean. Also, I was running out of time and finding it impossible to complete all three legs, but now they could cover the Rio to Paraguay trip on their own and I could cover the other two stages.

I went out to South America to do a quick reconnaissance trip of the routes, the support that Rory had put in place and the two drivers. I met the politician's son who seemed genuine and enthusiastic, but at that young age, naive. However, his bodyguard had been in some tight situations and he had a relaxed confidence that compensated for his charge's inexperience. I talked them through the journey, which was in my view the safest and most direct of all of them. I laid down my standard rules; travel only during the day when it was safest; do not shout about any aspect of the journey and, in particular, the details about the cargo they were carrying. There were to be no internet communications that would indicate where they were or where they were going; social networks in particular were to be avoided.

After my reconnaissance trip, I returned to England and had further background checks done on the two Brazilian drivers. Nothing came up, so I finally accepted their offer to cover the Brazilian leg of the journey. That was the first of many mistakes that I would make over the next few weeks.

On my last night before I took the morning flight to Santiago to begin the first stage of Rory's relief effort, I went to see Luisa and the children. She was a different woman; she was excited that Rory's last quest was to be fulfilled, her energy and strength had returned. Her eyes were bright and the darkness under them had gone, her dark hair had the shine of a Rolls Royce in a Belgravia car showroom and she wore a thin white cotton dress that seemed to shimmer almost over the quickened beat of her heart. Her mourning was at an end; it was not that she had forgotten Rory but she had taken on his dream. Over recent weeks she had thrown herself into the task of organizing all the documents required for the trip. I had taken on Rory's challenge, but it was Luisa who put the trucks on the road to complete her husband's final mission.

Her two boys had their mother's colouring; tanned oval fresh faces under jet black mops of hair. When you added Rory's grin and his piercing blue eyes to the mix, you had two heart-breakers for the women of the future to contend with. Luisa and I had a drink on the veranda of her house while the kids ran around playing, but occasionally they would stop and look over when they heard their father's name mentioned. Eduardo, the eldest boy, took the hand of his little brother, Raul, and they came over to sit on the steps of the veranda and listen to their mother and the pale Englishman tell incredible stories about their brave father. The mention of his name had an instant effect, as the smiles would appear and then disappear and a look of loss would take their place. I noticed many men that night, they would walk by looking at the beautiful woman on the veranda, but Rory would be a hard man to replace.

"He was a great man, a good man and I know he was never happier than when he was with you and the boys. He really was going to give up all his humanitarian convoys after this one you know; you and the kids were his life," I said. The three looked at each other and this seemed to confirm all that they had been wanting to hear.

"He admired you, you know?" Luisa said, changing the subject.

"He was cleverer than I thought." I replied.

Luisa laughed and countered, "Yes, he also warned me that on the surface you treat everything as a joke, but you care deeply."

"Yes, it's true, deep down I have a passion for my football team. Which reminds me; I brought with me two Chelsea football shirts for the boys."

"Yes, Rory also said you were impossible and that you never let anyone get close to you. He also said your team was how you say in English … shite?" That was hilarious as on the last word she mimicked her late husband's Scottish accent.

"Nonsense, the face of a beautiful woman with a liberal attitude to sex was a good way to scale my defensive wall." I said. "As for my team, I think he must have been referring to the contents of the Leeds United trophy cabinet."

I was determined that we would talk of Rory through the night with an air of humour. Rory would not have wanted me to leave Luisa and the boys in tears. He would have wanted them to be smiling and laughing at stories celebrating his life and, above all, his love for his family.

It was getting dark so I took my leave as I had to set off early. I kissed Luisa on the cheek and I promised her that I would get every single item of aid that Rory had secured through to its destination. I shook Rory's sons by the hand, in a big, manly, no nonsense way. As I went to turn away, I failed in my attempt not to leave Luisa in tears; she had grabbed her boys, which set them crying too. I walked away, contemplating that I had failed in my task to leave them smiling and wondered if I would also fail in the task ahead. I was in new terrain, working with people I didn't know, my preparation was not as thorough as it had been on previous journeys and, due to so little time remaining there was no contingency for any mistakes. Even with all of these doubts, I still had little idea of how bad things were going to be.

I covered Chile and Peru easily enough. The people I met in South America were fantastic and the main tension appeared to be the opinions held by the people of each country on the neighbouring country's football team. On the first trip, the Chileans were very European and that was reflected in their manner and style, and it could also be seen in their architecture. They were wealthier than their neighbours; the difference was highlighted by the Peruvians who viewed the Chileans, as one Peruvian nurse told me in her best English, as "up themselves", but it was all banter and she and her staff were delighted to receive the medicines and medical equipment.

The only problem for me was a lack of sleep, caused by having to complete the journey with no co-driver. After day three, it certainly came home to me that I was not a young man anymore: I ached all over and was desperate for sleep. On the journey, I knew that all was not going well with the second relief truck covering the Brazil-Paraguay leg. I hadn't received any responses to the frequent calls I made from roadside telephone boxes, whenever I could find one that worked. But each time I returned to my truck I reassured myself that if there was a serious problem, then they would have contacted me. Again I was wrong, but there was little I could have done from where I was, on the other side of the continent.

In Lima, I finally found out that all the cargo of the second truck was back in Brazil and that the truck was badly damaged. The politician's son was at his father's house under sedation, while his bodyguard was in hospital. My advice had been ignored. I should have guessed that the driver being a politician's son, his dad would have seen an opportunity for self-promotion. The truck had set off in a blaze of publicity, which delayed its start and then they broke my second instruction as the truck finally set off as darkness fell. The truck had then been involved in an attempted hijack as two trucks rammed it; they were trying to secure the cargo of legal drugs, which had been documented in a press release as being 'worth millions if it got into the wrong hands.'

The only thing that saved them and the cargo was that, after being forced off the road, the bodyguard was driving and when he was forced to stop, he stood his ground in a shootout. In the gunfight he took a bullet to his groin and lost half his penis and one testicle, but he continued to protect the cargo and the politician's son until the police arrived.

I caught the next available flight to Rio and went to visit the bodyguard in hospital, but he seemed happy enough as obviously guns were more his passion than women. Even when the attending priest said "All was God's will', he just smiled. I then went to the politician's house, but on his father's instructions the son refused to see me. His father ran out and threw a punch at me; I ducked and he disappeared into a ditch. 'Fuck him,' I thought, he wasn't getting my vote if that was his idea of a public engagement. In the meantime, I enlisted the help of some local mechanics in Rio to repair the damaged truck, which involved the big task of replacing its cracked chassis.

While the mechanics repaired the second truck in Rio, I caught a flight to Argentina and set off on the second aid trip from Buenos Aires. I still had not caught up on my sleep. As I travelled through Argentina on my way to Uruguay, it was easy to see that it was no longer the wealthiest nation in Latin America, while its neighbouring rival Brazil was prospering. Of course there was the usual football rivalry. This time it was not just about who had the superior national football team between the two nations, as South American football teams were frequent lifters of the Jules Rimet Trophy and therefore competed to be best in the entire world. I completely understood this tension and in fact I regretted that there was no such rivalry between England, Scotland, Wales and Ireland, as we were all crap.

The second stage of the journey again went without a hitch. An added bonus was that I had one day to replenish my energies in Montevideo. I made camp in the main market, sat on a stool by one of its huge meat-laden barbecue grills and ate the finest steaks I have ever eaten; I had three of them plus a small light salad.

I did not return to Buenos Aires, but got a flight back to Rio, now that the truck was repaired. We reloaded its cargo and though it was three days behind schedule, I was ready to go the next morning. It turned out that the politician had paid the mechanics extra to work around the clock and repair the truck, but he could not reconcile this with his rage at what had happened. I liked the contradictions between head and heart in the Latin American culture; Argentina in particular is one of the few countries where I've felt at home.

I made it to the Paraguayan border without any further problems, but there was a last minute change of plan. The medical supplies I had onboard had been due to be collected by a hospital vehicle, but because of the delay, the driver had been ordered to return with his vehicle back to Asunción. I now had to drive directly into Paraguay, which was a bit of problem, as I had no paperwork to enter the country.

I decided to call the Brazilian politician once more and ask for his help, knowing that even though he blamed me, he also blamed himself for what had happened to his son and for nearly losing all the aid. I was right in my assessment and, though he called me every name under the sun, he pulled all the strings he could and I got some papers and the identification documents to match them. I got into Paraguay by a rather unorthodox route, but everything was finally delivered.

The truck was emptied with the help of the hospital staff and those children from the hospital who could walk and carry the supplies. The supplies were obviously very much needed and the truck was unloaded in half the time it took to load it. I briefly said my goodbyes, as I had to get out quickly. All I had to do was get through the border, but I thought that that should not be a major problem; even if I was detained at the border I no longer had any cargo to lose. Also, though my exit papers were slightly suspicious, I had no goods to warrant any interest in keeping me for long. Once again I was wrong, but at least I was consistent.

At the Paraguayan border I was taken into a cell for questioning. It was one of those rare occasions when being English and therefore ignorant of any foreign language proved a perverse benefit. The lead interrogator was screaming and shouting at me; his face was bright scarlet like an Arsenal fan at the end of a typical season. For what seemed like hours he screamed and ranted in my face, while I just looked at him in an idiotic way with an innocent smile on face. I could do nothing but wait it out. In such situations a thousand thoughts go through your mind; I could be locked up with no one knowing where I was, I knew that I was without official documents and that what little money I had could disappear.

As more guards came in, I knew it would turn violent. I recalled the scene from David Lean's *Lawrence of Arabia* where the hero of the title falls into a similar situation and is then buggered by his jailer. All the guards were wearing wedding rings; I prayed that they were happily married and would hopefully settle for beating the crap out of me instead. For the first time in weeks I was right, and I was relieved that they settled for giving me a good old-fashioned kicking.

I was smashed against the cell wall with such fury that I felt my arm dislocate from my right shoulder socket. I was then held and punches rained down on my lower back. I have been roughed up a couple of times in some dangerous places, from Bosnia, to Romania, to Cricklewood but these were experts and I knew that the effects of the punishment were going to stay with me for some time. I was lifted back up on to my feet and then repeatedly punched in the kidneys, and then I was thrown down on to the floor of my cell. I knew I had to stay conscious or I would be finished; I needed to talk my way out of there. At one point, I was trying to get back on my feet and I couldn't recognise my features in the pool of my own blood on the floor. I remember thinking that my reflection was distorted. It wasn't – that was how I looked.

As always my only weapon was my wits. I kept asking for the British Consul, to no avail and then I changed tack and made repeated requests for a Catholic priest. I started reciting the Lord's Prayer and alternating it with a Hail Mary and then I sprinkled in a few shouts of John Paul II and held my hands in the air. You're probably thinking that finally, I had found God, you're wrong. I realised that I was in a Catholic country and I had more chance of getting out by calling on my 'very tenuous' Catholic rather than British credentials. Apart from wedding rings, I observed that three of the five guards wore crucifixes, two on chains and one as a tattoo on his blood-covered fist. I tried to forge a bond that any prisoner tries to make with his captors; mine was based on our mutual faith. I felt it that it wasn't the right time to mention the lapsed part.

The other two border guards were more in tune with my actual religious beliefs as I think they just got tired of my prayers. It was those two men who finally picked me up and threw me on to the grass out outside, as if they were emptying a slop bucket, which was ironic considering it was the same area they did this. Once they had walked back into the main building, I rolled on my back and looked up at the rising sun, as dawn was breaking. For those who have not had the experience, to find yourself on a shit-covered, scorched grass bank, with a mouth filled with your own blood, your vehicle nearby but stripped of all of your possessions, but with ones' arse intact it was a state of utter bliss considering what could have happened.

I couldn't move as my shoulder was killing me, and knowing that I was no longer of any interest to the border guards, I just lay still trying to conserve what energy I had. As I lay there I thought of my situation: I was again in the shit, in more ways than one, there was no knowing how I was going to get back and I was wondering if anyone would care. I treated myself to a little self-indulgence and I surveyed my life. Though Biddy loved me, she had the family and indeed her church if I didn't return. My Auntie Julia was now well into her eighties;

with her Alzheimer's she no longer remembered Little Johnny. I had no wife, no girlfriend, no children and I thought that even mad bastards like Rory and Loftus had made some commitment in their short but eventful lives. As I lay battered on that Paraguayan make-shift piss-stop, I thought perhaps it was time for me to try to settle into polite society. Then reason returned and I remembered that I had no social skills. That brought to an end to my bout of self-pity for another few decades.

I laughed at the absurdity of situation and proceeded to crawl up the bank, throwing up blood and vomit along the way until I finally got back on to the main road beside my truck and lifted myself into the cabin. Thankfully, in their haste to drag me into the cell, they had left the keys in the ignition and having taken one passport, my camera, the film containing my photos of the last few weeks and a small stash of money in the glove compartment, they didn't bother to look for my real passport and the majority of my money wrapped in an airtight bag secured by a string in the fuel tank. My watch and the US half-dollar that Kerry had given me for good luck many years before had been taken off me in the cell.

I returned the truck to Rio, but then headed to Buenos Aires; I really loved the city and I knew that it was a place where I could watch the world go by, while drinking numerous coffees and while my body repaired itself. Sitting in a square one lunchtime watching a dancers' tango for the tourists, I tried to work out why the border guards had bothered to give me a work-over. All I could think was that the border guards were enraged that they had missed the opportunity to seize my cargo to sell on the black market, rather than irritated at my lack of 'official' papers. Also, the fact that none of them could speak English and therefore they could not extract from me how I had got into the country must have added to their frustration, hence the violence.

I said goodbye to all those who had helped me in Argentina and I decided that I would definitely return, especially as I then heard that my adopted 'dad', Alex, was soon to settle there.

After my Paraguayan 'workout' a pirate DVD which is probably available on any stall only five minutes from the border on the Paraguayan side, the spasms in my back that used to frequently paralyze me were getting more frequent. My shoulder wasn't dislocated, but the tears to my shoulder muscles were, according to my doctor, irreparable damage and pain shoots through my right shoulder even today. When people see me flinch and ask why, I just say it is because I've overheard someone mention Arsenal or George Osborne. But Rory's legacy was achieved and the three cargos of aid ended up where they were supposed to. From the relief on the recipients' faces, I could see why Rory had risked everything to achieve it. On the plane, as was the ritual for any Englishman heading home and, though my kidneys continued to ache, I needed a few stiff drinks to relax. It was the end of a chapter; I knew that my relief work really was finally at an end.

When I got back to England, nothing had changed and all was as though I had not been away. After recent events that was the best therapy possible. So I went to have a drink in The Bobbie and I saw a mate of mine, Jockey, at the bar and I walked over to him.

"How's the love life?" I asked.

"Shite, I asked a girl last night in the pub if I could get into her pants."

"Yes, always best to adopt a slow build-up," I added.

"She said no, as she already had one arsehole in them," then added Jockey, "Bastards!", as he dejectedly returned to his pint of lager, complaining that I hadn't bought him one in weeks.

Jockey and I then watched England play Paraguay in a 2006 World Cup qualifier on the pub's television and no one celebrated more than I when we won. Jockey couldn't understand why I was so animated at the England victory, as I had never been a big fan of the national side. That night it was packed in the bar, but no one asked about my trip and I had no interest in relating my tale. I did know that it was good to be back with rogues like Jockey and that it was good to have tears of laughter roll down my cheeks again.

Perhaps the existence of the two worlds, the 'normal' one and the 'humanitarian' one, in their own way, kept me sane. No, seriously I am sane, though Stephanie or one of his psychiatric colleagues would say that proclaiming that I am sane is a sign that I'm not. Jockey had no interest in where I had been, unless women had been involved; it was a relief as I didn't want to talk about my little trips abroad; all I wanted was to forget about them for a while and relax with a beer and good laugh. In a way, I now understood my Uncle Paddy's reluctance to talk about the war. I had no need for meditation tapes or balancing polished stones on my head whilst listening to CDs of dolphins singing. Those remedies work for some, but not for me; after the trials and tribulations of South America, meeting Jockey 'earthed me' back to my normal world. In turn, my humanitarian work gave me a chance to do something, to try at least to leave the world a little better than it had been when I'd arrived.

Edmund Burke famously said, "Evil prevails when good men do not act." I am not what you would call a good man; I'm more of an awkward one who likes a challenge. Those who deserve to be called 'good men', and 'good women', are those who sacrifice so much for what they believe is right. Rory sacrificed his career, time with his new family and then finally his life. I could have written two books; one a serious highbrow novel, (well in theory I could), the other would be a series of humorous anecdotes of anarchic characters. My life is not one or the other, what I have learnt in one world has helped me to get through in the other. Without the skills I learnt in St Vesuvius, Warshalton and Trumpton I doubt I could have coped with Romania, Bosnia and South America. Perhaps it's the other way around.

Chapter 29 – *The Demise of Rogues*

Fifteen years after the war in the Balkans ended, I once again travelled to the region on my way to see Sam and her beautiful daughter, who is also my goddaughter, in Turkey. I travelled by train and decided to take a detour and visit Belgrade and see some old friends. That night in Belgrade I got through the border with no problems, so it appeared that I was no longer on any black list. Later, after seeing my old Serbian and Bosnian friends, I returned to the Serjia Hotel. In my room, I received a call from a very terrified receptionist.

"There are some... police downstairs wanting to speak to you ... please."

"Ok, where?" I asked.

"In the ... lobby," she said, barely getting the words out as her voice was trembling so much.

I went down to the lobby and there were two men seated on two separate sofas. They wore the standard uniform of intimidation in the Balkans; matching black leather jackets and shirts. They could have been police, secret police or local mafia, but with their relaxed demeanor I knew that they were plainclothes policemen; they were happy to be sitting in a public area that was open and exposed and they didn't have their chairs strategically placed with the backs against a wall. They sat stretched out like bored emperors sitting on their thrones. They sent out a clear message that this was their city, they were in charge and they made the law. Whatever they wanted, I was screwed and I'd better accept it. It was time to adjust the odds.

"A beer please," I asked the terrified waitress who reminded me of Bambi, though with fewer spots. She was trying to steady herself against the reception desk. She was too scared to move, until the largest, oldest and baldest of the two officers nodded to her that he approved of my order.

"Make it three Shiljivovitca's and three beers, Halva," I followed.

This would be interesting, I thought, if I knew one thing it was that the police, given the look of these two bloated officers, could always be sidetracked by a few beers.

"Why are you here?" asked the smaller of the two policemen.

"The receptionist called me," I said, I was determined to not make it easy for them.

"In Belgrade?" he demanded.

"Tourist," I smiled.

"There is nothing here to see, your governments blew shit out of my country."

"Not according to the brochure," I said. "I will contact trading standards when I get back home."

"You think the loss of life here is a joke?" he asked.

"No, but civilians suffer the world over for the abuses of their government." I replied.

"We know you."

"Ah, that's nice, but you now have me at a disadvantage, who are you?" I asked.

"We police."

"Wow, why didn't you say?" as I turned to the reception area and called out, "Waitress please take their drinks away, these men are on duty." But no one dared to come back to remove the drinks that had very quickly appeared in front of us.

The senior officer leant forward and spoke for the first time, "You brought illegal contraband into my country during the war."

"No, not your country, I brought medicine to Bosnian hospitals to treat children, from Serbia and Croatia, as well as Bosnia."

"You come back to gloat or maybe you spy on us again?" said the bigger man as he smashed his glass on to the glass table. This was a standard ploy to unsettle your subject; mind you, I noticed that he made sure he'd emptied his glass tumbler first. I replied "I am not a spy, no secret service would have me; I have too many vices and I'm therefore too easy to blackmail. You don't seem so fussy, do you have any vacancies?"

At that point I leant forward, an approach I learnt working at the card table and in the defendant's box. This was a device to take back control of the situation, by giving a signal that I was not defensive, that I was confident and unafraid. "Look I have no issue with the Serbian people, only with the genocide carried out by your leaders at the time. Now we have all moved on I hope, as this is a new century."

"You helped our enemies."

"No, I helped in a very small way the children of all three nations. I had nothing to do with the armies of any side, you and your armies can all go on an island and carry on beating the Balkan shit out of each other for all eternity for all I care."

"You spy!" the bigger officer screamed once more.

I remained calm, I had no choice, but it was a struggle to smother the fear in my voice as I spoke, "If a good-looking woman is stripping off opposite my hotel room tonight, I will have a quick look but that is as far as my spying goes."

The smaller police officer suddenly laughed aloud. Maybe the larger thug would have preferred a young man opposite his window. The laugh broke the mood, thank God for humour and all I learnt from rogues over the years. I needed to get them drunk. "*dopasti se*" I called over to the lesser spotted Bambi peeking out from behind the hotel column, ordering another round of drinks.

The drink continued to flow, and though I could hold my own after all the training I'd had over the years in the company of rogues, it took every ounce of concentration to keep my wits together beside those two seasoned drinkers. Six hours later they left. Bambi I think had leapt over a hedge hours ago, in fact the two policemen and I had been taking turns in helping ourselves to drink from behind the bar. I passed the still terrified receptionist and uttered a barely coherent, "Goodnight," as I staggered up the stairs and fell into bed.

My last thoughts on the evening were about the words of the larger policeman, "You come here to gloat." They had been on the losing side. In fact I came to the conclusion that, like me with the Iraq War, they did not support the conflict either–   certainly the smaller policeman expressed later that he was glad to raise his little girl in a time of peace. It did not matter, for whether you believe a war is just or not, it's human nature that you don't want to be on the losing side. We are lucky in England in that we cannot remember losing a war; but maybe that is why we get involved in so many of them.

I mentioned a few rules that I stuck to during those days delivering humanitarian aid. I stayed away from politicians and journalists, not because they were bad people, though most of the ones I met were, but because they had a separate agenda. Politicians would look for the publicity angle, which might put them in a favourable light. That was dependent on providing information to the media, thus potentially exposing the details of where we were going and the contents of our cargo. Journalists, for the same reason, though they were not out to publicise themselves, would release information and jeopardise our relief missions. Also, films or articles naming any local people who had helped our relief efforts might leave them vulnerable to reprisals later. For that reason, I will never release the real names of any person who assisted me. My last rule (for the same reason) was never to take photographs of where I was, who was with me or of those whom we helped. Some photos do exist, but they are mainly from Romania and 'official aid convoys' to Bosnia.

Frank Sinatra sang the immortal line in *My Way*, "Regrets I have had a few, but then again too few to mention." Well good luck to Frank; it was different in my case, as I screwed up so many times in relationships by saying the wrong things or, in my youth, by punching someone I shouldn't. The success of South America was due more to luck than anything else; it confirmed to me that I would be a liability to any future aid convoys. To those I know, who are still out there delivering aid to those who need it, I am sorry if you feel I betrayed you, but I don't regret it.

As for my benevolent rogues, not one truckload of the aid I have mentioned would have reached its destination without them. If playing by the rules was required, then others would have done the job earlier. It was when the rules barred the aid from reaching those who needed it, that I enlisted the help of characters who had a talent for 'calligraphy' (as one forger described his work) or had the know how to access fuel on a military airbase.

This book is meant to make you smile, in between bouts of revulsion. If there is a message amid my ramblings, then it is to give people a chance. There are no profound lessons to be taken from my scribbling, only that those who came from often dysfunctional backgrounds can beat the odds, despite being dealt a poor hand. If you have a position of power and would shut the door on someone because they were from a broken home; were raised on a notorious housing estate; were a pupil from a failing school; had a criminal record but served their time or because they would say toilet instead of lavatory, don't. If you have not met someone before, or encountered someone whose reputation precedes them, give them the benefit of doubt and leave judgment to those in white wigs, with suspender belts hidden beneath their robes. Granted it's not easy; just look at how quickly I judge the judges.

The beating I took in Paraguay took its toll on me, so I decided to take a real holiday for the first time in ten years to convalesce. I set off for the Greek islands hoping to relax for a week, but the pain in my back was getting worse and the spasms were so bad that when I had an attack I could barely move. It was like having two skewers plunged into your back three or four times a day, but not knowing when it was going to happen. Eventually, I decided to pay a visit to the nurse on board one of the biggest ships in the harbour. My nurse's name was Lorraine.

"Well, gorgeous, what's the verdict?" I asked her after my examination.

"You must have got a good going over recently, your kidneys look like they have been tenderized by an elephant and they're bleeding into you, Johnny," I looked up at her; no one had called this battered object simply Johnny for a long time.

"No, please don't hold back, I can take it."

She smiled but just shook her head in despair, as I slowly lifted myself off the bench in the surgery.

I added, "You don't know the half of it, our Chelsea manager Mourinho parted from my beloved Chelsea this morning. I may be fucked, but my team is even more so."

Lorraine organized my passage on her ship and she nursed me over those next few days, but little did we know that she was sicker than I was.

UNITED ROGUES

"I don't know what effect these men have on the enemy, but, God, they terrify me."

Arthur Wellesley, Duke of Wellington

Chapter 30 – *Trumpton-The Lunatics Have Taken Over the Asylum*

When I first began my relief work, I encountered another very unusual collection of rogues. These characters were from a working class estate, only a few miles away from mine, in Somers Town. If Warshalton provided a taste of London life south of the Thames, then Trumpton, as I call it, was its equivalent north of the river. This is the final piece of my story about benevolent rogues and I complete my tale by returning to the idiocy, the eccentricity and the humour that made all the humanitarian convoys possible.

I was introduced to Trumpton by Chelsea Mark, who had a job in an immigration centre in Manor House in North London. He introduced me to a porter who worked there. Numbnuts and we did not click at first: he was a sarcastic, cynical, cocky bastard and therefore, it goes without saying, he was an Arsenal fan. Yet, I grew to like this angry individual; it was through him that I met the characters who lived in Highbury Quadrant, in an area that I call Trumpton, where everyone grew up together and knew each other. If there was ever a sense of community in Somers Town, it has long since gone. Those people I did know have since moved away to try to secure a better quality of life, others are in jail or can only be contacted by employing the services of a medium.

I still live in Camden and I bump into former school friends occasionally, but usually the men are married and are embarrassed to talk about their childhood antics in front of their wives. The women I knew are pushing multiple prams around with chariot skills that *Ben Hur* could only have dreamt of. I bumped into an old girlfriend outside one of the most notorious drug dealing pubs in Camden a few years back.

"You're the first woman from St Vesuvius I have met that is not pushing a pram," I said.

"I lost custody of my three kids to their father, due to my drug and alcohol addiction," she replied and she then proceeded to burst into tears. She turned and entered the pub, which was doing a two-for-one promotion on spirits as a pre-lunchtime special.

The rogues of Trumpton in the north were like those of Warshalton in the south and I felt I had again stumbled across some undiscovered tribe which had not adopted the so-called civilized standards of polite society. Through Numbnuts I got to know the rest of his family, the Connellys. I got to know his brothers and sisters and, in particular, Surf, who became one of my best friends and is, like Numbnuts, a brother to me. Whereas Surf was a brother in the proper family sense of the word, Numbnuts was more the kind of brother that you would pay a monthly bribe to, to keep him out of the limelight should you ever become famous.

Surf and Numbnuts are unique amongst the rare breed of benevolent rogues, in that they are related. If you're familiar with that other rare group of individuals, Nobel prize winners, only once has the award gone to brothers, Jan and Nikolaas Tinbergen. One of the numerous differences with securing the accolade of 'benevolent rogue', and the only one that would matter to the Connelly brothers is that there is no cash prize. It's also an award that no one in their right mind would want; indeed if any benevolent rogue was found to be in their right mind, they would be stripped of the award. The Connelly brothers, though different in so many ways, share the core necessary traits of mischief and challenging everything. When I first met them, the identifiable signs were present, the electricity was there and that tension that immediately put you on your toes. Both, I found, had moral courage although Numbnuts would angrily deny that he had any saving graces whatsoever.

Numbnuts was once was discovered arguing with a homeless man outside a café and it was a commonly held belief that he had been having a go at the poor man. It turned out that Numbnuts had planned to treat himself to a big fried breakfast, but when he saw the man, he decided to forgo a proper breakfast and get a bacon roll for himself and the man. When the recipient of his generous offer said, "What, no sauce?" the conversation went downhill from there. Surf is a far more amiable character; unlike his brother he has social skills, but he too deserves the label 'benevolent' as he has helped me with my relief work over the years.

The Connellys accepted me as a friend; I got to know all the family very well over the years. I had heard of sibling rivalry but, as an only child, I was intrigued by it. There was the bond between them that you would expect – no outsider could say a word against any member of the family, but that did not stop wars breaking out between the siblings every now and again. The principal instigator of trouble was Numbnuts. If you had put him amongst a group of monks who had taken a lifetime vow of silence, within ten minutes they would break it to tell him to go and fuck himself.

MacConnelly, a pseudonym adopted because he had once been married to a Scottish woman, was the eldest and he kept his family and his professional life separate from his siblings. Surf was in the middle, but he adopted the mantle of patriarch of the family. He was also the entrepreneur of the family. There were three sisters, Ann, Stacy and the youngest, Colleen. There were no weaklings in the family and the women could knock the boys down if they got too far out of their prams. Theirs was an alien world to me, but a great world and I am glad that they let me become a part of it.

Numbnuts was the youngest of the brothers and unlike his two elder brothers, he was not successful in business; in fact, he had no ambitions in that department at all. His stated ambitions were to get drunk, meet women and have fun. He had a sense of humour that I understand, but it's almost as though he is an artist born in an age that does not appreciate his genius. When Numbnuts makes what I recognise as a humorous remark, most of the Trumptonites just look at him blankly. At a family funeral, he was in the lead car with his brothers and sisters. Colleen was crying and saying that most of her aunties and uncles had died and there was only one auntie in Ireland left. Unfortunately, in her grief she couldn't remember her name.

"Do you remember her name?" she asked Numbnuts.

"Next," replied Numbnuts and he kept a completely deadpan look on his face.

Surf, who does instantly get his brother's humour, told me later that he had had to keep his face pressed against the window of the hearse for fear that the rest of the occupants, who hadn't appreciated that remark, would see the tears of laughter streaming down his face. I too experienced this pained reaction to Numbnuts' acerbic humour. Numbnuts and I were walking down the Holloway Road one Saturday trying to negotiate the onslaught of various insurance salesmen outside the numerous supermarkets. As we passed by one supermarket, one such salesman jumped out in front of Numbnuts and pointed directly at him.

"Excuse me sir, have you had an accident in the last week?" he asked.

Without breaking his stride or into a smile, Numbnuts replied, "Yes, I shat myself last Thursday," and continued on his way, leaving the salesman with his mouth open.

We continued walking along 'the murder mile' as it was called by the local press (Holloway was a violent area then), looking for a pub in which to have a drink and watch the football. One pub would frequently throw Numbnuts out, as the barmaid hated the sight of him; we went in there. Numbnuts and the barmaid rekindled their instant mutual dislike. As I walked in, I could see the two eyeing up each other like neighbouring cats marking out their territories. The big hefty barmaid gave him her best scolding stare.

Numbnuts, without turning to me, said, "That reminds me, I must get a handle for my bucket."

I knew that she had heard him implying that she was a woman of somewhat loose virtue and as a result had a cavernous vagina. She stormed over to wipe our table and she swept all the spilt lager from previous customers on to him.

He then, in a very polite manner, enquired, "What time do you close?"

"Eleven o'clock," she replied curtly.

"And the pub?" replied Numbnuts.

If Numbnuts has only one saving grace, it is that he is consistent; he was barred before the players had come out to warm up for the lunchtime kick-off.

Numbnuts and I were invited by Surf and his wife Ann to join them for Christmas dinner, so I offered to give Numbnuts a lift on the back of my motorbike. I prepared for our journey and found my passenger the smallest motorbike helmet possible, just to watch him squeeze his disproportionately large head into it. I loaded up my panniers and a backpack, and gave Numbnuts the two extra bags of presents that I had bought for Surf and Ann's children. Numbnuts' contribution was a rumbling belly. My motorbike, even though I still had my Zephyr, could barely move under the weight of our luggage, but more so under the weight of Numbnuts. On the journey, all I needed to do was buy flowers for Ann, and though it was Christmas Day, I knew that there would be a flower seller outside Finchley Cemetery. As I returned to my bike carrying something the size of a Rhododendron bush, Numbnuts lifted his visor.

"Could we look any more bent?" he uttered.

I was still laughing at Numbnuts' comment as we drove along the streets; we were virtually riding on the back wheel due to my passenger's fat arse. My only fear was that children would run out and pile furniture on us when we stopped at traffic lights, perhaps thinking that we were a life-size game of Buckaroo.

My only regret in the time I have known the Connellys is that their mother died just as I got to know them and so I never had a chance to meet her. She, like all Irish mothers, was the head of the family. Even after her death, nearly twenty years ago, their mother has left a positive impact on all of them; despite their feuds, they will always support each other. Amongst the three brothers, opportunities to take advantage of each of the other two are never lost. This is usually at Numbnuts' expense as he often ends up doing the donkeywork for one of Surf's moneymaking ventures.

One day Surf had the idea to demolish the chimney of his house and send Numbnuts up in the bucket of a mobile crane that he had hired. Surf sat at the controls. Numbnuts was lifted sixty feet into the air with his nose throbbing red, which was always an indicator that he had had more than a sweet sherry on the night before. Numbnuts was knocking down the chimney with something smaller than a child's hammer. I doubted his hammer would have knocked in a drawing pin. I concluded that Surf wanted the chimney to remain and that his real motive was to leave his brother on the roof indefinitely using his glowing nose as a landing light to ensure his house would be seen by low flying aircraft after dark.

Chelsea Mark being a Chelsea fan and Numbnuts being an Arsenal fan, meant that they were constantly trying to get one over on each other. Unfortunately, my fellow Chelsea fan would often lose out in his battle of wits with Numbnuts. The reason for this was that Numbnuts was a complete cynic who trusted no one, whereas Chelsea Mark was completely gullible. I remember lifting my face to the smoke-stained ceiling of The Arch, when Chelsea Mark told me that he was going to do an insurance job on his car and that Numbnuts had offered to help. Numbnuts had advised Chelsea Mark, who was going on holiday, to leave his car on the estate opposite The Bobbie, his local pub. Numbnuts advised Chelsea Mark to leave the doors unlocked; promising that the car would be gone by the time Chelsea Mark returned from holiday. Chelsea Mark did as he was told only to find on his return that his car was exactly where he had left it, but it had been used by the customers of The Bobbie as a urinal for the whole week.

It was a great source of discomfort for H that Chelsea Mark was often held up to ridicule by the Arsenal-supporting residents of Trumpton, even though H did not have much time for Chelsea Mark.

"It's the principle of it!" declared H.

The people of Trumpton were not always to have it their own way. In the nineties, Arsenal relocated to a new stadium and Surf and the Arsenal-supporting Trumptonites, were invited to a farewell party at their Highbury ground. No doubt thinking that we would be a good mockery prop, they also invited Tommy, a Spurs fan, and me to join them. Being the only non-Arsenal invitees, we were the subject of a fair amount of abuse over the evening but we took it in good part, apart from Tommy referring to each Arsenal fan in turn as an "inbred banjo player". Later at night, the Arsenal contingent asked me to take a group photo of them. I duly obliged but I noticed that a rather drunk friend of theirs, Dale, was not part of the group. I dragged him over and positioned him resting on one knee to help him support himself at the front of the group. A few days Surf stormed up to me, holding the recently developed photographs of the evening in his hands. He seemed quite angry about something.

"You fucking knew that Dale had pissed down his leg and that is why you had him kneeling with the pissed leg of his trousers in full view, right in the centre of the picture!" Not for the first time, I thought to myself, 'Was ever a man more misunderstood?'

The Arsenal fans' devotion was as fervent as that which H, Wurzel, Amelia and I gave to Chelsea. At Chelsea, one would occasionally hear a voice over the tannoy congratulating a member of the crowd on having a baby. In turn, I knew of one of the Trumptonites who had left his girlfriend in childbirth in Scotland to get to a match down in London, in the belief that she would not notice he was missing. Arsenal fans are known for not making much noise, hence the 'Library' tag for their previous stadium, so he probably got away with it.

Numbnuts and Surf would tell me tales of growing up in Trumpton and stories about the people they grew up with, in particular Brains, Tommy, Snotbox and Jockey. Trumpton was a mirror image of the rough and ready working class environment that I grew up in. It was multicultural and all the better for it. St Vesuvius had had a large number of black pupils and so did their school, St Judas in Trumpton. Violence would break out but not in the way that Enoch Powell or the NF would have you believe; it was usually rebellion against authority and had nothing to do with colour. Ones' colour, race or religion was often used as source of humour rather than for malicious abuse and, even now, Numbnuts, Surf and I are often welcomed as "Fenian bastards," by Jockey, a Scotsman, because we are of Irish descent.

Trumpton had its own football team, and it held players of every colour and every size and so it was nicked named 'the allsorts'. Despite being probably the most multicultural team in London, before one big game the players were warned not to do anything that would aggravate racial tensions in the area. The reason for that was that the opposing team were mainly of Turkish origin and, the weekend before, the Turkish and Kurdish communities had had running battles in the nearby area of Green Lanes. Numbnuts therefore started his warm up by going up to the visiting supporters and shouted 'Three cheers for the ethnics'; he then kicked the ball into the crowd where it unfortunately hit a Turkish woman on the head. As is often the case, Numbnuts really did try to build bridges with the Turkish supporters and he was genuinely trying to welcome them, but being Numbnuts, he sparked off a riot. Racism had nothing to do with it – half the Trumpton team were black, two were Turkish and the rest were London Irish.

The game continued but more trouble was to follow. The opposing team's captain eventually had to go up to the referee and complain about the appalling bad language. The referee called Numbnuts over and told him to stop screaming profanities whenever his side was on the attack.

Numbnuts replied, "Well, you try and keep your cool when Brains runs into the offside position every time and keeps shouting for the fucking ball."

Brains was Trumpton's equivalent of Bear. During the game, Numbnuts would be surrounded by the opposition, with no back up offered from his fellow striker, Brains, who would steadfastly remain in an offside position. Rather than lose the ball, Numbnuts would have no choice but to pass the ball to Brains, which always resulted in the whistle being blown by the referee and a free kick being given to the opposite team. Brains would then castigate Numbnuts for not passing the ball sooner though he was the one at fault, sometimes standing between the goalkeeper and the goal.

Numbnuts would then erupt, usually warming up with, "You stupid big cocked Irish prick," and the opposing team would try to cover their ears to avoid hearing this dreadful abuse being hurled by one opponent to his fellow teammate.

Numbnuts would then be red carded by the referee for swearing at Brains who looked completely unaffected.

"That boy has a serious problem," was all Brains said as Numbnuts was manhandled away by both teams.

Brains was at the core of Trumpton, like a worm in a bad apple. You had to focus on him, for you never knew what he would say or do next. Brains was his surname and upon meeting him, I was reminded of the genius of Sir Isaac Newton. That was because Brains at once told me of his daily activity, which reminded me of the eminent scientist, who was known for having so many ideas in his head when he woke up that he would sit on the bed for hours trying to work out what to do first. Brains was similar as he told me that when he awoke, he would sit on his bed powerless to move for some time, as his mind would be in turmoil trying to fathom who he was and when he done what, where he was.

When Surf was a kid, he was hit by a car and the accident broke both his hips. This meant he had to remain in hospital for a few months with his legs in traction, which involved keeping his legs in the same position suspended by weights. His recovery was going well until Brains came to visit. On arrival, Brains set about trying to impress the nurses by showing off his muscles; he exercised by tugging and lifting the weights suspended at the end of Surf's bed. Surf's screams could be heard throughout the hospital, but Brains was oblivious to his cries and just carried on pumping iron and smiling at the nurses who were too stunned to scream.

Years later, Numbnuts was also confined to hospital with his jaw completely wired up after being knocked out. The list of suspects was as long as one of Surf's legs: girlfriends, pacifists, we never found out. Frustrated at not being able to pass comment on the large nurse who kept passing wind every time she tucked him in, Numbnuts was even angrier than usual. Surf mentioned that his brother needed cheering up, so Brains turned up with a large basket of fruit for his good friend. Brains had been warned that Numbnuts was being fed intravenously as his jaw was firmly wired shut, so he turned up with a basket of various fruit, including golden delicious apples he'd brought directly from the icebox. Numbnuts used sign language, mainly in the form of threatening and abusive hand gestures, to inform his friend that his selection was not appropriate, while trying to knock him out.

Surf commented later to Brains that perhaps that frozen fruit may have been too hard and too big for Numbnuts.

"I would have happily diced them and pressed them through his mouth brace," replied Brains, he then turned to me, "Baloo, he's never happy so I need to get him something else, any suggestions?"

"Coconuts are out of season," I said, "so I would suggest chewing gum as you can bend it through the wiring, but try not to press too hard on his chest with your knee."

Brains smiled; he then set off once more to help himself to items from *Waitanhour's* exotic fruit counter.

You could hardly describe Numbnuts as a victim and his revenge on Brains was swift and merciless. When they were in a café after the doctor had released Numbnuts from his facial birdcage, he challenged Brains to fit a whole *Wagon Wheel*, a very large chocolate biscuit, into his mouth.

"Only a true Irishman could get a four-inch chocolate biscuit into his mouth in one piece," declared Numbnuts.

Brains was always out to prove himself, especially when anyone challenged his Irish credentials. With the whole unbroken chocolate biscuit in his mouth, he looked like he had caught and swallowed a Frisbee.

"Now try and lift your top lip over the end of it, like only a true Irishman can," added Numbnuts.

Brains duly obliged.

"Now just close your teeth over the end," said Numbnuts, and Brains attempted to complete the last and final stage of the operation, which pushed the still complete Wagon Wheel down Brains' throat and he nearly asphyxiated himself.

"I haven't seen a pair of eyes sticking that far out of someone's head, since that big Russian landed on the beam, nuts first, in the last Olympics," said Numbnuts, looking down at his somewhat distressed friend choking on the floor of the café.

In Trumpton, everyone was trying to get one over on each other. I had to have my wits about me, particularly as I was an outsider and a rival supporter at that. Entering the world of Trumpton, was the equivalent of signing up to an intensive SAS training course to get back into shape, except this course was not for Her Majesty's elite forces on the front line, but for rogues. On this course, you tripped up your comrades and, if a friend stumbled and fell, you ran back and repeatedly jumped up and down on his head.

One time, Surf needed to move some very heavy building material. The Connellys would once again get Brains to do the job for free, by goading him to do it in a manner similar to the attempted murder by Wagon Wheel incident. Surf said to Brains that any decent Irishman would be able to lift the three tree stumps out of his garden without a shovel. Brains, who was never one to have his heritage undermined or manly prowess questioned, proceeded to pull the hundred-year-old tree trunk out of the ground by hand. Incredibly, after three days he accomplished the task, popping three vertebrae and nearly replacing his tonsils with his bollocks in the process. Once Brains was back on his feet, even if his head was still ninety-degrees to his knees, Surf asked him to dig a trench for some new electric cable to be fitted to his house. Surf, I think, was regretting his previous generosity in hiring a crane to get his brother on to his roof, when he could have got Brains up there for the price of a packet of biscuits. However, an alternative fee of a half a trumpet was agreed.

"Half a trumpet?" I asked, when Brains informed me of the deal. Brains was obviously still very pleased at securing, or so he thought, such a generous rate after completing a task which had involved removing a ton of rubble with a child's bucket and spade. Brains then asked me how much that was worth exactly and was surprised when I informed him that on the Trumpton Exchange Mechanism that was worth "Bugger all."

"What is a full trumpet worth?" he turned and asked Surf.

"Two bugger alls," replied his employer.

Brains it turned out believed that a Trumpet was cockney rhyming slang for two hundred pounds. I asked him why as I never heard this expression, and he said that that is what he thought the job was worth and when Surf said 'half a trumpet' then this must be that amount. Brains pride would never allow him to admit he didn't know something, a weakness that was taken full advantage of.

I think Brains got some payback in a way, as he had, without anyone's knowledge, laid a cable from the traffic lights outside Surf's house. The house only received light intermittently, according to how busy the crossing was. The fuse box would also blow regularly, and all the lights in the house would go out if an impatient pedestrian repeatedly pressed the button on the pedestrian crossing box to activate a red light and cross the road.

Growing up in the same flats as Brains' family was, according to the Connellys, 'an experience'; this is an Irish euphemism meaning 'left mentally scarred for life.' One story in Trumpton folklore was based on Brains' father, Billy, on the occasion that he lost his thumb in a frenzied axe-wielding incident when he was a kid. Billy was brought up on a farm in Ireland and his mother had ordered him and his sister to kill one of the farm's chickens for dinner. While little Billy was holding the chicken on a wooden block, his even smaller little sister aimed the man-sized axe at the chicken's neck, only to miss and cut her little brother's thumb off. The story went that the chicken had then picked up Billy's thumb and run off with it. The chicken and Billy's thumb were never seen again.

Thirty years later, Billy Brains summoned his three teenage sons to inform them that he was seriously ill. Two of the boys burst out crying, except for the youngest, Brains. Eventually, thinking the worst and that cancer was to take away their beloved father, Harry, the eldest, asked his father how long he had left to live. Billy Brains replied that with his kind of illness you could never know.

"Is it cancer, father?" asked the middle son, Shaun, still sobbing away.

"No, but it's just as bad. It's gout," replied their father.

Now the two eldest boys were bemused, but to the youngest son this was still a life-threatening situation.

"Is it of the heart Dad?" asked Brains.

"No, it's of my big toe," replied Billy Brains.

Brains, now a little puzzled, sat up and leant forward, "Your big toe?"

"Yes, I am afraid so, I could lose the entire toe."

Now the youngest son got quite animated and leant forward a little further.

"Perhaps," he said, now thinking he had discovered a cure for his father's 'fatal' illness. Meanwhile while his two eldest brothers were pressing themselves back into the sofa. Undaunted by this lack of support Brains bravely continued, "Perhaps we could go to Ireland and find your thumb and then the butchers could sew it back on to where your toe used to be?"

Brains was pleased with the ingenuity of his scheme; he sat back comfortably on the family sofa as his brothers disappeared. But then he too had to leap over the furniture, as Billy chased him out of the living room, out of the flat and right across Highbury Corner with his belt swinging violently in his hand. Everyone, when they heard the story was surprised, not because of what Brains had said, but that Billy was able to catch his youngest after a four-mile sprint.

Whenever the Brains family got together, there was always an opportunity for Brains to exert his superiority over his elder siblings. His brothers, as you have no doubt gathered, were in a perpetual state of terror. One Christmas, their mother Milly entered the room with the turkey only to find that Brains had tried out his new K-Tel home-cutting-hair kit on his two brothers. His brothers sat at the dinner table with whole clumps of hair lying next to them on the floor. Brains sat there with a dumb smile holding his present, which was basically a large comb with a now-blunted long razor running through it. Mrs Brains was in shock as she surveyed her sons; they looked like they had received a session of chemotherapy for Christmas. As her two eldest sons looked like turkeys that had plucked by an epileptic, Mrs Brains called for Billy Brains to get his belt. Brains leapt over the sofa and ran off across Finsbury Park chased by his father, plus belt. This was turning into the Brains' family outing, but it was good to see that that time they took the scenic route.

At school, Brains' best mate was Snotbox, whose name was the result of his nose being constantly bunged up with mucus and some sort of white powdery stuff – but more of his job later. He was one of the wildest Trumptonites and academically he was on a par with his best friend. When they were kids, the school's career advisor asked what their ambitions were for when they left school. 'Left' sounded so much better than expelled.

Snotbox was young and black, he said that his ambition was "Not to be beaten up by the police."

Brains replied in his usual inquisitive manner, and asked the careers advisor, "What do you mean by ambition?"

"Well, what do you wish to succeed in?" replied the careers advisor.

Brains understood, "To get out of here and not be put back again for another year," which was a good point as he was the oldest kid in the class apart from Snotbox.

Snotbox added, "Why worry? You'll be in here so long that by the time you leave you'll walk straight into a pension, man."

"I mean when you leave school," the career advisor interjected.

"Well, not get arrested when the police beat up Snotbox for being black," said Brains, acknowledging to his good friend that in such a circumstance he was on his own.

"No, no, no, what is your ultimate aim?" continued the frustrated careers advisor.

"I have no idea what you're talking about, but I do know what I want to do with my life," replied Brains.

"Yes, exactly, that is exactly it, what do you want to do with your life?" responded the flustered careers advisor.

"Die in a mass explosion," replied Brains.

"I always wanted to be a drug mule," added Snotbox.

The door slammed and later we heard that the careers advisor handed in his resignation, without a thought to his next job.

Throughout Brains' life many people have tried to get through to him, but the result was always the same; they soon head for the exit with the door slamming loudly behind them. Yet of all his classmates, I have every confidence that Brains and Snotbox will be the ones, amongst all the potential astronauts and pop singers, who will achieve their stated ambitions.

Brains' first job was as a butcher's assistant. As part of Brains' induction, they briefed him on health and safety. After he had confirmed that he understood all of the instructions given to him he signed a form stating that he was ready for work. He then applied a butcher's cleaver to a joint of meat for the first time. Blood spurted everywhere and Brains remembered one part of his health and safety induction.

"Can someone call me an ambulance?" he asked.

It was the shortest career in butchery history, but he had continued the family tradition of removing a part of his body with a sharp implement. The Japanese mafia remove a finger with a samurai sword as a punishment for failure. In Brains' case, it was a failure to listen during his health and safety induction.

As in Warshalton, Trumpton's main sporting activity was football, watching it or playing it. It's the sport of choice in poor English working class communities. Sometimes, I would join the Trumptonites for an Arsenal away game. I was always up for a good night out with the Kids, but to their annoyance rather than see the game, as a good Chelsea fan I would say I had to head to the nearest pub to monitor its paint drying. On one occasion, all the Trumptonites headed to Cardiff, as Arsenal was in the FA Cup Final being played at the Millennium Stadium while the new Wembley was being built. After the game, we hit the bars and clubs of the city. In the Walkabout pub, the Welsh fans gathered together singing songs implying that the visitors from London were young male prostitutes. When the Welsh choir eventually disbanded they left Brains still singing away in the middle. Later, we managed to get into a nightclub; all of us were drunk, Brains most of all. Almost immediately, a very large ginger-headed woman approached Brains. Her opening line was the usual flirtatious, subtle indication that she was interested in forming some kind of relationship, as you will hear in many nightclubs across our green and pleasant land.

"Come on, take me home and fuck me," she said.

Brains considered his response to this gracious offer and replied, "The only way you will get me to take you home and fuck you, is to throw me over your shoulder."

Later, at about three in the morning, we were placing our orders in the nearby kebab shop, when we were surprised by a large bang on the shop's window. We were as startled as it is possible to be after consuming a large amount of alcohol and looked outside, only to see the large ginger girl with Brains up on her back; he was still banging on the window, wearing a big smile and sticking both thumbs up in the air.

The next morning, we all sat on the mini-bus which would take us back to London. Snotbox welcomed his best friend aboard

"Did you take any precautions, man?" he asked.

Brains responded to his friend's natural concern, "Yes, I kept my wallet under the mattress all the time."

Numbnuts turned his face away and slowly but deliberately started to bang his forehead against the window. Anthropologists have recorded similar symptoms exhibited by depressed gorillas in captivity.

Tommy Spitfire, like me, chose his football team irrespective of peer pressure. Even though Tottenham was a North London team, Trumpton was pure Arsenal territory. It would have been an easier life choice if he had come out as gay and turned up in The Bobbie dressed as Judy Garland, but Tommy walked his own path and no one stood in his way. When the North London rivals played each other, you would find Tommy in a Highbury pub wearing a white Spurs top and standing right in the middle of a sea of red Arsenal shirts. Anyone who challenged him, and many did, would need back up to take on Mr Spitfire. John Donne said "No man is an island," and that may be so, but in all the years I've known Tommy, I never saw anyone in a rush to build a bridge over to his little sanctuary.

Like Brains, I had heard of Tommy before our first encounter. In particular, I'd heard how handy he was with his fists. He had a girlfriend who lived on a particularly rough estate on the Isle of Dogs in East London. She asked him if he would help her with her shopping one Saturday, as she had been accosted a few times on the estate. On their return, she was ahead of Tommy and had to negotiate three guys at the base of her stairs. They were jostling anyone entering the tower block, particularly vulnerable young women, children and pensioners. This was Tommy's first visit to the estate and when she got to the second floor, she decided to look over the balcony to make sure Tommy was OK. She needn't have worried. He had already entered the building and her three assailants were laid out cold on the grass by entrance.

If Brains was similar to Bear, then Tommy reminded me of H in that they both had that kind of facial expression that said 'Go on do your worst, but you will only get one shot.' Tommy was no taller than I was, but he had that presence, like H, that said he could handle himself. However, unlike H, you got a warning if someone had pushed Tommy too far. Like Dirty Harry's "Go ahead, punk, make my day", if Tommy said, "Oh I get it!" you knew that the police and an ambulance would be arriving shortly. Yet both H and Tommy, by their sheer presence, stopped most confrontations erupting into violence and once an opponent backed off, to their credit, neither Tommy or H ever pursued them.

Surf, Tommy's best friend, told me that Tommy had always had that presence even as a young man. Surf told me the story of when they were employed at their first job in a screw factory in Islington. Surf was given a half-inch-thick steel rod to carry across the workshop to be cut and turned into the little screws that secure the bells on children's bicycles. Meanwhile, Tommy was given solid eight-inch-thick steel pipes to turn into rivets for shipbuilding. Both were the same height and build, but Tommy had that sort of presence that he could lift a ship in dry dock and slip on a propeller single-handed.

I don't think I met anyone in Trumpton who did not have a Tommy story about people who had crossed him and come off the worse, but the ones I remember were those based on his roguish nature, his quick wits and ability to take advantage of an opportunity.

He once received a commendation for bravery when he worked in a bank; he had tried to apprehend a robber and pursued him as the armed bandit ran off with sacks of money from the bank. Tommy received a three hundred pound cheque from the bank for his bravery. Naturally, there is a little bit more to the story than the official version. When the robber pointed his gun at Tommy, Tommy was not intimidated but as the money was not his, he just handed the sacks of cash over to the bank robbers. Tommy did notice that the robber had dropped a bag in his rush to get out of the bank. The police told Tommy later, that the thief admitted to them that when he threatened Tommy with a gun, his blank look unnerved him. Upon seeing the robber drop the bag, Tommy leapt over the counter and grabbed the bag on the way out. Tommy ran off in the opposition direction to the bank robbers and called his friend Tricky on his mobile. Tricky turned up five minutes later in his Vauxhall Viva and Tommy threw the bag containing five thousand pounds in cash on to the back seat. Later, Tommy did time for incidents like that one, but it never bothered him, for like Barker's Norman Stanley

Fletcher, doing time was an "occupational hazard" in Trumpton.

I had better explain the Baloo reference. To the Arsenal and Spurs contingent in North London I am known as Baloo, while my South London colleagues refer to me as Wombat. Baloo resulted from when I first met them all; I was in a disco and decided to take a break and lie down on a chaise longue for a quick nap. They thought that it was hilarious that I could just relax in a disco and go to sleep. Numbnuts christened me Baloo as I reminded him of the easy going character from The Jungle Book – not having a care in world. Brains gets very confused by my name and over the years Baloo evolved into Blue, as in my team's colour; on occasion Baboon cropped up for no discernible reason that I can think of (perhaps from the expression 'hung-like') and sometimes BaBa when his brain is fully occupied trying to remove the foil off a pot noodle.

Chapter 31 – *If Darwin Had Found Jockey on the Galapagos Islands?*

There were two local pubs in Trumpton, the oldest of which was The Bobbie, where a fight was the only time the furniture would move and the carpet would be hoovered. The other pub was The Snow Plough, where drugs outsold alcohol. Even when the toilets were broken, you would still find at least three crack heads in each cubicle at a time. One manager asked me how he could make money; I suggested that he abandon selling alcohol and just install a pay-in turnstile to use the toilets. Every manager of The Snow Plough had some form of breakdown following some failed initiative to change the pub's image.

One new manager thought a barbecue would be a good idea. Brains applied for the job on the basis that he ate burgers twice a day and that his kids were never short of a (free) toy. His credentials were impeccable in comparison to the only other candidate, a vegan, so he was given an apron, a brand new barbecue, a supply of meat and fifty cases of lager to sell. Later in the evening, the new owner ran out of the pub to tell Brains that the heat of the barbecue was setting off the fire sprinklers inside. He found Brains blind-drunk with melted plastic caked on to his bare hairy chest; the barbecue had caught fire and was spitting molten plastic in all directions. Standing there, stuffed from eating burnt meat all afternoon, Brains was muttering that he couldn't extinguish the fire having burnt all the knobs off. Brains then collapsed on to a bed of bread rolls.

Later, Brains left the fire brigade to deal with the now-out-of-control fire and walked into the rival pub, The Bobbie for, as he said, "a well-deserved drink."

He went up to Snotbox, "I saw one of your old girlfriends today caught in the fire."

"Which one?" asked Snotbox.

"The one with the false eye"

"Nar, don't know her," retorted Snotbox.

"The one you went to Skinhead's wedding with," persisted Brains.

"Nar, I don't know who you mean," replied Snotbox.

"The one with the big head of red hair and the big shoes," replied a very angry Brains.

"Nar, no idea who you mean," replied his nasally bunged-up friend.

Then Brains angrily shouted out, "She looks like a fucking clown!"

"Oh, Monica," responded the enlightened Snotbox.

Chelsea Mark entered and interrupted the conversation, "Women are wasted on you lot; if only they knew that mine was two-foot long when I fold it in half."

Chelsea Mark then announced that he had a new girlfriend; she was from Czechoslovakia and was the barmaid from the Gooners pub on the other side of Trumpton. Brains clicked his heals and did a Nazi salute while humming the *Marseilles*. As always, no one said anything, it was hard to know where to even start to correct him.

Brains, Tommy and Numbnuts worked together for a tarmacing firm, and a large part of their work involved putting in electric generators and laying adjoining electric cable. One day, as happens on many days actually, Brains called in sick and Numbnuts was left with a new co-worker. As they laid down the last cable of the day and shovelled a foot of soil on top of it, the new guy proceeded to lay down plastic yellow tarpaulin sheets on top of that.

"What's that for?" asked a bemused Numbnuts.

"It's the law, so that when the next guy comes along and digs here, he knows that a foot below is a mains cable, otherwise he would be shot thirty foot into the air."

Numbnuts backed away slowly; Brains had never told him that when he had first been put to work with him over a year earlier. Later, Numbnuts flew into a rage when Brains returned to work.

"You lunatic, do you realise over the next ten years there will be Irishmen putting metal shovels through electric cables and being shot out of holes across London like jumping-jacks?"

Brains responded, "Well we're working across all of the south-east, but the main thing is, please don't worry about me, I always remember the places I've worked before, so I will be fine."

Brains interpreted anyone having a go at him as suffering from some kind of anxiety; their shouting was an expression of their concern for him and, when he did register that someone was angry, it must really be directed at someone else. Numbnuts resigned himself to being a criminal accessory in what was soon to become the largest depletion of Irishmen since the potato famine.

That was not the end of the conversation, as Brains added, "You have been looking exhausted when you have got in each night this week: is that new fellow working you too hard?"

"How the fuck do you know how tired I have been? You've been sat on your arse nursing a chipped fingernail all week," retorted Numbnuts.

"And dry skin, you know, you should apply a moisturiser when you get out of the bath," added Brains.

And at that point, Numbnuts spewed all his lager across the table. "When I get out of the bath? What! Can you see through the window …?" Numbnuts stopped in the middle of his sentence; suddenly realising that that might indeed be the case, as, though Brains lived a mile away, his flat looked directly on to Numbnuts' flat.

"I am the proud owner of a recently acquired," (a euphemism for stolen) "telescope. There's no point in going to work when you have something like that to let you sit back and gaze at the world," he said proudly

Again and not for the first time in his life, Brains had done what no one else could do; he left Numbnuts speechless.

I loved hearing the daily stories of my friends working together as I had done years earlier with H and Bear; it was pure cabaret. At the end of one job, Brains took the lorry on a shortcut through a nearby park and it sank immediately into the wet grass. Numbnuts simply got out of the vehicle and walked off to get a bacon sandwich. In the meantime, Brains was running around in circles and making a madcap effort to dislodge the twenty-ton vehicle by jamming a hand shovel under the back tyre. When, with all his considerable might it still refused to budge, he jumped on it, only for it to break. The handle flew up, hitting him on the head and knocking him out cold. He later regained consciousness, as the tomato sauce dripped down from Numbnuts' bacon sandwich on to his head. Brains then returned to consciousness to hear some welcoming words.

"How are you still alive you thick prick," I left out the question mark as the comment was always rhetorical, and Numbnuts was not interested in an answer. Also, Brains thought he was having a go at someone else.

Later, Surf challenged his brother's lack of concern for Brains, though not because he cared overly much for Brains; it was just an excuse to continue the lifelong argument with his brother.

"Are you not concerned that he could have caused himself a serious injury trying to lift a twenty-ton truck with a hand shovel?" asked Surf.

"Not overly," said Numbnuts and we looked at Surf at this point, as it seemed to Tommy and me to be a strange question for Surf to ask.

"What about the cables the company told him to go back to dig up and apply the safety tarpaulin? He probably has a shovel in one now and thirty-thousand volts going through his brain at this moment," continued Surf.

"He's done that twice this year and survived, so who really is the expert on all this?" added Tommy who also worked for the same company.

Surf continued challenging all of us, "You make it sound like it's his ambition?"

"Actually, that was his ambition at school," I added

Surf ignored me and returned to the subject of castigating his brother along with his best friend.

"Look, you two should look after him; you're a team." Surf insisted.

"So you think I'm part of a team with a man who has an IQ of less than the number of fingers I am holding up?" Numbnuts held up one middle finger to his brother.

"And this morning, how could you run off when Brains started to dig in the same place he had laid a cable last year?" demanded Surf.

"So you're saying that no matter what he does I should stick with him through thick and thin?" Numbnuts charged his brother.

"Yes, exactly, no matter what," added Surf.

Numbnuts then offered a scenario, "Well, let's just say you were laying landmines with Brains." Surf nodded and his brother continued. "You laid the landmine and covered it with earth while Brains watched you do all the work. You stand back once the job is done. What would you do next?" demanded Numbnuts.

"I would run like fuck," replied Surf, for though he loved an argument with Numbnuts, he was no fool and we all knew that Brains' next action would be to step forward and tread the mine in.

I hailed our waitress over, "Another round of bacon sandwiches please, and do you sell walnuts? We may have to pay a visit to a slightly-fried friend in hospital"

Later that same month Brains had another accident, but it didn't involve twenty-thousand volts going through his head. As an innovator in the world of stupidity, on this occasion he managed to get his fingers caught under a huge 'York stone' paving slab, as he was laying pavement for a new pedestrian precinct in London's Oxford Street. In an attempt to free himself, he somehow managed to get a shovel under the other end of the paving stone, which he was trying to push up with his leg. The shovel then snapped. By the way, Brains always sold the new shovels provided by his company and replaced them with cheap rejects from the repair shop where Snotbox worked. Just at that moment, a very a large woman walked on the slab and crushed Brains' fingers even more.

"You fat cow," shouted the crushed Brains.

The woman then came back and jumped on the aforementioned paving slab, which explained why he was readmitted to hospital.

Later that same week, Brains was angry that Numbnuts was laughing at his crushed and bandaged fingers. In a fit of temper, he kicked a can of paraffin at Numbnuts, ignoring the smouldering cigarette in his mouth at the time. Brains' leg caught fire and got stuck in the blazing bucket of paraffin. After extracting his friend from the blazing bucket, Numbnuts told me later that his ribs were hurting so much from laughing that he had had to take the rest of the day off, go home and lie down.

Brains proudly announced in the work yard one morning that he had found a laptop. At that point, Brains' entry into the electronic age came to an abrupt halt.

"How do you open it?" he asked the crowd that had gathered in the yard.

"You might as well strap it to a broom handle and use it as a shovel," declared Tommy.

Brains knew that this was not its stated purpose, but it was certainly the most viable option.

Once, when Brains had no place to live due to his having upset his poor girlfriend Linda, Tommy agreed to put him up for a while. It made sense as they both worked together. That is, it made sense until the end of their first day living together; Tommy went to jump in the works' vehicle that Brains was driving home. But as Tommy was approaching the vehicle, Brains locked the door and out of the rolled-down window, informed his new flat-mate that it was against company policy to have anyone uninsured in the vehicle.

"You vacuous fuckwit, I work with you." Tommy said.

"Sorry, but after six pm, you're no longer on shift, so I can't let you in," replied Brains.

"But it's only two o'clock in the afternoon!" cried Tommy.

"Sorry, but I wrote myself up for overtime tonight; on my clock it is six pm, so technically we are no longer working." Brains replied.

"But it's not six and what you have done is fraud!"

"Ah fraud, that comes under a different section, my concern is insurance cover," said Brains as he drove off, thinking about the extra four hours overtime he would receive.

It was not the first time that those two would clash, but each time Tommy raged at Brains, it went straight over Brains' head. However, Brains did love getting a reaction from Tommy.

The company they worked for often had to carry out frequent re-training classes for its staff – usually due to some incident caused by Brains. Each vehicle was required to be fitted with a tachograph to monitor that drivers were taking their appropriate breaks, thus limiting road accidents. The boss found out that Brains had been using the cards for the tachograph device as flying saucers, which he would flick out of the window when sitting in traffic jams to ward off boredom. All the company staff were summoned for re-training, much to Tommy's fury and Numbnuts' delight.

Numbnuts later told me that, "It was another opportunity for Brains to make a complete prick of himself and for Tommy to blow a bollock with him."

Numbnuts took great delight in telling me that the training course went as follows:

"Good morning, I wish to provide a refresher course on the proper use of your vehicle's tachograph," said the instructor, engaging with his audience for the first time.

Brains leant towards Numbnuts and in a low voice asked him what a tachograph was. Numbnuts responded and leant towards his friend's ear.

"You woodenhead." Numbnuts whispered.

"Thank you, that's what I thought," replied Brains, who was determined to concentrate on the instructor's every word.

"Firstly, you write your initials, rather than your full name on this cylindrical paper, the tachograph insert, to certify that you are the driver of the vehicle," began the instructor.

Brains put his hand up, as he had a question, "With no due respect, what happens if your initials are your name?"

"No one's name is actually their initials, the initials are the first letters of each name that someone has," the instructor replied.

Brains pondered this and put his hand up again, "What about JP who manages The Mayflower?"

"I think you will find that his initials are the first letters of his Christian name and surname," replied the instructor, once again believing that he had closed the matter.

Brains' hand shot up once more, "So you know him?"

The instructor now left the room, another in a long line of Brains' ex-teachers and careers advisers.

Tommy walked up to Brains and shouted directly into his face, "How fucking stupid are you?"

Brains sat wondering what Tommy was screaming at and why everyone was leaving so soon as the class hadn't started yet.

Meanwhile, Numbnuts decided he needed a break from Brains and he walked into the main office to ask the bosses if he could work somewhere else for a week.

"Before I go fucking mad and strangle the fuckwit," was his exact phrasing.

He was put with Murphy (who was notorious for his swearing) for one week to dig a hole in which to put a new generator at the base of a mansion in a wealthy part of West London. It was different to tarmacing, but for Numbnuts it was like going into therapy.

Numbnuts later told me that everything went well, as the kindly woman of the house would walk down to the end of garden each day at lunchtime to bring him tea and sandwiches. Numbnuts said it was just the break he needed; Murphy was off sick and he enjoyed the peace and quiet, as well as the civility of Lady Grace. On the last day, Murphy returned to work to join Numbnuts.

"Murphy, don't fucking swear when Lady Grace comes down with lunch," said Numbnuts to his colleague.

"Fuck no, I never swear in front of a lady," declared Murphy.

"Just don't, she's been very good to me all week, so none of your shit, bollocks, cunt, fuck or arseholes." Numbnuts persisted.

"You can trust me," said Murphy sincerely.

At lunchtime Lady Grace arrived at their place of work for the last time, as the hole was now very deep and the generator's installers were arriving to begin the installation on the following Monday.

"Good morning boys, here is your tea and your sandwiches," said Lady Grace looking down and smiling at the two workmen in the hole.

"Thank you, this is my colleague Murphy," said Numbnuts standing on Murphy's foot and whispering, "Just nod you prick." Murphy dutifully nodded.

If everything had ended there and Lady Grace had said goodbye and waved before turning back towards the house, then all would have been fine. But, just then she turned to Murphy.

"I'm so sorry about the rain and you must be awfully cold, my man," she said.

"Yes, it's awfully cuntish," replied Murphy, with great sincerity.

Numbnuts moved forward to catch the poor old woman, as she tottered and nearly collapsed into the hole.

Brains' life ambition may have been to explode and distribute himself over a wide area, but his daily goal was to get to the pub as quickly as possible. So when he had to take an exam to keep his job as a result of some new government legislation, his answers were based solely on that premise. He showed me the questions and his responses, which he had had to provide to his boss the next morning. They were multiple-choice questions, but with only two suggested answers and I saw straight away that all the correct answers were *a*. You would have to be as thick as a brick to fail; I worried for Brains. I read through the twenty questions, which were clearly drawn up by some comedian in the company but with the objective to make sure that no one could possibly get a question wrong.

Brains answered *b* to every question and he asked me what I thought. I read the first question, 'If you were handed a club hammer and you saw that its head was loose would you,

   a) Refuse to use it, inform your supervisor that the tool was dangerous and fill out an accident report? or

   b) Use it as quickly as possible and shout out to those in the vicinity that the metal head may fly off and kill someone at any moment?'

I informed him that, though I was not an expert on these things, I believed he would secure a one hundred percent score, which was correct in that every answer he'd given was wrong. He did add that he was puzzled as to why he had to warn anyone. He had wanted to write 'Fuck 'em,' but he wasn't sure of the spelling, plus there wasn't enough room in the tick box.

There was more to life than the Kids trying to avoid work; there were cultural pursuits too. I met Brains one night after going to the cinema to see the film the *X-men*.

"I think you're showing your ignorance there,' he said, 'as you'll find it is called the *Twelve Men*," Brains corrected me. As always, he was wrong on so many levels, it was impossible to know where to begin. He also had views on any number of topics, as he confided to Numbnuts and me at a bar and without any prompting.

"Homosexuality is disgusting, perverse and depraved," he said and he thought about this for a moment and added, "Yet, strangely beautiful."

One morning I got a telephone call from Tommy.

"Fuck off," were Tommy's first words to me.

This was an unusual greeting, even for Tommy, but I replied, "I'm fine Tommy, thanks for asking, how are you?"

"Fancy a beer later for the England game and watch that overpaid bunch of dick-pullers get beaten by Croatia again?" he asked.

"Yep, I will give Surf, Numbnuts, Snotbox and Brains a call and head to The Church," I said. The Church was a big pub in North London's Muswell Hill.

As seasoned watchers of our national football team, we knew that despite the usual fevered press hype building up to the game, our national team would fail to achieve, so it was best to share the pain.

"I don't care where you fucking go, just fuck off," added Tommy.

"Have you got Tourette's Tommy?" I asked.

"Brains fell out of the window again last night," he continued.

"Christ, I hope he didn't hurt anyone?" I asked.

"Just get the fuck out of my garden!" shouted Tommy.

"Is that what Linda," (Brains' girlfriend), "said to Brains?" I asked in my attempt to ascertain the facts about Brains' descent on to his patio.

"Fortunately there was no one there to break his fall, so no one got hurt; Brains landed on his head."

"Thank God for that," I said on hearing this good news.

"Just get to fuck, you mad old cow," cried Tommy down the telephone.

"He took her response badly then?"

"No, you fuck off you mad cow," he told me angrily.

"Tommy the swearing I can take but cow is out of order, as I have never found my feminine side," I replied.

Later that evening in The Church, Tommy explained that while he was on the phone to me, some woman had climbed over his garden wall and proceeded to walk through his front room in an attempt to use it as a shortcut to God knows where. It seemed that it was not unusual for Tommy's flat to be used as a shortcut by some of the local residents of the mental institute across the road. There must be a map for the 'Care in the Community Ramblers' Club' somewhere with a big red public access path marked on it, going straight through Tommy's house.

That night in The Church, Numbnuts, Brains and Snotbox joined us just as England appeared to score the first goal. Surf was returning from the bar with a full tray carrying eight pints of lager when the goal went in, and in the excitement, Surf threw the whole tray of drinks into the air. The lager landed on us, just as the referee disallowed Sol Campbell's goal. We just looked at Surf, as we stood there in our lager-soaked clothes having now finished our first pint of the evening without it touching our lips. Surf was not easily deterred and returned to the bar while declaring that he thought Campbell would get a hat trick tonight and keep shooting at goal.

"Oh, bollocks," said Tommy, who had little faith in the Arsenal defender and in a voice of resignation he turned and said to me, "We might as well watch the game smoking a pipe and set the fire sprinklers off."

My girlfriend Angela and her friend Louise met Tommy and me in a pub later that night in Islington and we ended up back at Tommy's place. Tommy and Louise stayed downstairs as Angela and I took the upstairs room. The next morning when we emerged, we discovered Tommy lying face down with Dfor, his pit-bull Mastiff, asleep on his back. Dfor was a good guard dog and he was making sure that his master wasn't kidnapped by ramblers during the night. Tommy stirred and gradual opened his eyes.

"Ah fuck, Louise is gone," he muttered.

"Maybe she is down the jewellery selecting a wedding ring," I suggested.

"You can fuck off too and I'm not talking to a nut-nut in my garden this time."

"He's a very angry man," I said to Angela with a big smile, just as Louise walked into the living room holding a small black insect between her fingers tips.

"I just found this in my knickers, what she it?" she asked.

"It's only a dog-flea, but just go over to Dfor and place it in his ear and hopefully he won't have noticed it's missing," I added.

Louse was shocked and I could see that Angela was trying very hard not to burst out laughing. I turned to Angela and repeated my standard lines on these occasions.

"I can't live in a world without love," as we looked at Tommy and Dfor now both asleep and snoring to their little hearts' content.

Later Louise told Angela and me that Tommy gave her a lift in his truck back to the station, but the poor man had had to stop every now and then to get out and throw up. As an aside, Tommy did 'The Knowledge', which is the term for those who want to be a black cab driver and learn London's roads; but being Tommy, he did this in his own inimitable way, driving a ten-ton vehicle from work. If you ever find yourself walking the narrow lanes of Hampstead, you may notice the lack of metal bollards. Usually budding cab drivers did 'The Knowledge' riding a moped.

Tommy is a rogue in the true sense of the word. One evening, in a mini-cab on his way back from a birthday party, his friend Tricky got into an argument with the driver about something. Tricky was paralytic, so Tommy just sat back and let the two of them carry on their argument. Tricky then suddenly jumped out of the mini-cab and ran off into the distance. He pulled the mini-cab's rear view mirror off the windscreen in the process. The cab was an illegal one and not in the best of shape, so the whole windscreen came off with the mirror. The driver was not hurt, but he was certainly not happy. Tommy continued to sit in the back seat, waiting for the driver to get going again to take him home. Instead of moving, the driver turned on Tommy and demanded that he paid for the damage his friend had caused. Tommy refused to pay, so he got out and slowly started walking while the driver ran alongside him demanding full payment and calling the police on his mobile phone. Tommy continued to amble along in his usual nonchalant manner, despite having an irate mini-cab driver screaming in his face, a non-driveable vehicle behind him and a police car pulling up abruptly on the pavement in front of him, from which two policemen got out.

"What is the name and address of your friend who damaged this man's car?" demanded the police sergeant.

"I did not know him, we just shared a cab, as we were going in the same direction," replied Tommy, unruffled.

"You have to pay for the damage to the car," the police sergeant continued.

"No. I didn't know him, so I am not paying for something that has nothing to do with me. I was just sitting quietly in the back of the cab when all this happened." Tommy replied.

The driver continued screaming at Tommy, "You pay, you pay!"

The police sergeant could tell by Tommy's deadpan responses that this was going nowhere, so he tried a different tack on what he believed was firmer ground.

"OK, forget the other guy and the damage, the driver says you owe him seven pounds for the fare so just give him that to shut him up."

"No, I don't," replied Tommy, "my fare is only three pound and fifty-pence, as I was only half the ride."

The two policemen were starting to despair and the driver was by now on his knees, banging his fists against the pavement. Tommy just stood there completely unfazed by any of it, until the second police officer decided to assist his colleague.

"Ok, just pay the man the three pounds fifty." he said.

"No," said Tommy. "It was three fifty to my door, and I had to walk the last bit."

Both policemen were exasperated by now and were virtually pleading with Tommy to pay the man, so that they could be back in the station in time for breakfast.

The second policeman continued, "OK, just give the man something."

Tommy then turned to the mini-cab driver who was still on his knees banging his head against the pavement.

"Here's two pounds and forget the tip," said Tommy.

The driver and the policemen were beaten men. Then Tommy turned and walked at his usual slow pace into his house, which to the bemusement of the policemen and the driver was right where they were standing in full view of the mini-cab.

That summer being a World Cup year, Tommy had a barbecue in his garden; as you will learn, this four-yearly event would come around far too soon for some. There was always plenty of food at the end of Tommy's barbecues, due to Dfor's hairs being all over the grill; he would often sleep on it, as Tommy's flat had no heating. Tommy's garden had a low wall and led directly to the street, which explained, to some extent, why it was viewed by some as a public right of way. Tommy, after consuming a few beers and a burger with Dfor, and as Surf and I declared that we weren't hungry, sat back in his deckchair facing the main road.

"Amazing," said Tommy, "people walk up and down this street twenty-four hours a day and anyone could easily step over into my garden and just help themselves to any of this."

Surf and I surveyed the 'any of this' around us and, if you included the contents of the front room, it might reach five or six pounds on eBay, and that was including the furniture as tinder for a bonfire. Tommy continued, "You know what? I have never been burgled." Surf and I looked at Dfor, who was chewing on the metal tripod of the barbecue. The mastiff played an active role in rites of passage for those born in Trumpton, he had bitten a piece out of most of them. Surf gave me a knowing glance, indicating that you would have to be stark raving mad to enter Tommy's garden. We were the only visitors Tommy ever received, except for the inhabitants of the building across the road whose inhabitants were mentally ill.

You probably have worked out that Dfor, stood for D for Dog; Tommy calls things what they are. Tommy loved Dfor, so when Dfor became ill to the point that it would be doing the poor creature a favour to put him to sleep, he couldn't bear the thought of it. No one would raise the possibility with Tommy, knowing that the messenger would have his lights turned out quicker than Tommy's beloved hound. However, eventually Tommy accepted that Dfor had to go and after the vet had put the dog to sleep, Tommy called us to meet him for a drink and break the news.

"Dfor, I will really miss him," said Tommy sadly; they had fought and bitten a lot of people between them over the years.

"Well D for Dead," now said Numbnuts.

"More likely S for Stuffed," added Brains.

Tommy exploded, "Oh I get it, take the piss out of my loss you fat bastards! I'll have you two stuffed and then I will order a bigger mantelpiece to mount your two fat arses on it!"

Tommy, his red face now inches from theirs, continued to rage but Numbnuts and Brains just smiled, knowing they had roused Tommy out of his gloom. I looked at my two friends proudly; their job was done. That is what friends are for.

Tommy did not have too many long-term relationships, though that was not always his fault. When I first met Tommy, he had just ended a relationship with Edward Scissorhands: that's a name I should explain. Scissorhands had a bit of a temper; when Tommy popped out to see Tricky for a quick beer one Saturday lunchtime, he returned to the house later than expected - Sunday morning. Scissorhands was slightly irritable. When Tommy got back to the flat they shared on the Trumpton estate, she had already smashed the windows and torn up the carpets.

As he explained to me later, he respected her right to express herself and so he had decided to sit down next to the wall of their block and wait until she had calmed down. Unfortunately, he fell asleep. When he woke up, he found himself surrounded by his clothes all of which had been shredded with even his shoes cut into numerous pieces. When he entered through the broken door to their flat, he found everything he owned had been destroyed;

pictures, books, even his mobile phone had been sawn into sections and stuffed in the toilet bowl. Hence, her nickname and a story that was a great source of amusement for the Kids for years after the event.

Women were always drawn to Tommy, as he was rough and ready; what the tabloids would call a 'man's man' in that he would not compromise for anyone. I know many women who could describe everything they hate in a man and chances are you would end up with an identikit of Tommy. Yet, I never heard a bad word about him from any of the women who knew him. In an ever-confused world with changing values and moral compasses with arrows missing, women knew exactly where they were with Tommy. Like H, Tommy is brutally honest, but in a world that minds its Ps and Qs, those two are as rare as Christmas cards on Bin Laden's doorstep. Tommy also appealed to basic primeval female tendencies: he was protective, he was as solid as a shithouse at a summer curry festival and he was an ideal mate if you wanted to ensure that you and your children would never want for anything or fear anyone.

Tommy also has charisma, being very funny in a dry deadpan way. I once observed him chatting to a woman and complaining about his 'farmers', taking from the cockney rhyming slang 'Farmer Giles' as a name for haemorrhoids. She stayed with him all evening, laughing at the top of her voice. Now that's charisma.

Men, I think, are simple to understand. We are driven by the pursuit of women; we relish a beer with our mates and talking about nothing of consequence; we are territorial and strive to have power over our 'virtual kingdom' or as much power as our woman will let us believe we have. Women are more complex, they want attention, but don't want their men to get under their feet; they want dependability but they also want excitement and to be pleasantly surprised; they want family yet fiercely protect their independence; they talk through every configuration of a question and then any solutions; they want to be seen as more than their physical attributes but want their man to worship every curve of their body; they want to discuss everything with their man, but rarely seek an honest answer. Then, once they have chosen the man of their dreams, the process of changing them begins. I think this is one of the reasons I love the company of women more than men: you never know what will happen next, well at least I never do.

Tommy once took us all on a pub-crawl on Upper Street to celebrate Tottenham beating Arsenal. The fact that all of his friends (excepting me) were supporters of the beaten team and that the street was in Arsenal territory, made no difference whatsoever as all the 'unusual', but usual suspects came out. After leaving the fourth pub, the Kids embarked on what was some kind of random assault course with Brains leaping every piece of street furniture along the high street, while Tommy attempted to jump up on the bar of the canopy outside each shop, like a drunken gorilla swinging through the jungle. The first step of his journey along the restaurant canopies of Islington had a bit of a false start. The flaw in his plan was that when he jumped up to grab the bar of the canopy, rather than securing his weight and propelling him to the next bar, it bent down to his level. While the diners looked around to see Tommy standing there, feet firmly nailed to the ground and his hands in the air holding the bent canopy frame, a very demure French waitress came out of the restaurant.

In her best Queen's English she shouted over to our friend. "Get down off that, you fat cunt."

Tommy let go, but the canopy stayed in its new position. In the meantime, Brains continued leaping over the numerous display boards enticing customers into the restaurants. At that moment, he was distracted by the abuse Tommy was getting and did not realise that he was leapfrogging a display board far higher and wider than the previous ones. This resulted in him catching his testicles on the first corner of the hard, wooden display; he bounced along until he collapsed face down on the pavement holding his swollen private parts. Tommy was a little happier on seeing that Brains had seemingly castrated himself and was on the short-list for the annual Darwin Awards. Brains, though, somehow survived, so technically, he was disqualified from winning, but he certainly deserved to be highly commended by judges that year.

Due to his sudden disability, Brains crawled into an empty cab in the hope that it would take him to the next bar at the end of the road, but he left the door open only for a passing bus to rip it clean off.

"Oh my God! Did you get a number?" cried the cab driver returning to his vehicle.

"Yes, a 43," replied Brains who still expected a lift up the road.

As you may have noticed, it takes a lot to get a Trumpton passenger out of a taxi: if it had been on fire, I still think Brains would had sat and wondered where the driver was. By the time we reached the end of the street, Brains had caught up to us and got out of the three-door cab, only for both him and Tommy to be refused entry into a late night bar, The Redoubtable Knob, as it was couples only. The two returned ten minutes later dressed in old women's clothing that they had liberated from some black bags left outside the charity shop across the road. The bar was very liberal in how it defined a 'couple', as two large transvestites in Laura Ashley hand-me-downs and Doc Martin boots appeared not to be a problem.

Warshalton was also like Trumpton, in that violence was not really part of the agenda, in fact the only item on the agenda was having fun. Unfortunately Tommy's reputation and his steadfast refusal to compromise or even keep his head down, meant that wherever he went, trouble would follow. Tommy asked me to meet him for a drink in a pub on the Holloway Road in North London to watch our teams play, but it was busy with Arsenal fans after Arsenal had played earlier and lost. I walked in to find Tommy already at the bar.

"Best to keep our heads down, this could get ugly," I said.

"Too late," said Tommy "it's going to kick off at any moment."

"How do you know?" I asked.

"Well you see that mean looking bastard over by the bar? He is going to lead a charge at us any moment now." I then noted the big stocky character at the other side of the bar; he had one long eyebrow which crossed the full length of his forehead above a closed black eye and he was growling directly at us.

"Yes and his mates look a bit handy; particularly the tall ugly one, she looks a bit nasty. I wish we had Brains or Chelsea Mark here to chat her up." Though I had no reason to doubt him, I asked, "Why do you think it will kick off?"

"Well I was coming back from the toilet and he said he was going to kill me." Tommy replied.

"Ah that's a shame, as it looks like the window cleaner washed the windows for nothing," I said, guessing that Tommy hadn't then asked them all to take part in a group hug.

"Well, I'm not one for long conversations, so I threw my pint over the lot of 'em." he said.

"Well, I hope I get a hospital bed with a window view," I said just as Cyclops and his gang ran at us. We traded punches until the bouncers dragged the two of us outside, holding back the rest of the pub who tried to follow.

We then came under a shower of bottles, as some of eyebrow's crew had called more of their gang, who were waiting for us across the road and were throwing bottles and glasses at us. They ran across the road and steamed, fists flying, into us. I managed to knock down the first two or three until I went down. Tommy was still trading punches but he too was eventually knocked to the ground by the sheer number of them and then the mob piled into us feet first. The bouncers were terrified, but they must have believed we would be murdered, as they came out and somehow dragged us back into the bar. In the meantime, more bottles rained down on the pub, some coming through the windows.

"I think my ribs are broken." I said to Tommy, as I tried to straighten myself up at the bar.

Tommy added, "Yer, me too' and with that and, to the amazement of everyone, he then bent over the counter and asked the bar staff cowering underneath it, if they were still serving? This was followed by a little squeak from one of the terrified barmaids, which must have sounded like a yes to Tommy who then said,

"Two lagers please, love."

I had heard many stories of Jockey, another resident of Trumpton, prior to meeting him. I first met him in a bar in the West End, when I brought Tracy along to meet my new friends from Trumpton. Tracy was complaining about a film she had been dragged along to see that afternoon, when the Scottish man brought along by Numbnuts joined in the conversation.

"Is that the one about two posh blokes trying to give a fat bird one up the arse?" asked Jockey.

"Err. Yes, that is the one, it got an Oscar for best screenplay," Tracy added, not the least bit put out by the Scotsman's question.

"Quite right, they don't make 'em like that anymore. Bastards," sighed Jockey.

Tracy, had just lost her job, so shortly she began wondering aloud what to do next.

"Sex telephone call-lines, you make over four hundred pounds a day, it has great prospects, plus you meet interesting people," added Jockey.

"Interesting people?" asked Tracy.

"Well, I'm fucking interesting," added Jockey, appearing a little put out by her lack of enthusiasm.

Thinking he had ingratiated himself with Tracy, Jockey then asked if she fancied sex behind the dustbins. I don't think I had ever seen her laugh so much in all the years I had known her, and it took her some time to compose herself before she could finally answer.

"NO!"

Later in the Blues Club in London's Oxford Street, I saw Jockey dancing with a six-foot blonde and as I have yet to describe him, I should mention that this meant a woman twice his size. At the same time, I bumped into a public school colleague from work, who was known even to his educated work colleagues as Toff-boy.

"Jonathan," he called out.

Jesus I thought, he can't even say Johnny or John.

"How are you, Jonathan?" Toff-boy continued, and at that point Jockey barged between the two of us towards the bar.

"Do you know that appalling fellow?" continued Toff-boy.

"Well we have not been properly introduced" I said, which was true as Jockey had no interest in me and hadn't even taken his eyes off Tracy's breasts until the big blonde arrived on the scene.

Just then, Jockey pushed through us again carrying a bottle of champagne.

"Out of my way Baloo and Chinless," he grumbled, "she's getting a bit of Scottish loving tonight."

Toff-boy realised that the girl Jockey was in pursuit of, was in fact his sister; if he had had a chin it would have started to wobble. I hoped it would be the beginning of a loving relationship, as that would be one wedding I would go to just for the speeches alone.

I was soon to learn that my new friend Jockey was a very excitable, lovable, little Scotsman who was not only short but also appeared to have no neck. I thought that if I was his best man for his prospective nuptials with Chinless' sister, my main job would be to put my toe up his arse to make his head pop out from his collar and put a tie on him. The little bundle of testosterone was obsessed with sex, and I have no doubt he would have used horseshit as a deodorant if he had read in a magazine that it would attract women. Of all the rogues I have described, Jockey was the most sexually obsessed, even more so that Bertie. Twenty-four hours a day, sex was all that occupied his mind.

"Even my dreams won't give me a break," he once told me.

"You must have sheets like popadoms," I noted.

No matter what the conversation, he could only view it from a sexual angle.

Numbnuts once commented on to my recent move to Highgate. "A bit of a change from Camden, not much violent crime there, I expect, unless somebody puts their elbow on your almond croissant."

"It has its moments." I countered. "It all kicked off last week in the Soya Bean Coffee Shop, when a deluxe twin-baby pram rear ended another. On the theme of violence, the conversation then turned to self-defence.

"I keep a baseball bat by the side of my bed, in case anyone breaks in," said Slugger another good friend from Trumpton, who then turned to his old schoolmate, "What about you Jockey, what do you keep by the side of the bed for emergencies?"

"A box of tissues," added Jockey, who I think had a different perspective of what an emergency was.

Jockey was not discriminatory; indeed he was the most egalitarian of lovers, caring little for a woman's shape, size or education – even to the point of disregarding whether his lovers were able to speak English, or speak at all for that matter or were even in the middle of psychotic episode.

Numbnuts remarked in his usual sensitive manner, that this led to a blurring of his work and social life as Jockey was a 'hod carrier by day', a brick carrier on a building site and a 'hog carrier by night'. Numbnuts had a talent for exaggeration: we all doubted that Jockey could physically lift any of his women.

At work one day, Jockey had a fit of paralysis while he was climbing a ladder carrying fifty bricks on his building hod. The sudden convulsion resulted in him shooting his arms almost out of their shoulder sockets, as he then fell off the ladder and landed head-first on the ground. Some animals have an instinctive reaction when falling. Cats for example, at least in folklore, adjust in mid-air and land on their paws and Trumptonites like Jockey and Brains are similarly able to alter their bodies. The impact is greatly reduced by their rotating their bodies in free-fall and landing on their heads. Jockey broke his collarbone and was off work for a month.

When he had recovered, he returned to work to find that the two-storey building he had fallen off now had an extra six levels. Though he noticed this, he failed to notice the new lift shaft on the top floor, which he then immediately fell down while carrying his first hod of the day full of wet plaster. Fortunately, a metal bar stretched across the open shaft on the fourth floor and it broke his fall; bringing him to an abrupt halt, with a testicle on either side now a good six inches apart. He then, as there was no other way out, had to crawl crab-like back up the shaft. After an hour, his red sweating face popped up from the top of the lift-shaft. The plasterer was off his head on cocaine and very angry due to the lack of wet plaster.

"More muck, your Scotch bastard!" he shouted at Jockey.

Normally, Jockey would challenge the Scotch remark explaining that it was a reference to the drink not to the people of Scotland, but as he was hanging on for dear life, this was not the time. He remained there for another hour with just his red round head and fingers on show, as if in homage to the graffiti image 'Kilroy was here.'

The love of Jockey's life was a girl whom Numbnuts referred to as Fathead, due, unsurprisingly to her disproportionately sized head. Seeing how angry Jockey became when Numbnuts first mentioned the nickname for his girlfriend, Numbnuts tried to placate his good friend and added that with the amount she ate, she would be sure to grow into it.

Jockey is one of the unluckiest men I know and Surf once commented that if Jockey was on the Titanic and had somehow survived that ordeal, he would have climbed on board the Marie Celeste. It's not just that things just go wrong; it's that when they do, it's also his bad judgement that when he wants to share his troubles with friends, the friends he chose are us. For instance, a few years later, when his relationship ended Jockey told us despondently, "That she's run off with black Toby," before he broke down in tears. He was on his twelfth pint of lager and all he had eaten all day was a 'Connelly three course meal'; a delicacy comprising of a packet of cheese and onion crisps, a packet of salted and another of dry-roasted peanuts all freshly opened and piled together into a heap on the bar mat.

"Fucking right, man, she is a good looking woman so what would she be doing with small-cocked white trash?" said Snotbox, who was also a good friend from school.

"How can I attract her back?" said a very despondent Jockey.

"Don't bother, man, as your mum used to say, once you've had black you don't go back," added Snotbox.

"Well you could cover yourself in black paint like the guy in *The League of Gentlemen*," suggested Numbnuts.

Jockey gave him a glaring look. Numbnuts got the message and went back to his pint, but added, "Well, at least she's getting some decent cock now."

Things did take a turn for the better later that week, when Jockey met a woman during another late drinking session.

"Pal, she was stunning, the most beautiful woman I have ever seen in my life, a goddess pal, I love her," he boasted to Numbnuts and me the next day.

"You were smashed out of your little Scottish skull, but didn't you meet your goddess at lunchtime today?" asked Numbnuts.

"I'm fucken devastated, in daylight she looked like a fifty-year-old peacock."

"That's the male" I noted.

I just received a torrent of Scottish abuse for my concern, ended in the traditional manner with "Bastard!"

Jockey was depressed, until he suddenly announced, "Fuck it pal, I've had enough, I'm off on a *pool* holiday."

"Great, chilling out in the sun by a swimming pool sounds like just what you need after a long spring, summer, autumn and winter of sexual frustration," I replied.

"No, a *pool* holiday," he responded angrily.

"Playing pool all day with a few beers, without a care or a girlfriend in the world, excellent idea," I replied.

"No, a *pool* holiday," he continued growing apoplectic with rage. "A fucken *pool* holiday, a *pool* holiday, trying to *pool* women, I'm going on the *pool,* you Bastard!"

"Oh, a pull holiday," I said finally grasping what my good friend was saying.

Brains walked in the door and Jockey greeted another old school friend, "Another relationship ended, my life has more twists and turns than an Agatha Marples murder mystery."

"You try reading a Catherine Cockson horror, it's not even in English," but more of why Brains held strong views on the mispronounced author's literary output later.

I decided to ignore the Trumpton book review club, as I was unfamiliar with the works of Marples and Cockson and instead I tried to cheer up our little Scottish experiment. I told him of the success that my good friend Bear had on Internet dating. Perhaps I stretched the meaning of the word 'success' but Jockey decided that he would give it a go and he asked my advice on the text to go on his dating profile. I wrote something out on a piece of toilet paper and passed it to him.

"SCOTTISH SEX-CASE SEEKS PULSE AND SPINE FOR LOVING RELATIONSHIP. Yes, that would do," he said, content with my offering and confirming to me that he was illiterate.

If Darwin had lived today and went to the Gorbels rather than the Galapagos Islands, the Origin of the Species would be an entirely different book; there would be no mention of evolution just a warning that Dodos should smarten up or they would find themselves going the way of Jockey.

Sometimes in life, my parallel worlds collide. Bourgeois phoned me up to say that he wanted to catch up over a drink. Fine I said, but I was meeting up with some of my Trumpton

Kids, so I invited him to join us. Bourgeois recommended a 'quiet pub', called The Happy Traveller in Camden. As Tommy, Surf, Numbnuts and I entered the pub, we were met by two women lashing out at each other with their stilettos in hand.

"I like it here," said Tommy.

Random fights broke out all evening, Tommy saw an opportunity between the missiles and moved in on three Welsh women who were sitting terrified in the corner of the bar. Tommy invited the Welsh party to join us in the The Arch for a late drink, since the police had turned up in full-force in a futile attempt to stop a riot taking place. The three Welsh women willingly agreed and together we broke through the police line as they were trying to stave off the barrage of chairs.

In The Arch, Tommy appeared to be doing well with one of the Welsh contingent, though Bourgeois was a little nervous; he was noting that The Arch had a bit of a reputation.

"For what, letting its customers out alive?" asked a mystified Surf, who was still thinking about Bourgeois' recommendation of a 'quiet pub.'

In the background, I could see that the 'potman', the slang name for the person who offered to collect the empty glasses in pubs in return for a free drink, was getting on very well with the Welsh woman whom Tommy fancied.

After a while, Bourgeois turned to me and said, "Righten, in all the years I worked with you, H and Bear, I never told you what I thought of the three of you."

"Feel free, if a High Court judge can express an opinion, why not you?" I replied.

Over his shoulder, I could see the Welsh woman now scribbling her telephone number on a bar mat and handing it to the potman. I returned to the conversation.

"Well, you three were the ...," but before Bourgeois could finish his sentence, the potman came flying between the two of us after Tommy landed a left hook right on his chin.

The whole pub erupted into a massive brawl with Tommy landing punches on the bouncers and the bar staff. As we defended ourselves, I could see Bourgeois in the distance being carried off in what was known as 'The Arch Combo special'; a headlock sealed with a French kiss. It was being expertly delivered by what looked like a sixty-year-old woman, who was dragging him into a waiting taxi. That was the last I ever saw of him.

The Arch was the second pub that evening to be surrounded by the police as they were making a futile attempt to close it. Many of the policemen who arrived came straight from The Happy Traveller and once again, I could see from their faces that they did not exude any confidence in trying to get control of the outbreak of anarchy. I greeted them with a hello, as some of them were old friends now. One traumatised policemen told me that he was leaving the force, as he was still recovering from the battering they had taken on New Year's Eve in the Pigs' Knuckle ballroom across the road. I had heard that that night had been especially violent; order was only restored using a relay involving thirteen police vehicles to arrest the entire occupants of the dance hall.

We had lost Surf and Numbnuts and outside I turned to Tommy and said I had no interest in getting involved in any more fights.

"Yes, I try to avoid trouble myself," replied Tommy. "Anyway fancy a drink in The Mayflower?"

I gave Tommy I dubious look, but shrugged my shoulders, "Why not? It might liven up a rather dull evening."

That night was they were having a karaoke evening in The Mayflower. A duo was on stage and from a distance, it looked like one of them was playing an instrument, which was an unusual accompaniment for a Karaoke I thought, until on closer inspection I realised it did not have strings. The singer belted out the lyrics from U2's, *New Year's Day*, while his musical accompanist played air-banjo.

The next act was a beautiful black woman with a very large afro who got up on stage and proceeded to give a rendition of the Gloria Gaynor hit *I Will Survive.* She had a fantastic voice and sang with great passion but halfway through her song, somehow she managed to part her hair in the middle. It parted sideways. I then realised that water was leaking from the ceiling and dripping on to her head while she was holding an electric microphone plugged in at the mains. Her audience looked on with horror and then dived in all directions for cover. The singer's eyes were firmly shut, she was still oblivious to the danger she was in. It was only when she finished her song and opened her eyes, noting that the clapping was fairly muted due to the cramped conditions now experienced by her audience that she realised. By that point, I was laughing so hard that I did not notice the man next to me had a Rottweiler, which bit me in the leg having been startled by my laughter.

JP, the manager, said that if we wanted to stay on for a drink later, the bar upstairs was doing a Hide and Seek theme night. He said that it was only open until six in the morning but we were still very welcome to go up. Tommy and I went upstairs; we were confronted by the sight of around a hundred Asian-looking men wearing turbans and lipstick.

"The thing about JP is that I can only understand every third word he says," I said to Tommy, as we both read the poster on the wall advertising in big letters 'Hide a Gay Sikh Night.'

JP, as always, carried out an equal opportunities policy and catered for all minority groups, if drink was bought. It seems that The Mayflower was the one pub outside the West End to allow gays of different races and creeds to have secret events.

"I don't care what a person's race or sexual orientation is, but you're more right-wing than me so I take it you're off?" I said to Tommy.

"Fuck that bollocks, it sells beer and it's open until dawn," replied Tommy as he ordered from an Indian man in a blonde wig and high-heels, "four pints love."

The gay Sikhs just ignored us while we propped up the bar, but I did hear one turbaned customer say to his partner as he passed us, "God, they're enough to turn you to women."

I told Tommy the story of my Uncle Paddy's experience of the war and he told me of his father and his brothers fight against the Nazis, and we agreed that we were thankful that their sacrifice gave us the freedoms we have today. I did qualify this, noting that that their sacrifice was not simply for gay Asians to sing in a chorus Erasure's pop-hit, *Respect*. However, I said to Tommy that it was an example of how all sections of society have the right to express themselves in our country, whereas in most parts of the world they might be executed by their governments if discovered.

"And us along with 'em. Stoned to death for being gay, we'd never live it down," added Tommy, forgetting that we wouldn't be alive to live it down.

We were still propping up the bar and, no doubt due to meeting Bourgeois that night, Tommy asked me how my friends from Warshalton and Trumpton, the Kids *en masse* would react in the event of a war if they were called up to fight for their country.

I thought of their various traits and said, "You and H are similar in many respects. You talk only when you have something to say, neither of you has time for bullshit and I know that when the chips were down you would stand together in the trenches."

"And Bear?" asked Tommy who, like other Trumptonites, never quite believed that Bear was real, even after they had met him.

"He would tuck his balls up his arse, buy a dress, marry a Swiss lunatic and declare himself to be a resident of a neutral country."

"Chelsea Mark?" Tommy persisted.

"He would urge us all to declare war on both sides and fight for a Socialist state, while booking himself a flight to Switzerland with Bear dressed as one of his bridesmaids."

"Brains and Jockey?"

"Brains would shoot a flare into the air to give away our position to the enemy in an attempt to try and trade his life for ours." I suggested. "Jockey too, but he would trade all of our lives just for a quick knee-trembler with a large female in the enemy ranks."

"Snotbox?" asked Tommy.

"Well, he would say that all wars where white man's wars, so he would tell the military to go and fuck themselves, not that they would understand a word. Afghanistan would be different though, as he would join the Taliban to protect the opium fields."

"Numbnuts?"

"Well, he would suffer the fate of many officers in a time of war and would be murdered by his own men. However, rather than being shot for being an incompetent officer, he would antagonise both sides who would call a truce to their hostilities, form a temporary alliance and train their guns on him at the same time."

"Surf?"

"Having sold the location of Numbnuts to both sides, he would later dig the bullets out of his brother and secure a good price on the metals market. He would then use the money to treat the rest of us to beer in memory of his fallen brother, while crying his eyes out at his loss. That is why Surf is the one I would want to stand beside in the trenches."

"Absolutely," agreed Tommy.

RAMPAGING ROGUES

"Go and never darken my towels again."

Groucho Marx.

Chapter 32 – *The League of Extraordinary Lunatics.*

The Connellys had another branch of the family living in Doncaster, who Numbnuts referred to as 'northern monkeys'. Their two cousins, Daniel and his younger brother Dillon, were the equivalent of Surf and Numbnuts. Daniel was like Surf, in that he always found an angle and knew how to seize an opportunity to make a pound. But Daniel was a socialist, for as an ex-coal miner he had seen how the mining communities of the north had been decimated under Thatcher. In this respect he was very different to the entrepreneurial Surf who was a staunch capitalist. Dillon, on the other hand, was like Numbnuts, who was no good with money and just wanted to have a good time. However, there was one difference, and again I think this reflected the demise of the mining industry around Doncaster. Community meant a lot to Dillon and his targets were more carefully selected, usually soft southern bastards like me. Numbnuts would abuse anyone.

Things never seemed to go right for Dillon, which meant I was drawn to him, as I wanted to be there when things went wrong. This was for the humour I could extract from the situation and fortunately, with Dillon I never had far to travel, in turn he had a brilliant quick northern wit. The Connelly humour was there and I remember when I first met the brothers, Uncle Charlie, who was well into his eighties came over, threw his arms around his nephews.

"How do you know these wankers?" he asked me.

Next to him was a little man, whom they referred to as Charlie's apprentice; Charlie often ran out of drinking buddies, as they frequently dropped dead in their attempts to keep up with his drinking. I didn't hold out much hope for the young kid swaying next to him, who though he was only in his early sixties, he seemed already to be walking towards the light.

As an example of how things tend to go a bit awry for Dillon, I'll tell you about his birthday when we found ourselves in a packed and vibrant pub in York after a day at the York races. The pub was playing seventies music, mostly songs that I'd hated when they had first been released, let alone when they were resurrected. There was at least a great atmosphere as always after a day at the races. Also, we had come up on top after a few wins on the day, except for Numbnuts and Dillon who had lost the lot, which made it even better; plus the pub was full of women, the lager was cheap and it had a pool table. So what could go wrong? Well, everything really. After an hour of drinking and excellent banter in the bar, the head bouncer came up to me to tell me that he had had to throw my "little grey-haired mate" out.

"Come on it's his birthday, it can't be that bad, what has he done?" I asked.

"Indecency," the lumbering, but friendly ox of a bouncer replied.

"Granted he is a scruffy individual, but we will tidy him up and I will buy a belt for his trousers."

"No, it's worse than that," said the bouncer.

"Christ, what are you doing letting sheep in here?" I replied, feigning outrage.

"Worse than that," he continued though he was now laughing and trying not to. "I was checking the toilets and I found him taking a photo of his dick with his mobile phone."

"Maybe he was just letting his mum know he was OK, they're like that up north you know," interjected Numbnuts. It was typical of Numbnuts to seize the opportunity to try to get us all killed.

"You're joking, where is he?" I asked the friendly ox.

The ox pointed out of the window to our little grey-haired badger friend, standing in a doorway across the road and trying to keep out of the rain since he didn't have a coat. Well, we all had to leave the bar and so we stormed across the road and challenged the birthday boy for the reason for our exit.

"You prick, what are you doing taking photos of your knob?" Numbnuts demanded.

"Well, I don't know, I was pissed, so it seemed a good idea to send a photo of my cock to my girlfriend," answered Dillon.

"I'll replace the 'Dillon wedding day' date entry in my diary with 'Dillon jail visit,'" I replied.

We managed to find another club that would let us all in. To my surprise, as we entered, a young woman suddenly ran towards me with her arms open only for the bouncer to rugby tackle her to the ground and then with his colleagues drag her back into the bar by her ankles.

"Looks like I am starting from scratch again. Is it gone midnight?" I said to Numbnuts.

"You were close there, Baloo, it was one of those rare occasions when a woman has drunk so much she is beyond caring."

"Ah, thanks, Numbnuts and I must apologise if I ever gave you the slightest impression that I give a flying fuck what you think. Any fear of you exciting a woman?" I asked.

"Well my last girlfriend nearly collapsed with delight when I fell into a workman's hole."

"Please God, tell me you're referring to one he had dug?"

After the club, we headed off to the Chinese takeaway across the road. In the queue, I was immediately accosted by a small old woman, who spoke so softly that I had to bend my head down towards to hear her.

"Do you like anal?" she then shouted in my ear.

"What?" I asked.

"I like anal," she continued.

"I'm very pleased for you, but sorry love I don't. Anyone of this assortment standing behind me would be up for it," I said, pointing to Chelsea Mark and Numbnuts.

"Bollocks," replied Numbnuts, though he did keep his hand in the air along with Chelsea Mark's.

Then another elderly woman came over and took the one beside me by the arm, "Come on mum, time to get you home," she said, then turned to me. "Sorry, duck she does get a little over-excited when she goes out."

"What a place, I do love it here," said Numbnuts, as Chelsea Mark moved in for a takeaway mother and daughter combo.

Dillon and I are good friends, but you would never know it by our conversation, which has always been pure unadulterated abuse. He is a graphic artist by trade, but a few years ago, he decided to say bugger it, and dropped out of the rat race to look after his dad who was dying. He also intended to live off the state and play golf. I actually respect him for making this life choice, not because (as he said) it was a personal statement saying there is more to life than work, but because he could be a contrary bastard, and it was only a matter of time before his work colleagues killed him.

On one of my visits to Doncaster, Dillon verbally laid into me over my hotel breakfast following a heavy night on the town.

"You sick bastard!" he shouted across the breakfast bar.

"I'm fine, thanks for asking," I replied cheerfully.

"That poor dumb girl you were chatting up!"

"Personally, I think she was very lucky to meet me." I replied.

"I mean dumb, as in she can't talk."

"If you mean Susan," I said, "the term is profoundly deaf, unless she once went out with you, then you would have the right to call her dumb."

"Poor girl, being with you means she has a drink problem as well," Dillon continued.

"Look, Susan has the same needs as any woman and she is entitled to the same pain and suffering that anyone else experiences as a result of knowing me."

"Well, she has an advantage over the rest of us, in that she doesn't have to listen to you, you Cockney prick!" Dillon yelled.

"I have no argument there," I added. I had to admit that she was indeed fortunate.

"How the bollocks did you chat her up? I was watching you and I couldn't work out how you two were communicating."

"You helped and you played your part perfectly." I informed him.

"What?"

"I was communicating to Susan, that you were my mentally-retarded brother."

What followed was some of the most appalling language I have ever encountered. Dillon was not a morning person, I thought.

Later that night in a bar in town, Susan came over to me. She asked me through sign language, in the sense that she had to point and exaggerate the movement of her lips, how my brother was.

"I have to stay close when he is like this, in case he stabs himself in the head with his fork again." I replied.

"You will burn in hell, if the devil can take the competition!" shouted Dillon across from the bar. He was then warned by the bouncer to behave himself or he would be thrown out again. I was puzzled by that, as we had never been in that particular pub before, perhaps Dillon has a photo album packed with pictures of his penis in various locations.

It had been difficult to interact with Susan, as I am not good-looking, I have no fashion sense or even a car; all I ever had was the ability to talk. Susan did, for some reason, find me funny and she soon got in on the joke that there was nothing mentally wrong with Dillon. Susan and I had a very brief relationship: it was a short, sweet, fun encounter. Susan was looking for someone for the long-term and, though we liked each other, being at different ends of the country, and unable to converse freely, meant that it ended before it really began.

Skinhead was getting married to a very beautiful Scottish girl, Irena, and we all headed to Blackpool to celebrate his excellent fortune. Skinhead was originally from Trumpton, but he had moved to London's East End, where he blended into his new environment, being very loud and bald. The weekend in Blackpool resulted in the largest gathering of Trumptonites since they had attended school together, and for me it was an excellent opportunity finally to meet some of the more notorious residents whom I had heard of and was looking forward to meeting.

It was also the afternoon of the World Cup quarter-final game between England and Argentina, so we were desperate to get to the pub before kick-off. We just made it, lobbed our bags into our hotel's reception and went straight to the adjoining pub. Surf was drunk already, as he had consumed eight cans of larger on the train and he was overwhelmed with the excitement of the build up to the game. All through the first half of the game he did not stop dancing; he was jumping up and down giving ear-piercing whistles every time the England team went on the attack. Sibling rivalry was ever-present and Numbnuts was keen to press the advantage, now that his brother was drunk. Surf leapt into the air whenever an England player touched the ball.

"Look at that stupid prick of a brother of mine," Numbnuts continued to tell everyone, relishing his brother's drunken antics.

At the same time, I noticed from my position across the bar, that every time Surf jumped up and down, the vibration gradually moved a foot-high concrete model of a lighthouse that was positioned on the edge of a shelf above Numbnuts. Just then, England scored and the pub went berserk but no one more so than Surf, who leapt into the air and then landed with an almighty thud on the pub's wooden floor. Just then as Numbnuts shouted across at his brother, "You stupid prick!" The vibration finally toppled the lighthouse off the shelf and hit him right in the middle of his head. His feet disappeared from under him, as the weight coupled with the shock knocked him back against the bar.

Quickly, Numbnuts tried to compose himself in the hope that no one else had seen what had happened, but he could see that I had had a clear view. Numbnuts' pride would not let him acknowledge that his brother, who was as drunk as a high court judge, had somehow got the better of him; he refused to acknowledge the pain he was in. Though still crying with laughter I decided to look away, knowing that Numbnuts would only let his guard down if he knew no one was looking. I was still able to keeping him in sight as I sat looking up at the panoramic mirror positioned above the bar. I listened to Chelsea Mark informing the barmaid that he was a multi-millionaire and that he had just moored his yacht up by Blackpool Tower. Now Numbnuts, thinking that no one was looking, seized the chance to clasp his head in his hands.

He bent double and let out a muffled, "Jesus Christ!"

When Numbnuts straightened back up, I was there to welcome his return with a big smile and a little wave.

As the first surge of Londoners drank at the bar after England won the match, the next projectile of London vomit was thrown up on Blackpool's doorstep as the train from Euston pulled in at the station. Fifteen minutes later, another wave of luggage was lobbed into the hotel reception area. Snotbox entered the bar and I then had my first ever sight of Dale, sober. I then learnt that Dale really could start a fight in a room on his own. Dale walked straight up to the pool table, where Slugger was just about to win the game by potting a very simple black ball balanced over the pocket.

Dale picked up the black ball and dropped it in the pocket, went up to Slugger and said, "You Fenian bastard!" He then he turned to me, who he must have remembered and said, "I'll stick the cue up your arse, you Chelsea prick."

Slugger came over to me, "Well now you have finally met Dale sober,"

"Yes, I like both versions." I said.

Dancing Spice then turned up with an inflatable Frankenstein; none of the others batted an eyelid, as it was accepted that he was a total nutjob. You may have seen him on TV whenever Arsenal made it to a FA Cup Final, if you can remember that far back. He always turned for his team's games dressed as Elvis, for some unfathomable reason. For the rest of the weekend, Dancing Spice got in and out of his hotel room through the window, wearing infrared glasses, but again no one seemed surprised by this apart from me.

Jockey then leapt off the train onto the main platform; he was like a horny tartan Tasmanian devil, just released after twenty years of solitary confinement. Once he had thrown his case on top of the growing pile in the hotel's reception, he joined the growing throng in the bar. I knew that the night would be a wild one, as it was only three in the afternoon and the newly-arrived guests at our hotel were already negotiating their way over a mountain of cases and trying to locate reception.

Later we found our rooms, got cleaned up and headed into town. We found a club and all the others were inside, but I was trying to help Snotbox who had already been refused entrance.

"Is it because I'm black?" he demanded to know of the bouncer at the door who was barring his entry.

"No, and in case you hadn't noticed I'm black too, but you can't come in, in those jeans," decreed the bouncer.

Snotbox then went back to his hotel to change and returned to the club an hour later. The bouncer tried to explain that jeans were not allowed to a bemused and confused Snotbox.

"But you said 'not in those jeans' and these are not the same jeans, look I have a hole in the arse of these ones," pleaded Snotbox, turning around to provide evidence to the bouncer. The bouncer couldn't believe what an idiot he had in front of him, he started laughing, gave up trying to explain the club's policy to Snotbox and let him in.

I went up to Surf at the bar, "I'll get a round in so we can toast England's victory."

"Christ! Did England win?" Surf cried out and then said we should get drunk to celebrate.

Chelsea Mark walked into the club, after having disappeared with a woman he had met earlier in the evening. He then proudly told us what a wonderful woman she was, that it was the best sex he had ever had and that he was in love.

"Wallet gone?" I asked.

"Christ yes, can someone buy me a drink?" was his reply.

Jockey joined us and immediately attempted to chat up a young woman at the bar.

"What do you do for a living?" he asked her.

"I'm a barmaid," she replied as she put the pint he had just ordered on the bar in front of him.

Slugger was the next to get married and he asked a few of us to go for a weekend in Dublin as his stag do. Dublin was a popular destination for stag weekends: groups would land and drink themselves stupid for two days. It wasn't that we needed drink; we were in stupid mode right from the start. Me, Surf, Numbnuts and Fat Bloke had agreed to meet Snotbox in The Bobbie on the Friday morning, as he had offered to drive us to Stansted Airport. Snotbox was late and turned up straight from work, wearing a white hat and white overalls that were covered in blood.

"I didn't know you were a butcher?" I asked.

"I'm not, I'm a baker," replied a confused Snotbox.

"He cut off one of his bollocks last week, with a cookie cutter," added Numbnuts.

"Fuck off," replied Snotbox, who appeared more bunged up than usual that morning. Later we found out he had accidentally snorted baking soda out of the wrong bag.

The two-hour journey to the airport took us four hours for many reasons. Firstly, Snotbox drove to Ilford to collect his case for the weekend, but he had forgotten that he and his girlfriend had moved to Romford a month earlier. Secondly, every time we pulled up at traffic lights, Surf, who was sitting in front next to Snotbox, would apply the handbrake. This meant that each time Snotbox would try to pull away, the back wheels would spin like a dragster on the starting line. Snotbox never once realised that Surf did this every time. We sat at the lights, the tyres spinning around, smelling burning rubber and surrounded by black smoke, until Surf released the handbrake.

"Fucking hell man, I don't know my own strength," replied Snotbox.

When we arrived at the airport car park with completely bald tyres on the car, Snotbox turned to Surf.

"I have no fucking tax man, so I need a place to park this out of the way for the weekend so one will notice."

"No problem, just park over there in the no-tax-disc parking space,' said Surf.

I looked over; the sign over the parking bay read: 'HERTZ STAFF PARKING ONLY – CLAMPING IN OPERATION 24 HOURS A DAY'

"Ah well, Snotbox is another star graduate from Trumpton's educational system," I said to Surf as Snotbox parked his car under the sign.

Surf replied, "No, he can read alright, he's just as blind as a bat." He then turned to his friend, "I will leave a note in the empty tax-holder to say 'tax applied for.'"

"Thanks man," said Snotbox, as Surf slipped a note into the tax-holder with the words 'FUCK OFF' written in big letters.

"Don't leave your driving licence in the glove compartment, in case you need something to light your joint with on the plane," I added, as Snotbox, unlike the rest of us, loved to smoke a joint when he wasn't stuffing powder up his beak.

"Licence? I don't have a licence man, if I did they'd only take it away from me," replied Snotbox, looking at me as if I was mad.

We finally managed to wrestle a joint out of Snotbox's hands and get him on the plane. He just didn't care and he told me once, 'Fuck the legal system man, drugs didn't do me no harm.' It was well known that he always had a spliff on him in the defendant's box, where he had appeared on many occasions for possession.

"Why are you reading a paper that only has Irish news in it?" Snotbox asked the passenger next to him.

"It's an Irish paper," the bemused passenger replied.

"A bit weird to be reading a paper about Ireland man, why do that?" continued Snotbox.

"This plane is going to Ireland," replied the passenger, who was getting increasingly nervous.

"Fuck me man, are we?" that was news to Snotbox.

When we arrived, Skinhead was already in our hotel and on the telephone to the other guests.

"'Tis is the management, yer haven't finished yer breakfast, so get yer fucken arses back down here now and eat your *mate*," demanded Skinhead in a terrible O'rish accent. While we were unpacking our cases, you could hear terrified guests running up and down the stairs. They were trying to keep the manager happy and avoid being thrown out with their bags.

I showered and then set off to meet the others in a nearby Dublin pub. Whenever we head off *en masse*, as we did that afternoon into Dublin's Temple Bar area, it's not long before we split up into little groups engaging with the locals. That was due usually to getting sidetracked by women as the others moved on, or if some of the larger group liked a particular bar and stayed put.

That evening, Snotbox and another friend, Duggie, were ahead of us, as both were fast drinkers and were keen to explore this new town. Then the trouble began when the barman refused to serve them, making it clear that chatting up Irish girls in his Dublin bar would not be tolerated.

"Is it because we are black?" Snotbox asked the barman.

"Yes," replied the barman.

Snotbox turned to Duggie, "He's got a point, man. We are black."

"Thick racist idiot," replied Duggie.

Snotbox turned to the barman, "He's got a point man, as you are a thick racist idiot."

Three bouncers now joined the barman; surrounding Snotbox and Duggie, they were very keen to teach the two black men a lesson. At that point, Surf and I walked in. We sized up the situation quickly, I moved into the centre of the group to join our two friends, giving Surf a chance to send a text to the others that Snotbox and Duggie were in trouble. Surf then joined us in the centre of the circle formed by the barmen and bouncers. More barmen joined the group surrounding us, but we weren't going anywhere.

Then more of our group descended on the pub, with Slugger entering first with Brains. Now Slugger's nickname was not ironic, he is a good man and he does not seek trouble, but God help you if you start it. Brains too was not a little lad though he would disappear at the first sign of trouble quicker than shit off a shovel. The staff of the pub really started to worry.

Brains added to their anxiety when he called out to his best mate Snotbox, "Don't worry, I'll jump in and save you but only after they've buggered your little black arse off."

"You'll be up there too, you fat bent honkey!" Snotbox shouted back to his friend.

Though still outnumbered, the balance suddenly turned in our favour when JP, the manager from The Mayflower and his equally handy mate Luke pounded through the doors.

"Is it a scrape they're fucken wanting?" asked JP, keen to become the worst kind of customer a pub could have: well, he was on holiday after all.

Now that the odds had changed, though we were still outnumbered about three to two, the bouncers and bar staff backed off and kept their heads down for the rest of the session. They knew that if they forcibly removed us we would rip the bar out of the floor and take it with us. The women Snotbox and Duggie had met were delighted to join us when we set off to the next bar. We separated again into various groups and met up from time to time as the night went on.

Over the years, I have heard Brains and Snotbox deride each other with racial abuse; it's a personal dialogue they have between them. However, if any remark is made to my friends in a malicious sense, the person making it had better have an army behind them. When I look back at that afternoon, the locals must have been perplexed to see born and bred Irishmen, such as JP and Luke, standing alongside what they would call English guys, like Surf and Slugger, to fight alongside their black mates Snotbox and Duggie. The dialogue between the Kids is blistering, often cruel to an observer, but when one of them is in trouble their friends will emerge, even Jockey and Brains, if only to catch a glimpse of a woman in uniform.

Perhaps the funeral of the Connelly father in Dublin sums up the fusion of the Irish love of the 'craic' and London's edginess. I hadn't known their father but by the time of his death, I had gotten to know all of the brothers and sisters. I went to the funeral with Tommy, who had grown up with the family.

It was an unusual funeral: the coffin was transported through Dublin's busiest street, O'Connell Street, by horse-drawn carriage. We all walked slowly behind the hearse, as the funeral procession made its way slowly through Dublin's city centre.

I turned to Numbnuts and in a low, respectful voice said, "He must have loved horses?"

"Not really, but he fucking hated Dublin, so he just wanted to clog the streets up and stop the traffic," replied Numbnuts in his naturally loud and aggressive voice.

"You must have been your father's favourite," I replied in a low voice.

At the graveside, once the coffin had been lowered into the grave, it is the Catholic practice for each mourner to pick up a handful of soil and gently throw it on to the coffin. I was distracted from this part of the ceremony by Tommy, who fancied one of the nieces of the deceased. Tommy is naturally more London than the rest of us and I think even I was slightly surprised later that night, when everyone was sitting around the table, reminiscing about the deceased, and Tommy made the comment, "Christ, this is boring."

Suddenly, there was a huge crash and I just caught sight of someone who had run up to the grave and thrown something large on to the coffin. Colleen, the youngest

Connelly sister, informed me that it was her dad's best friend, Gerald Less. Colleen explained that Gerald could not bear to be at the service and see the body of his good friend interned, but at the last minute, he had run to the cemetery to throw soil on the coffin as a gesture of farewell. In his haste, whatever he picked up was more like a paving slab than a handful of earth, hence the loud bang as it landed on the coffin.

Amidst all the confusion, Surf turned to me, wiping a tear from his eye with his handkerchief.

"I'm glad I did not go for the cheap plywood," he said.

For this reason, I believe we are collectively London Irish. Perhaps it's just the black humour, as Irish comedy in my experience is usually respectful, but with the added London influence we will go for the jugular. We are a mixture of two very distinct cultures, encompassing I believe the best of their traits. Others I am sure believe we are mutants, encompassing the worst of both.

Chapter 33 – *The Empire Strikes Back*

Surf worked for an American company and he was asked if he wanted to go to their annual St Patrick's Day party on 3rd Avenue, and told that he could bring a few guests if he liked. The largest expedition in Trumpton's history was soon assembled. It also coincided with Tommy's birthday. When one first meets Surf, you meet a seemingly well-adjusted man, happily married to Ann, with two great kids and a well-paid responsible job. But he is a rogue and a benevolent one too, as he raised funds on numerous occasions to finance my medical aid convoys. Here he was, bringing along the Kids to his employer's high-profile social event on the other side of the Atlantic. Surf cannot fight his nature, like me he cannot help but mix different characters together to revel in the mayhem that will undoubtedly ensue.

Everyone knew that the casualties would be high, many a liver would be lost, all would at some stage lose the ability to see and many an open hole in the sidewalk would fill with toppling bodies originally from North London. Even veterans from battles long past came out of retirement or cold storage. In the case of Tom the Bomb and his best friend Don, they were both were in their seventies and turned up to set forth on one last voyage to the Americas. It would be messy for sure and with Tommy, Brains, Jockey and Snotbox among our number it would certainly be ugly.

Tom the Bomb and Don were the first problem; due to their age, getting their travel insurance was near impossible. In the end, I managed to secure their cover through the ASHA supermarket website.

"What do I need to do now, son?" Tom the Bomb asked me.

"All you need to do is take Don, go down to the supermarket and get your heads bar-coded, so that you can be scanned by the funeral director in America if you drop dead," I replied.

"Fair enough, son," he said and then he took another drink from his pint, "You little shit-pot."

On our foreign excursions it was the ritual that I got to the airport first to engage with the airline booking staff and gently prepare them for the shock of the rest coming around the corner. I got talking to the nice woman manning the British Airlines desk.

"How many are in your party?" she asked.

"Well the first wave should arrive any moment," I said.

She was laughing just as the first wave crashed up against her desk; our little Scottish experiment, who was already drunk, toppled head first into it.

I watched Numbnuts and Tom the Bomb drag Jockey out of the boarding desk by his arms. It looked like a further instalment of the *Rocky* films, number seventeen in this case, as Numbnuts had been on the booze all night and looked like a punch-drunk prize fighter who never had reached his prime. Jockey looked like the stereotypical bloated coach in his corner who had just accidentally taken a punch for his fighter. Tom the Bomb in his big Panama hat made Don King look more respectable than the Archbishop of Canterbury. The woman at the check-in shook her head when Jockey tried to turn on his charm and ask for an upgrade even though there was an imprint of his head in the front panel of her booking station.

"I would downgrade you if I could," she replied in a very professional and courteous manner.

It didn't help that when she asked the oldest gentleman in our party for his name, he replied, "Tom the Bomb and I'm ready to go."

I asked her for a feedback form to acknowledge the excellent service she had provided, the excellent service being for those in Executive Class who were spared the pleasure of our company. She laughed and let Jockey give her a hug; we are quite a huggable lot once the fear subsides. All this was before security was stepped up after 9/11; if we did that now all four of us would be stuffed and mounted ready to be exhibited in the George Bush Junior Museum of Terrorists in Bum Fuck, Idaho.

At the security check-in point, I told Jockey that there was a special entrance for drunken Scottish passengers and pointed him to the luggage portal next to the metal-detector. My good friend then waddled away to try and squeeze through it, but he got wedged between a handbag and a cello case. How he got out of this country and into the United States exposes a major flaw in our security system; incredibly, not one member of airline security envisaged our little drunken experiment as a major threat. Every official, excepting one on the American side, just shook their heads and laughed as he zigzagged his way across the Atlantic. Al-Qaeda could bring the entire western world to its knees by simply opening a recruitment office in Glasgow. An offer of free beer and travel mixed with the following helpful advice from their travel agent would set the enterprise on the desired track:

"Now here is the bottle of scotch and the package I want you to drop off to a friend of mine who lives in a big white house in Washington. Did I tell you that my sister is sweet on you? Two tickets will be waiting for you to take her to a Rangers' game when you get back. Don't forget Allah's ... err, sorry mum's the word."

Then the birthday boy appeared, looking like he had been shot out of a cannon. He had started off with us, but we'd lost him when he got to the airport as he'd had to buy some underpants. He was still drunk but, after a night in the cells, he didn't want to disgrace his family and he made sure he would be wearing clean underwear if he was knocked over by a Mack truck. Tommy stood there with a small battered case, which looked like a make-up bag he had just retrieved from a lost property office after years of neglect. I should mention that the night before we had enjoyed a little gathering at The Mayflower. I was going out with Kathleen at the time and she decided to come along, which was great, but unusual. Although she liked a good laugh and a few beers with her male friends, my Kids were in another league entirely. Kathleen had always maintained a discrete distance so the only ones she had met up until then had been the Connellys. Kathleen, though she was too much of a woman to say it, realised upon meeting Brains for the first time that his reputation was well earned and without him even opening his mouth she turned to me.

"His insurance will not cover what you and Numbnuts are going to do to him," she said.

What was wonderful about Kathleen was that, rather than keeping Brains at an even greater distance of a mile or two; she became quite protective and of her own volition she decided to sit next to him. Perhaps it's like a mother's maternal instincts on discovering an abandoned child playing on a railway line.

"I can't get me cock up," were Brains' first words to Kathleen.

"Oh dear, I'm sorry to hear that," she replied politely and without irony, no doubt wishing that she had kept her maternal instincts to herself.

"No, I can't get it up anymore!" stressed a dejected Brains. "Your Baloo and Numbnuts told me about Viagra, that's why I am going to America; mind you, I hear it's some idiot's birthday."

Kathleen looked straight at me; Numbnuts and I looked away, attempting to avoid her scowl. Meanwhile Tommy was looking a little bit miffed at having overheard Brains' birthday greeting.

Brains continued, "Numbnuts told me to stick the Viagra capsules down my 'jap's eye' that's until there's enough inside to make it hard, that's how it works you know. Geology is just amazing isn't it?" Brains then put me in Kathleen's line of fire, "Baloo warned me that whatever happens, I've never to swallow it because with my metabolism, I would end up with a stiff neck, leading to brain damage from my head knocking against low ceilings."

Kathleen rounded on me and Numbnuts, "He will hurt himself resulting in complete renal shutdown, so you'd better tell him the truth now you pair of idiots!"

Kathleen then disappeared to the ladies. Now that we had received our orders, Numbnuts and I proceeded to clarify the instructions previously given to our flaccid colleague. Kathleen returned and was pleased, when Brains turned to her.

"Thanks, Susan, for sorting out the Viagra thing." he said.

"So you know what to do now?" she asked, ignoring her new name.

"Yes, Lisa," replied a very smug Brains, "Baloo told me to stick the whole box up my arse."

For the rest of the evening Brains sought my advice on a number of matters, ranging from what language Americans speak, to how to operate the hand activated dryer in the pub toilet. I can barely look after myself, let alone anyone else, yet earlier in my life I had had to look after Bear. I knew that I would have to keep a watching brief on Jockey and now I found myself looking after Brains. It would be nice if one day I had to mentor a piano-playing protégée, but at this late stage of my life I've only had these three geniuses to my name. Brains' reliance on me for reassurance and guidance over that weekend was a burden that Numbnuts said he was happy to share. That, in part, explains Kathleen's last comment before I left, "The US President is giving his State of the Union address to the Senate tomorrow; I will murder you both if Brains streaks naked past him."

Kathleen and I decided to leave after our quiet drink and we left the others to order their tenth round. Kathleen was not speaking to me, even though I offered to go back and tell Brains that he could take the Viagra suppositories out of the box before insertion.

The next morning the mini-bus arrived with everyone on board.

"Morning, anyone had breakfast?" I asked.

"Me, I had a cooked one made for me," replied Tommy.

"What's open at this time of the morning?" I asked, then I realised that only in Her Majesty's custody would you receive beans and a sausage at five in the morning.

Surf then told me of the events that took place after Kathleen and I had left The Mayflower. They had all gone to The Arch, and Tricky, Tommy's mate, had joined them.

Tommy had been welcomed by the bouncer with the standard greeting, "Can't let you in tonight Tommy, you were a handful last time."

The definition of 'a handful' was that it had taken five bouncers to drag him out the night before. Tommy then gave his traditional response of "Bollocks," as he walked straight though the two bouncers standing in front of him. That was not the start of the trouble; that occurred a few hours later when Tommy hailed a mini-cab and he was quoted a fair of thirty pounds to get home, even though his house was only a mile down the hill. Tommy was never one to get involved in protracted negotiations, so he proceeded to tear the bonnet off the mini-cab.

After his arrest, Tommy was asked if there was any reason why he should not be locked up for a month and he replied, "I have to get a flight to New York in a few hours and if I miss it, I won't be happy." The local police knew that that meant they might as well have a football team of Millwall fans in the cells. He asked if they could give him a lift home to collect his bag. They did.

On our way to the airport, I asked Tommy why he was known to some in Trumpton as 'Bungalow Bill'. Tommy then informed me that the nickname first came about because the actress Joan Collins had had a big boyfriend who was nicknamed Bungalow Bill. Tommy added that he was given the name because of their similar build and because both were supposedly quite large in the 'bedroom department'. Brains then made his own contribution and said that he thought the nickname was due to Tommy having nothing up top. Tommy gave Brains a stare, which made me think that as both of them were knocking sparks off each other already, the trip would not be without incident.

Fat Bloke and Numbnuts dragged Snotbox and Brains out of the airport's business lounge where they had each drunk a bottle of champagne purchased on Surf's business club card. The commotion that followed in getting them seated was therefore to be expected. When the Chief Steward asked all of the passengers to pay attention to the safety instructions, Numbnuts turned to me.

He added his own guidelines, "Never wank your cock with a steak knife in your hand," he said with a nod and a knowing look in the direction of Brains, implying that he knew someone who had.

Tommy, meanwhile, was still drunk. The stewardess noticed that he had a litre-bottle of Smirnoff Vodka tucked into the side of his seat.

"I'm sorry sir, but you can't consume duty-free on the airplane," she said, as Tommy lifted the empty bottle of vodka and gave her a big stupid drunken smile.

I had my own issues. The Hasidic Jewish gentlemen sitting behind me had hailed the stewardess to complain that because of his religious beliefs he could not sit next to a female and therefore he wanted the woman beside him to be moved. The woman, who I learnt later was also Jewish, went ballistic and called him every name I knew and few more that I noted down and still use on occasion. Whatever the scriptures say, their interpreters over the centuries (men to my limited knowledge) have taken the opportunity to subjugate women to cover their own inadequacies. I once encountered a devotee of the god-squad on a train carriage. He, without giving me any warning, started to read to me from his bible in an attempt to persuade me that women could not be ordained as priests. I decided not to engage in conversation, as it was like trying to have a sensible discussion with a Nazi or a member of the flat earth society.

I decided to assist the distressed Jewish gentleman and as I was sitting in front of him, I offered to swop seats with him. The stewardess was relieved that I had resolved the situation. The woman had been swearing so loudly, that the rest of the passengers on the plane had been diverted from looking at Brains and Snotbox trying to fasten their seatbelts by tying them into a knot. However, as with all my plans, it had a slight flaw. My seat was in the middle of the row and the gentleman who was to take my seat needed my drunken friend Brains to get up from his aisle seat to let him in. As I took my seat, I thought it would be an interesting exchange and indeed it was.

The Jewish gentleman ended up shouting at Brains to let him in, but Brains had finished off another bottle of champagne and he was oblivious to the man's protestations.

Another stewardess then informed the Jewish gentleman that he could not stand in the aisle as the plane was ready to take off. I just smiled at him and made it clear that I was not moving. He really started losing his temper at that stage, which resulted in him being forced to take a seat next to a large black woman. This wonderful lady, I was to learn, was called Annie, she informed him immediately, "that if he gave her any shit she would mount his head."

Gillian, the woman who had originally been the subject of his request to be moved, told me that she was Jewish but she couldn't take sexist crap dressed up as religion. Gillian was a very funny woman and, while her boyfriend sitting by the window kept himself to himself, we, along with Annie, got a little party of our own going. Actually, Gillian's boyfriend did aim a look of disapproval at us once we got stuck into a bottle of champagne.

He soon went back to his paper when Gillian turned to him and said, "Why couldn't you stand up for me and be a man like him?" she pointed at me and continued, "When are you going to grow a pair of balls?"

Meanwhile, amongst my compatriots the problems continued to mount. The stewardess dealing with Brains and Snotbox was particularly stressed.

"These brown sugar cubes are stale," Brains informed her.

"Sir, you have put croutons in your tea," she replied, with considerable patience.

Snotbox was also having trouble trying to turn on his in-flight television. He called her over, "Babe, this remote control for the TV isn't working!"

"Sir, I think you should give the little boy next to you his Transformer toy back," said the stewardess.

"Ah, shit, sorry, kid, here you go," said Snotbox, handing Megatron back to the kid while being watched by his opened-mouthed mother.

Things did not get any better when we arrived at JFK airport in New York. Jockey was technically what we call absolutely bollocksed by this stage, having consumed copious amounts of free whiskey on the plane. I had to carry our little Scottish experiment into the arrivals hall by his collar, as one would hold a toddler by a walking harness. For those not familiar with procedure at US immigration, only one person at a time is allowed to step beyond the white line up to the immigration officer's desk. As the official waved our little Scottish friend forward, he looked at me panic-stricken at having to negotiate those ten steps all on his own.

"Go on, it will be fine," I said, but he just looked up at me like a little kid on his first day at school. "No, go on, he won't bite," I said, though like any parent I lied. The US official looked like he might eat our little Scottish experiment without salt. "Go on, I will buy you a pint and a wee sharpener on the other side," I insisted, just like any doting parent bribing their offspring to go through the school gates with the promise that their favourite dessert would be waiting when they get home.

Jockey gradually put one foot uncertainly in front the other. It felt like my charge had regressed to his baby years and was taking his first steps on his own. He staggered a bit and at one stage veered off into the wrong queue, but we were all so proud of him when after five minutes he had covered the ten yards. All of us broke into applause.

"I'm so proud of our little Scottish experiment, mother," I said, turning to Numbnuts.

"I think I am going to cry, I do hope they apply a glove to his hairy bottom," Numbnuts replied, wiping away an invisible tear.

"Put your left index finger on the ID pad sir," ordered the US immigration official.

Our little Scottish experiment was completely confused by which finger was his index finger. What confidence he had had after his trek evaporated. To the disbelief of the US Immigration officer, he put his little finger on the screen. The officer had to repeat his request three more times until finally Jockey put the correct digit on the screen. The US Immigration officer then asked him to repeat the process for his right hand. Jockey turned to look at us for reassurance, as he hesitantly lifted his right hand. With a combination of pride and piss-take, Numbnuts and I hugged each other following that achievement. It still took him a further four attempts to find his index finger.

The far-from-amused official continued, "Now, sir, can you look into the camera?"

Jockey's confidence returned following his recent success in finding his right hand and he fearlessly leant forward, grabbed the flex-stand which contained the tiny camera and pulled it towards his eye to look directly into it.

After a minute, Jockey shook his head and said, "I cannut see anything pal."

Without the bat of an eyelid the US Immigration Officer called over to his colleagues on the intercom, "This must be the dumbest idiot we have ever let in the country."

Jockey turned to us and gave us a big smile and the thumbs up sign, as the only part that mattered to him was that he was going to be admitted into the country.

However, whilst we all celebrated, Brains was despondent. He was the unanimous holder of the 'dumbest idiot' title in England and had lost the world title to the new kid before he had even had a chance to enter the contest. As for my good friend Bear, he has never been to the US, so technically he is what the American boxing fraternity would call a title contender.

After getting through customs we all piled into cabs, with Brains jumping in through the open window of the one Tommy and I were in. He was very excited at seeing the Chrysler Building for the first time.

"Look that's the building that Godzilla ate!" he shouted.

Once again I had nothing to say. Only Brains could utter a few words that left you with so much to correct that it was impossible to know where to begin. I offered our Russian driver a few bottles from the hundreds I had stock-piled for us for the weekend; I had taken them from the gallery on the plane. Probably not a good idea; when we shot through the toll bridge Tommy suddenly sobered up.

"Don't give the Cossack any more drink!" he screamed, "he's more rat-arsed than I am."

When we got to our hotel, we all fell out on the pavement and were then picked up by our Cossack who gave us all a big hug; he was even more delighted when we gave him ten more miniatures.

Tommy was not so generous with his tip, "If you ever offer to drive me again," he said, "I'll fucking dropkick you back to Siberia." Our drunken Cossack gave him a big drunken hug too and then he fell back into his cab facedown and went to sleep.

Surf is the cleverest of the Kids, so while I had taken care of booking the flights, he had taken charge of the hotel bookings. On arrival at the hotel the receptionist confirmed our bookings for the weekend of the seventeenth of February, the slight problem being that we were there to celebrate St Patrick's Day which was in March. I have never known Numbnuts to be as happy as he was at that moment. He launched into a tirade, mocking the stupidity of his brother for getting the dates for the celebration of our ancestral patron wrong by an entire month. It was as if he had waited his whole life for that moment.

"You stupid prick and you're supposed to be the brains of the operation." Numbnuts said, with great relish.

I had a back-up and secured the Carling Arms Hotel; it certainly met our requirements, though Jockey was later so drunk that he returned to our original hotel to argue with the receptionist for a key to his room. The Carling is legendary in New York for having been decorated by various artists during the drug-crazed sixties. Every room had a theme and I found myself residing in the 'Cunnilingus Suite'. I managed to get the one room with four single beds as every other room had only two doubles. As the others with me were Tommy, Brains and Snotbox, the thought of sharing a bed with any of them was like playing Russian roulette with every chamber containing a bullet. I threw myself on to the bed nearest the door and decided to grab a quick hour's sleep and try not to think about what sounds of the wild would shortly wake me up.

I was soon awoken by Snotbox, who must have been watching some hard core porn on the cable television, as he was shouting out, "Go on, fucken give it to her, big boy." As I had no hope of getting back to sleep, I lifted myself up on my elbows to see that he was watching a funeral scene from the TV soap *Dynasty*.

After our arrival in New York, I knew I needed a break from the others before the serious drinking began. The only sanctuary was the bathroom, so I bolted the door and drew a long hot bath. As I eased myself into it, Tommy broke the door in and parked his backside on the toilet. I will spare you the details, but it wasn't only the broken lock on the toilet door that indicated that he was in a hurry.

This was probably the most degrading position I had ever been in, in my entire life. I immediately jumped naked out of the bath, wrapped a towel around myself and set off on a quest for sanctuary. I finally found a shower cubicle down in the hall by the reception. I stripped off to have a wash and get some peace for the first time in twenty-four hours. Then there was a banging on the door. I wrapped a towel back around my waist, which is more than can be said for Brains who was standing in front of me naked.

He was breathless, "Baloo, I just ran back from Times Square."

"Naked?" I asked, feigning surprise.

"No, even I'm not that stupid, I'm only naked now," he said just as two mortified Korean girls were trying to squeeze themselves and two large suitcases past him in the narrow public corridor.

He continued, "I was all sweat from running so I stripped off, but I wanted to tell you I changed my American Express Traveller's Cheques in the only place that would cash them in the whole of New York."

"You're telling me that only one place in the whole of New York, the capitalist capital of the world, at midday on a Thursday, would cash your Traveller's Cheques? And this was an office three miles from here in Times Square. Let me guess, who told you this?"

"Numbnuts. He said it only opened for ten minutes each month, so I nearly collapsed a lung running there. Anyway I will wait here until you're finished, so that I can have a shower," said Brains.

Fortunately I had a book, so it was good to finally relax for a few hours and immerse myself in it, though I was interrupted every five minutes by the screaming of people who had turned into the corridor and caught sight of Brains in his natural state.

It was now time to get out and take over a bar in New York, so we all headed to the street to hail a couple of yellow cabs. Numbnuts was first to get a cab but said he had to go somewhere on his own.

Numbnuts gave the driver his instructions, "Quick get me to the best 'titty bar' in town."

He had unfortunately picked a Sikh who was probably the most devotedly religious cab-driver in New York and he replied, "I know no titty bar."

Numbnuts jumped out, "I guess I'm back with the animals."

It had been a while since I had been to New York and I was bemused to find that a number of the inhabitants of Manhattan had been struck by some form of facial paralysis. I asked myself a number of questions – had visitors from another planet landed and adopted our form? They hadn't got the mouth right and had just glued it on at the last moment. Or could it be simply that the rich, as they seemed to be the ones most afflicted, had been shot with some form of tranquilizing dart and therefore could not express revulsion later that night when Numbnuts broke wind in a salubrious, though crowded, bar? This scary affliction later reached our shores; the medical term is Botoxed, it has harmed more people than E. coli and is becoming as prevalent as malaria. I cannot watch an episode of CSI, without thinking it's a science-fiction programme about alien invasion and wondering which part of a character's arse is now wielded to their nose.

The first bar we hit was The Rodeo: a favorite haunt of mine. It was ten in the morning, so we were its first customers. The bartender was a young, fresh-faced college student and it turned out that this was the first day of her new summer job, so she would welcome a nice easy crowd. We were not it.

Brains was first to the bar, "Do you have a fire-extinguisher in the building?" he asked, resulting in the barmaid having her first panic attack of the morning.

As the morning moved into the afternoon, Brains turned to me, "I think she likes me."

"Why do you think that?" I asked.

"She hasn't called the police."

"Marry her." I advised.

Brains continued his conversation with the barmaid, who really did seem to like the clowns from that little island trouble across the Atlantic.

"What's that flag?" he asked her.

"That is the Lone Star of Texas."

"Why have you got that big picture of Woody Allen up on the wall?"

"That is John Wayne and he starred in the film, *The Alamo,* based on the fight for Texan independence," she told him.

"And the reindeer antlers?" he persisted.

"They are from the Texas Longhorn."

Brains nodded his head in approval. "Yes, they are a great band," he said, referring to the Scottish pop group.

Two elderly women walked in; if their make-up had been expertly applied they might have passed for being in their late fifties, however it looked like they had slapped it on each other at point-blank range with a paint-gun. Tommy was drunk again; with the amount of alcohol in his body, it only took a few drinks to top him up and he started to chat up on the shorter of the two women who had liberally splattered herself with vibrant crimson lipstick. Brains in turn hit on her blonde friend whose Botoxed face made her look like a display model in the shop front of a pawnbrokers; she was adorned with pearls the size of billiard balls and earrings almost the size of bicycle wheels.

The beers were flowing, with the barmaid, whose name was Vicky, throwing in free pitchers for good measure.

Tommy turned to me, "Our two birds have been in the toilets downstairs for a while."

I put the flat of my hand up to his face and said "Our! Don't speak to me again today."

Surf turned to me as I brushed off every attempt Tommy made to engage me in conversation and said "You are the bravest man alive."

"Birds?" added Brains who had taken over the conversation with Tommy, "I must have missed them, but the Joan girls, Collins and Rivers, have been downstairs for a while."

"Fuck off, I'm serious, do you think they're OK?" replied an anxious Tommy.

"Could be anything, the stair-lift might be broken, or they're calling their grandchildren," added Brains helpfully.

"Maybe she is stuck down there, after she shat her false leg down the pan?" added Numbnuts.

"Oh, I get it," replied Tommy.

When Tommy said those words it was his standard warning that it was about to kick off, but in our case it was his acknowledgement that he was the subject of the piss-take and he accepted it with bad grace.

We headed off to the The Old Peculiar Bar in Bleecker Street, again another favorite of mine. It was early evening and the bar was packed. Numbnuts tried to get the waitress' attention by bouncing peanuts off her head, but his actions, like our banter, seemed not to cause offence.

"You guys!" was her response, followed by four pitchers of beer landing on our table with such force that lager was dripping off all of our noses.

I gave her a big tip and apologised for Numbnuts' behaviour.

"No problem, we have pricks in America too and if he does it again I will have his fat hairy arse kicked," she said.

Numbnuts, who has the ears, but not as Brains says the 'trunk' of an elephant, whispered in my ear, "I love her. How did she know my arse is hairy?"

I was chatting to a woman for most of the evening and she asked if I had any mates for her girlfriends.

"Certainly," I said, "but my colleagues Brains and Tommy are still in the last bar we visited, helping some pensioners find their teeth. Surf is already sold I'm afraid, but we do have a large selection freshly-imported beef from England available in all shapes and sizes, so feel free to browse."

She looked at Fat Bloke, one of the Briggs family, one of three boys and three girls who never argued and always supported each other so how they got into Trumpton is a still a mystery today. Snotbox was available, but she had missed her chance with Numbnuts, who was already engaged in deep and meaningful conversation with another college girl in the bar.

"I was raised on a pig-farm," the woman informed him.

"How did you get out?" replied Numbnuts.

She simply ignored this or more likely it went over her head, believing that no one would be so rude and continued, "What do you like with your drink?"

"A straw," replied Numbnuts.

"Do you know where I am from?" she continued.

"I would say you're from New York, probably the West Side."

"Wow, how did you know that?" she asked.

"Well, look at your beak." Numbnuts just can't himself, even though he really fancied the woman. Brains pulled up outside in a limousine.

"How come you didn't get a yellow cab like I told you?" I asked.

"I didn't know what they looked like," he replied and moved straight on to the bar and immediately engaged another woman in conservation.

"My parents are very rich, I have my own plane, you know," she informed Brains.

"Me too," he replied.

"Really, what type?" she enquired.

"Mechano," he replied immediately and as usual with great pride.

"You look like a bit of a hulk, do you do sports?" she asked.

"Yes, I'm an international ironing-board surfing champion," he said.

My only surprise that evening was that in a country where it is the right of any individual to carry arms, it was unbelievable that half the Kids were not shot on the spot.

The nickname Tom the Bomb was left over from his days in the ring. Tom the Bomb had been a boxer and very good one at that; though he was well into his seventies he was a big man, still very fit and a real East London showman. We all loved Tom the Bomb and New York loved him too. In every bar we entered, Tom the Bomb was an instant hit; they loved his deep gravelly voice and the boxing moves that he would display for the 'ladies'. In each bar he would seek out the most beautiful woman and lead her on to the dance floor for a tango. He was quick-witted too. A bouncer, a mean Turkish man, was a little jealous of the attention that Tom the Bomb was receiving in a bar on our first night; he came up to us and started to show off his martial art skills. Tom the Bomb, an ex-bouncer, asked him if he was prepared for anything that might happen.

"Yer man, I have body-armour too, stab me here or here," said the bouncer, as he pointed to his chest and arms, "and I would still not go down."

At that point Tom the Bomb made his hand into a gun shape and stuck his forefinger to the bouncer's forehead and said 'Bang!' The bouncer smiled a little but as the blood drained from his face, he returned to his place of safety at the door. Only occasionally would the bouncer surreptitiously look over as the evening continued and Tom was increasingly surrounded by his American fans.

After all the stress of our arrival, and a quiet nap on the bathroom floor during which he cuddled up with our hotel toilet, Jockey was back in action and ready to take on Manhattan. Jockey has developed some amazing wooing techniques over the years. To impress a crowd of young women in a New York bar that night, Jockey performed a very bizarre mating ritual. It involved him lifting his leg and quickly slapping his hands together around about his ankle, though never quite making contact. I was relieved when he explained to me that he was dancing; to me he looked like he was miming an elderly person trying to apply soap to their legs in the shower, but hesitating in case they lost their balance and fell arse over tit out of the bath.

Later, in a bar on the plush Upper East Side, Jockey suddenly appeared with what looked like a toilet roll wrapped around his head and demanded that someone tied a balloon to his nose. Naturally, we all suspected Numbnuts of being behind this latest wooing initiative, but for once Numbnuts looked as puzzled as the rest of us. Incredibly Jockey managed to persuade a voluptuous Mexican woman out on to the dance-floor with his display. Jockey got no further; she grabbed his balloon and disappeared into the night with it.

The next day, all was suddenly wonderful in the life of Jockey; he had no hangover, the sun was shining in New York and he had an abundance of pornographic text messages from his Mexican balloon thief. These messages were very intimate and of a private nature, so he proudly showed them to us all.

'HI, JOCKEY YOU SEX-GOD, I'VE BEING THINKING OF YOU ALL NIGHT,' was one of the more discernible messages.

The texts continued to arrive into the evening, which kept our little Scottish friend in a state of sexual excitement all day long. The level of his arousal was indicated by the various shades of crimson that his face would become depending on the explicit nature of the text message. Later that night, he reminded me of a red traffic light, flickering violently before its bulb blew.

"Look at this, she is so hot for me, I am in here, pal!" Jockey would cry as each new text arrived and his face became a darker shade of red.

The evening of text filth ranged from, 'I NEED YOU NOW BIG BOY. WHERE ARE YOU?' to 'THINKING OF YOU SO MUCH, THAT I CAN'T GET MY LEGS TOGETHER TO GET IN THE CAB TO COME TO YOU.'

If you had hit Jockey on the head with a teaspoon at that precise moment, he would have taken out the bar in an explosion of testosterone. It could have saved the makers of the film *Inception* millions on pyrotechnics, all they needed was a sexually-aroused Scotsman, a simple kitchen utensil and an escape route.

After eight hours of texting, his balloon thief was on her way to meet him, but his demeanor suddenly changed.

"What's up, Jockey, has the cab crashed as her legs straddled the cab driver's head?" asked Numbnuts.

We pressed him to show us the latest text, as up till then he had paraded around the entire bar proudly showing off each message to every customer. I think it affected the digestion of a young couple eating a nearby table, when he read out to them that a delightful Mexican lady wished to 'PUMMELL THE COCK OFF HIM.' At last, he showed us the latest text with a look of complete dejection; it read, 'JOCKEY, NEARLY THERE AND I CAN'T WAIT TO BOUNCE MY BALLS OFF YOUR CHIN.' The tears were pouring from everyone's eyes, none more so than Numbnuts and I. No words could add to the moment, even Numbnuts just waved a despondent Jockey away, as he was in pain from hanging off the bar and bent over from laughing. We stood around our confused Scottish friend, repeatedly bent double with laughter or turning to support ourselves on the bar as our tears continued to fall.

At that point, I received another text in line with the many I had received that evening from Jockey. Our sexually-aroused friend did not know that I had deleted my name from his phone while he was asleep. When I texted him he did not realise that it was me sending the messages; pretending to be his voluptuous horny balloon thief in the yellow cab. I instantly straightened up when I read his reply.

'ERR. OK BUT NO KISSING'.

At that point I showed Jockey's response to the others as I had done all night. Tom the Bomb, Don, MacConnelly and I, were all looking at each other, only to return to our doubled-over positions as more tears of laughter descended on to the floor. Eventually, I managed to straighten up; I had realised that regarding the content of Jockey's latest text that was perhaps not the safest position to be in. Jockey had no idea of why we were laughing, but he responded in his standard manner.

"Bastards!" he shouted.

Fear then got the better of Jockey; he dragged us out of the bar, before his beloved Mexican balloon thief cart-wheeled through the door. In the next bar, he was in a total panic for a different reason; he was all worked up and running out of time to entice a woman into bed. His scatter-gun approach of throwing in every conceivable line in to impress a woman got somewhat confused in translation. He advanced on a Columbian woman at the bar and after he bought her a bottle of champagne she asked him where he was from.

"Glasgow, London," he replied.

The room I shared with Jockey gave off static electric shocks from its metal surfaces. This was due to the electrical power system running under many Manhattan buildings, which in some places was not earthed sufficiently. Naturally, I did not wish to receive such jolts so each morning I would wake up my good friend by firing my shoe at his head, in the manner of an Arab member of the audience during a George Bush Junior speech. This produced in him a knee-jerk reaction making him jump up and run to the toilet to relieve the night's full bladder. On entering the bathroom he was able to absorb all of the electric static charges, which meant that I was able to safely touch any metal object for the next few minutes. This would carry on all morning, as goldfish had greater long-term memory capacity than Jockey; he kept touching every metal object and crying out "bastard!" every minute. I would lay there in bed, drinking my coffee and watching my friend repeatedly receiving electric shocks.

After at least thirteen "bastards!" it was time to go for breakfast; I would ask Jockey to press the button to call the lift, which raised his total to fourteen. One morning, when the doors opened we were met by the smiling faces of three elderly black women, wearing their beautiful flowery Sunday dresses and wide brimmed hats. They told us that they were heading to a service at their annual Baptist revival gathering in the church opposite the hotel. As I chatted cordially to the three women, I asked Jockey to press the button for the ground floor. The women, I think, spent longer in church that morning; praying for forgiveness after the appalling language that flowed from my Scottish companion as he easily passed a baker's dozen of "bastards" by the time we had reached the ground floor.

This was the big day, the St. Patrick's Day party. This was the main event and Surf had managed to get us all on the guest list. I learnt that he got us in by convincing them that he required nine baggage carriers to join him; it really shows that he is one of the most inventive entrepreneurs in the country.

Jockey had already blown all of his fuses and he was wiped out for the big day. He did make a special guest appearance later that afternoon. The lift arrived on the forty-fifth floor where the party was, to unveil our Scottish friend dressed all in black and wearing sunglasses as he edged his way around the room. His head was clearly still spinning and he only paused once to fall backwards into the broom cupboard. He picked himself up and continued his tour of the party's walls; apart from his other ailments he was suffering from a bout of vertigo that did not bode well for someone whose job was carrying bricks up ladders. When his tour was complete he got back into the lift and left without saying a word to his hosts let alone to us. Numbnuts did manage to rush over and slip a very fat pig's knuckle into Jockey's hand just before the lift doors closed. All we then heard was 'Bastard!' and our Scottish experiment proceeded to throw up all over himself on receipt of the traditional Fenian delicacy.

Our hosts were delighted to have us and made us all very welcome, though when people arrived they thought that Tom the Bomb, the consummate showman, was the host. A prospective congressman was making a tour of the major St Patrick's Day parties held along Sixth Avenue and Surf introduced him to everyone when he turned up at our party to 'press the flesh'. The prospective congressman introduced himself to Tom the Bomb, thinking that it was his party. Tom the Bomb was dressed in three-piece Saville Row suit, a long black Crombie coat, red silk woven tie, all topped off with a black Panama hat. Tom was always a dapper dresser and in turn he was impressed with the politician's 'threads', so he asked the prospective congressman how he made his money, then Tom turned to me and said, "Well, I am Jewish my boy".

The prospective congressman was happy to oblige his 'host.'

"Well I own a number of successful publishing houses," he said.

"That's how I made my money, publishing," replied Tom.

The prospective congressman was impressed at meeting a successful entrepreneur in the same field and asked, "Did the latest fall in the stock market affect trading?"

"No," replied Tom the Bomb, "But five years for forgery didn't help."

Rather than scaring him off, the congressman was even more endeared to him, perhaps viewing him as a future candidate for political office.

Tom the Bomb held court throughout the day, telling more stories, as his audience grew to love him. They loved his accent, which was familiar to his American audience following the release of Guy Ritchie's film *Lock, Stock and Two Smoking Barrels*. Eventually Tom the Bomb grabbed the microphone and sang some classic Sinatra songs, followed by Don who delightfully crooned a few Dean Martin tunes. Once they had finished, the duo sang New York, New York. I believe that, if they had called a snap poll that day, Tom the Bomb would have beaten the prospective congressman to the Senate.

We ended up in the Black Bull Tavern on Third Avenue; everyone was pretty well drunk by then and many of our group had already staggered back to our hotel. Kendra, our barmaid, came over and asked Fat Bloke and me why we were still propping up the bar.

"We can't go to bed," I said, "as our sleeping arrangements have changed and we have to share a room with him." I pointed to Brains who was off his face and running up and down between the pool tables.

"Oh, I can see why that would be a problem," replied Kendra sympathetically.

The room arrangements always altered, usually the first ones back grabbed the best beds. So our punishment was self-inflicted and we were stuck with Brains

Fat Bloke added, "You see Kendra, Brains is operated by a rabbit on a bicycle who resides inside his brain; when his normal functions closed down about two hours ago, the rabbit took the controls and he is coordinating his runs at this moment. He will be like that all night until he knocks himself out."

"What is your name?" Kendra asked Brains as he ran straight into the bar. Brains couldn't provide an answer.

"The Rabbit must be on a carrot break and it's left our good friend on autopilot. Perhaps ask him again in an hour?" I suggested, as Kendra appeared to be enjoying the idiotic company of three brain-dead English guys in her empty bar.

"Have a drink on me, you crazy English bastards," said Kendra, as she poured a few stiff ones into three shot glasses.

"You're quite a woman, Kendra," I said as Fat Bloke, Kendra and I toasted the best barmaid in New York. Brains just then knocked himself out by running directly into the juke-box behind us.

When it was daybreak, we left the bar and I queued up in the diner next door, to buy some juice and a bagel, as none of us had eaten more than breakfast on the day before. Brains was pestering a tramp who was trying to sleep in a doorway for a cigarette and Fat Bloke was chatting up the waitress.

The woman on the cash till welcomed everyone with a pleasant greeting ranging from, "You look good today," to "I like your tie," and "Good choice, that pie is delicious." Then she looked up at me, "Good God! What happened to your hair?" she cried.

"It's been a long night," I answered with a broad smile.

When we returned to the hotel, we were swept back out on to the sidewalk by the rest of our friends. They were determined that as we were flying back to England that afternoon, we should all have a last New York breakfast together. We entered a diner across the road from our hotel.

The young waitress came up to our table. "Morning guys, what can I get you?" she asked.

"What's your name?" asked Tommy.

"Polly."

"Well Polly, put the kettle on."

I wanted to spend my last remaining hours walking around the streets of my favourite city, second only to London. I sat for a few hours by the Hudson River, not knowing that a few years later the Twin Towers would be erased from Manhattan's skyline for good. Of all the global and national calamities over the last thirty years, 9/11 is the only event that I can remember where not one of us cracked a joke. We were all rogues, but we loved New York and its people had been friends to us all.

It was nearly midday when I headed back to the hotel. I entered Tommy's room to see how the birthday boy was and found him sitting on the bed with the others standing around by the walls. I knew I had missed something.

"Tommy you don't look like the happiest of birthday boys." I said.

"I've lost my money and passport, I feel like shit and I am a year older, all I have to my name is a well-packed suitcase." Tommy said.

At that point, I realised why the others were standing well back against the walls, like

Jockey had been on the day before, as if waiting for an earthquake to arrive. Behind Tommy and without his knowledge, Brains looked like the Indian god *Shiva* with his arms firing Tommy's clothes all over the place as he unpacked Tommy's suitcase. Tommy saw my look of horror and then he turned to find Brains carrying out his rendition of the Indian goddess' dance of the seven pairs of un-ironed underpants.

"What the fuck are you doing?" screamed Tommy.

"Looking for your passport," replied Brains.

Tommy exploded. The torrent of abuse he launched at Brains focused on his grey matter having the same texture as a certain farm yard animal's manure. When Tommy finished and collapsed back on to his bed, the room fell silent. After a few minutes, Brains had obviously appraised his friend's situation having looked under the last remaining pair of underpants in his bag and having come to an assessment.

"Well you're fucked then, aren't you?" he said.

"I'm fucked …" was the beginning of a new tirade by Tommy, which was now focused on Brains' ability to entertain the entire Greek naval fleet, which had docked on the Hudson River that weekend, in one evening sitting. It thought it was unlikely that Brains could complete even half of the positions Tommy described but, knowing Brains, I believed he would give it a good go.

Tommy kept exploding at Brains all the way back to England and Brains just looked at him blankly. When we landed at Heathrow, Tommy discovered that his bag had been lost in transit. Tommy had to join a queue at lost property; he had a look of thunder on his face and we all knew that one wrong word would set him off again.

"Don't forget to describe the porn mags you had in it! That will help them find your make-up bag!" Brains shouted across the terminal.

"Why don't you just fuck off and …," well you can guess what Tommy shouted back, which was littered with more colourful expletives and a final suggestion that Brains might like to try inserting his head up his own rectum. After a further thirty minutes of Tommy screaming abuse across the baggage hall at Brains, he finally stopped, exhausted.

At that point Brains turned to Numbnuts standing beside him, "He doesn't like you much, does he?"

Numbnuts was speechless and just shook his head and said to Brains, "How fucking dumb are you?" which Brains naturally assumed was directed at Tommy and turning to his friend across the hall he waited for him to answer.

My love for the US remains undiminished, so when Numbnuts asked if I wanted to go with him to Denver for his last trip to the US before he left England for good, I could hardly say no. We went to meet some mates of his who worked in an elite ski-resort in Aspen, Colorado. First Numbnuts and I stopped off in Chicago to meet Surf and stay at his millionaire boss Bobby's house.

That night we headed out for pizza and at the table behind us one very loud diner was telling everyone within earshot that all illegal Mexicans should be kicked out of the country immediately.

Numbnuts turned around to him and said, "Hi Big Man, I'll tell the waiter to cancel your pizza."

"Why?" replied the puzzled Big Man.

"Well who the fuck do you think is making it?" replied Numbnuts to Big Man who went very quiet.

Surf turned to me, "You may be on the cuddly left, but my brother is a member of the unhuggable section."

Numbnuts, Surf and I are very different characters, but we share a dislike of bullies and bigots and we will never forget that our families were immigrants too.

After a very heavy session that night, our host Bobby addressed two of his hung-over

English guests at breakfast the next morning.

"Baloo, Numbnuts, I'm going to barbecue a chicken tomorrow, do you have chicken in England?" he asked.

Numbnuts lifted his drink-sodden head to view our host, whose knowledge of England appeared to be based on watching Charles Dickens' *Oliver* on TV over Christmas.

Barely able to get the words passed his dehydrated lips, Numbnuts replied, "Is it like lard?"

The next morning, Numbnuts and I landed in Denver to be greeted by the TV headlines that a football player for the Denver Broncos had been suspended for 'a helmet to helmet clash.'

"If it were gay pride week he would have got away with it," observed Numbnuts.

American phraseology continued to confuse us; the next item was a review of a new teenage coming-of-age movie, as the film reviewer said, he would give the lead actress two thumbs up.

"Poor woman, her eyes are big enough as it is," I noted, as we finished our beers and then headed off to Aspen.

Numbnuts was reunited with his old friend Monty, who had been working in the ski-resort for fifteen years along with another friend called Shep and a few Brits. There was no work back home where Monty lived in the North East and if it were available, it would be mind-numbingly boring work such as packing boxes. His hometown was one of the poorest in the country.

He told me, "Last time I went home, some bloke tried to sell me a knocked off fucking Mars bar in the bookies. One fucking Mars bar, not even a box of them."

Numbnuts had gone to Aspen many years earlier, with his girlfriend at the time Delia, who was a Trumpton girl. Delia is a great woman and has a mischievous roguish humour. She once told me that her earliest memory of Trumpton was a cold arse, due to her mum stealing a frozen chicken from the local supermarket and hiding it under her as she sat in her pram.

"I still get shivers, whenever I see chicken on the menu," she had said.

"You must get the same feeling when Numbnuts strips to unveil what *Blackadder* once termed "the last turkey in the shop," I said.

"Think of a starving sparrow and you've got it," laughed Delia.

"No please, you have it," I replied unselfishly.

Monty had secured a job in the town and advertised his talents on a sign outside his new place of work, 'ASPEN'S BEST ILLEGAL BARMAN.' When I read this I thought it might be a short appointment, as two policemen were also reading it. Shep was looking for work, but the problem was he had not spoken to an American in all his years in the United States and that somewhat restricted his prospects.

"What? I'm not talking to those fucking robots," he explained to me when I asked him why he wouldn't speak to Americans. When an American would talk to him, Shep's face would contort; he'd pretend to be mentally deficit and he would tilt his head to the left with drool running from the side of his mouth straining out loudly, "Huhh."

We joined Monty for his evening shift and experienced a clash of cultures, style, language, well everything really. The locals loved Monty. His constant angry expletives due to the gout in his foot which constantly plagued him resulted in various facial contortions. Those plus his impenetrable accent, was as one American told me like watching a cabaret based on the noises of various trapped animals. I surveyed the bar and learnt that the big guy in the Stetson hat getting the drinks in was one of a group of Texans packing the bar that night to savour the entertainment offered by Monty, who was well known in town.

"A CC with soda and ice," bellowed the Texan.

This was met with a clearing of the throat and an "Aghhhh!!!" from Monty, as he hobbled over. Monty was not up on his cocktails and so I had to translate the Texan's bar orders all evening. We had a good night and all of us had a good laugh with the

Texan and his friends in the bar, with the exception of Shep who just sat on his stool, arms folded and his chin up in protest at our fraternising with the enemy. During the evening, two girls came into the bar and started kissing each other. I asked Shep if he had any views on the subject of homosexuality.

"I have nothing against lesbians, though they are Satan's daughters," he replied, but whether that was due to their being gay or American I never found out.

DAMAGED ROGUES

"I am ready to meet my Maker.

Whether my Maker is prepared for the great ordeal of meeting me is another matter."

Sir Winston Churchill

Chapter 34 – *Was Ever a Man More Misunderstood?*

I have no idea what I'm going to do when I grow up. Will my life continue to be one of fun and adventure? You won't be surprised to hear that I had never asked myself this question, but what was surprising was that one of the Kids had and the even bigger surprise was that it was Brains. It happened one time in a bar in Islington when I turned up alone, while most of the others were part of a couple, apart from Numbnuts obviously who had even fewer social skills that me. Brains shook his head disappointedly when he saw me and decided to offer me some advice based on his learned experience.

He began, "Baloo, you have to get your life in order, look – I have three screaming kids."

The opening line of his lecture caught the attention of the other Trumptonites who all turned to listen, particularly his girlfriend Linda. Until that point, she thought that the two he had with her were his only children. I imagine he had forgotten to tell her about his weekend away in Liverpool that had happened before he knew her, fifteen years earlier.

Brains continued, "I have my own rented accommodation crammed to the roof with shit-filled nappies."

Again, this also begged one or two questions from Linda, as the flat was in her name and again his depiction didn't seem to me such a big sell. On the point that it was Linda's flat, she was claiming housing benefit for herself and their children, Jennifer and Luke, on the basis that she was no longer in contact with their father. But as the father of her children was actually living with her, it was important that the authorities never discovered this. One day a man from the council rang the doorbell to check that Linda was the sole adult occupier. Linda viewed him through the observation eye on the door and signalled to Brains that he should hide. Once Brains had disappeared, Linda allowed the council official to enter, saying that she had nothing to hide and he could go into any room he liked, but to leave the wardrobes in the bedroom alone as they had just been sprayed for moths. Linda thought this would cover Brains who had run into the bedroom and would be crouched in the wardrobe; a natural hiding place for any normal person. Her mistake was expecting Brains to do something normal. Linda and the council official entered the main bedroom, where they found Brains sitting on the bed completely naked, except for Linda's glasses which he was wearing as he was pretending to read a book. Later Brains angrily updated us on this encounter: he now had to move out of the flat or Linda's housing benefit would be stopped. The suspicions of the council official could only have been raised further, as the sixteen stone naked tarmacer was reading a book upside down.

Brains then asked me, "Who the fuck is Catherine Cockson anyway?" I thought there was no point in correcting his pronunciation; I doubted he would ever go in the library to take out any of her books.

Brains continued with his lecture on where my life was going wrong, or as he termed it; "Why I was on living on shit-street."

"You should look at me," he said, "I have a top job as a qualified civil engineer."

Again, this was slightly disinguous, as it was more the case that he laid the top layer of tarmac on the roads but this was a mere detail in the mind of Brains. He then turned his attention to the other aspects of my "pathetic life," as he called it. This was a real departure from his previous high regard for me. The high regard had occurred when I was on one of my convoys and Surf told me that Brains had been asked by some regulars in The Bobbie about what aid I was delivering to various hotspots? Brains, it seems, was quite taken about being asked and very proud to be associated with someone who was carrying out such philanthropic deeds. Though to be honest, 'philanthropic' is not a term Brains would have used spelt or even pronounced correctly.

"He delivers beer and cigarettes to underage kids who can't be served in their local shops." Brains said. "Also, he delivers guns so they can fight off anyone trying to nick 'em!" he had announced proudly, much to Surf's amusement. To Brains, there is no greater act of human kindness a man can perform, than to deliver cigarettes and alcohol to the poor undernourished children of the ghettos, for in his mind that is real deprivation.

Brains was now in his stride in his analysis of my pathetic life, "And look at you, Baloo, it is sad for a man of my achievements to look down on the poor miserable failure that you are. I take no pleasure in telling you this." That was strange, because it looked like he was enjoying it immensely.

"You're so right, my good friend, but I have so many failings I wouldn't know where to begin," I agreed.

"I can help you with that." Brains suggested. Somehow, I thought he would. "You live alone, so you have to cook, clean and even dress yourself, I have Linda, and as her 'pride and joy' I don't have to worry about any of that bollocks."

Just behind Brains, I could see that Surf's wife, Ann, was trying to persuade Linda to put the bottle back down on the bar, the bottle she was aiming at her 'pride and joy', while probably still wondering where the third child had come from. 'Pride and Joy' was how Brains said Linda referred to him, though he did later acknowledge that he had never actually heard Linda use the term.

"I have watched you over the years with beautiful, educated women who must bore you to tears, wanting to talk about things apart from football and sex," he added before he moved on to the area of my pathetic love life.

"You're so right, my good friend, politics, world affairs, the environment, family, you have no idea what nonsense they will come out with next," I said as I shared a raised eyebrow with Ann and Linda. The two appeared distracted and were scanning the bar; perhaps they were both looking for something more lethal than a bottle to batter him with.

He continued, offering some examples of my failures over the years to assist me in identifying where I had gone wrong.

"Kathleen you see, she was sophisticated and well brought up; she did not fit in at all,

– now Angela, she is one of us," he added in full flow, as he helped himself to the drink of the man sitting next to him.

In one sentence, he had been able to insult two of my ex-girlfriends and infer that I along with everyone else present was stupid; including himself.

"That man is the sharpest tool in his kid's play set," whispered Numbnuts in my other ear.

Brains continued, "Look at your job, now how sad is that? You have to travel around the world to get work each day, whereas I only have to wait outside the tube station in the rain at five am each morning to get a day's work."

I think he was referring to my job, developing energy plants, which entailed working in different countries, all expenses paid, but again these were mere details.

"You have your own house and no mortgage. Good God, Baloo, do you think a man of your age should have a mortgage?"

At that moment, Numbnuts tried to interject, but I raised my hand to stop him.

"Don't mock Brains, he has some valid points here." I said.

Brains then turned to his friend Numbnuts, "As for you, you're no longer any good at mockery either."

As I have said, only Brains could strike Numbnuts dumb, but that night, with Numbnuts having no job, no girlfriend and recently no home which meant that he was sleeping on my couch, Brains had disarmed Numbnuts of the only thing he had left in life: ridiculing Brains.

There was a serious point to Brains' ravings. He had a family and I had made no personal commitments in life whatsoever. Brains loved his kids and would recite to us the rhymes he would recite to them, one of his favourites, which Luke once recited at school, was 'The Rabbit and the Monkey.' Brains would sing it to his little boy each night it went:

'The Rabbit and the Monkey were sitting on the Grass.

The Monkey got bored and stuck his finger up the Rabbit's arse.'

Brains continued, "What I enjoy most is when my daughter tucks me up in bed and then reads to me before I go to sleep."

"Ah, the innocence of fatherhood," sighed Numbnuts, though he was still smarting at having been denied any outlet for abuse.

I remember when Linda gave birth to Luke; – later that night in The Bobbie Brains had related to me his ambitions for his new son.

"Ah, looking at him I can see that my seven-pound boy will excel at boxing in the nursery and by the time he is at school he will be an amateur boxer and win the Olympic gold, then he will go on to lift the British heavy-weight title and by twenty-one the world title."

I posed a question, "What if the first punch he ever receives to his head renders him a lifeless cabbage?"

"Ah well, that's boxing!" replied Brains, as he jumped off his stool and grabbed Tom the Bomb's seventy-year old girlfriend in a playful headlock.

A few months later, I met Brains and fatherhood seemed to have turned his thoughts to providing his family with some financial stability. "I have moved on from my ambition to die in a massive explosion," he said.

This should be good I thought, perhaps this time his ambition may involve a future.

"My ambition is to stand bollock-naked in my own lesbian bar and let angry 'strap-ons' beat me with a stick for ten minutes or stick various implements up my arse."

I looked at him bemused, which he noticed and he then qualified his previous statement with the words, "For tips of course." He stressed that a financial return was a core component of his business plan.

Later on, he confessed that if that didn't work, his alternative plan was to marry a rich woman and, as he put it, "Keep himself in the 'custard' I aspired to." For once I would try to correct his dessert remark later, but by then I gave up as he had moved the conversation on to the 'opinion fields of the Taliban.'

As his children grew up, Brains would often relate, how through no fault of his own, he would often find himself in the doghouse.

"Was ever a man more misunderstood?" I asked sympathetically.

"Well, I was sleeping on the sofa, looking after the kids." Brains said.

"See, not your fault," added Numbnuts

"Quite right, but then Luke was waving at me from out in the garden as I was dozing off – I just waved back," explained Brains.

"Seems harmless, what could go wrong?" I added, knowing full well that of course it had gone wrong.

"Well, I didn't notice that when Luke was waving at me, he had half a brick in his hand. When I waved he thought I wanted him to throw it to me," continued Brains.

"I would have done the same," added Jockey, which I did not doubt for a moment, as only the day before he had knocked his cat, Tiddles, off the window ledge when he opened the window to let it in – it had dropped six floors.

"I take it the glass window to the patio was closed?" I asked.

"You've been to university Baloo, so of course you've figured it out. The whole fucking thing exploded when the brick came through it and I was covered in glass lying naked on the settee. How am I supposed to sleep and perform my babysitting duties with that racket going on?" concluded Brains, though I was now worried about the naked part of his tale. He added that he had waited until Linda returned from doing the shopping, before taking up her coat to cover his nakedness and walking passed her in the direction of The Bobbie. He added that he'd informed her that he had done his bit and now it was her turn to do her parental duties.

It is not easy raising a family, even for someone as accomplished as Brains is in the parenting department. Often children have stupid accidents and the parents blame themselves; mind you, Brains never did. For example, one day Luke drank a bottle of disinfectant.

I asked, "Don't all these products have child-resistant safety caps?"

Brains replied, "True, but I had to get Linda to remove all of the safety caps, as I can't work out how to open them."

Fortunately, Brains buys everything from the Poundstore, so no harm was done to poor Luke as such products are ninety percent water.

That night we saw that Brains, in terms of intellect, was a colossus amongst us pigmies. Just as my lesson came to a close, Numbnuts sighed, took a drink from his beer and turned to Brains.

"A better man would have taken his own life by now." Numbnuts said.

"A bit harsh, but Baloo you should consider it," Brains said, turning to me.

Numbnuts shook his head in disbelief. I then summoned up the courage and indeed the gall to ask this great man what his plans were now, particularly as he appeared to have no money and he had not bought a drink all night.

"I'm off," he said, "I just remembered there is a can of Kestrel lager in the cupboard; I hid it from Linda six Christmases ago." He then turned on his heels and ran out of the door before Linda returned from the toilet. I smiled as I watched him through the window – one of Trumpton's greatest success stories was scarpering off down the road in search of a can of low-alcohol lager well past its sell-by date.

Chapter 35 – *Rule Britannia*

When I came home from my South American odyssey, I was shattered, and the beating I'd had in Paraguay didn't help. I entered my flat, dropped my remaining possessions on the floor and collapsed on to my bed. I was exhausted to the point of being overtired; as I couldn't sleep I just lay there and listened to the radio. BBC Radio Four were interviewing a late-night TV announcer who was retiring after fifty years of providing the voiceover between programmes aimed at an audience of insomniacs and alcoholics. The interview was as dull as the programmes he had introduced, until the interviewer asked the retiring announcer if he had had any embarrassing moments over his half a century of tranquilising, my term not his, the nation. The radio went quiet as the retiring announcer contemplated for a while and then after much thought he replied.

"Well there was one time when I had to introduce a programme which was Jack Hargreaves' *The Old Country*," he said, "I was a bit slow and the next programme started just before I finished the sentence and I lost the last syllable."

I fell into a deep sleep with a smile on my face; now I now knew I was back in the country that I loved.

How do we get away with inserting the Great before Britain? Personally, I think it's hilarious that this little island puts two fingers up to the world, thinking nothing of imposing a veto in Europe or telling the world where it went wrong on the economy. Inhabitants of other countries may say that they are indifferent to us, but, despite this being a small little island, I have no doubt they have an opinion on us.

We strut our stuff around the world, though our colonial empire is gone, our natural reserves are depleted and we rely on a bunch of self-serving financiers in the city to have our country's interests in mind. Even with a superpower like the United States, 'we' invented the term 'special relationship' which always bewilders Americans, as they believe in the capitalist system and have no interest in skewing the market for our little nation if they can get a better deal elsewhere. That is how a capitalist system works after all. Recently, our politicians have used the term 'special relationship' to describe our links with Russia, China and India. When our politicians visit the heads of states of these super powers and talk of our 'special relationship' they must think 'What? Who said that?' We also call ourselves the mother of the Commonwealth; those are mostly countries that fought to be independent of us, so it's perverse to think that countries such as South Africa and Zimbabwe look to us as some kind of mother state. I don't have access to the private list in the Foreign Office of the 'The countries we couldn't give a toss about', but I think through a process of elimination it must be quite small.

I love that we punch above our weight, but we are special in that no other country in the world would have the nerve to declare that they are special. But based on our illustrious history, we have a real basis for saying it. As the greatest colonial power from Elizabeth I to Queen Victoria, we may have introduced slavery to the world, instigated the first concentration camps for the Boers in South Africa, but we should rightly be proud, when you look at the impact we have had on the world in terms of language, the arts, infrastructure and inventions such as Watt's steam engine which propelled the Industrial Revolution, leading to the modern world that we live in today. For that reason, I believe we are not just a rogue nation but also a benevolent one. During World War II, Churchill brought us back to the superpower table, but our lack of influence in restructuring Europe during the post-war period showed our impotence, when Truman and Stalin put their pens on the map and drew up the world's new boundaries.

I see something else that is special about this country. Maybe all nations have a strain of eccentricity underneath the civilised veneer, but for me though you can't beat England for the number and quality of eccentrics and lunatics. We also have a tremendous ability to not only take the piss out of others, but also ourselves, because we have the confidence to know that it really doesn't matter that much.

As for our island neighbours, the Welsh used to take themselves far too seriously, and were often very defensive in the belief that everyone was taking the piss out of them. They were right of course. It's not easy though, when even your most famous sons, like Even Dylan Thomas, said 'Land of my Fathers. My fathers can have it'. Much has changed though and *Gavin and Stacy* is evidence that they having lightened up.

The Scots have no such insecurities, and exhibit a fierce pride in their nation. The growth of Scottish nationalism has resulted in the sales of twee Scottish programmes, like *Monarch of the Glen,* beyond their borders, but they have no fear of sending themselves up; all you have to do is look at the TV comedy *Rab. C. Nesbit*, which proudly declares that you can make fun of the Scots if you like, but no one can do it better than the Scots.

Across the seas, Ireland produced Father Ted and Australia gave us Sir Les. All have their critics at home, fretting about the fragility of their country's image, but this says more about the critics' insecurities than it does about Rab, Sir Les and Ted.

Jeremy Paxman's book *The English* asserts that the English identity is based on what we are not, in that the English are not Celts, but one characteristic that defines being English is that confidence we have in being English; we believe we have nothing to prove to the world. This manifests itself in the smugness and arrogance that our Celtic neighbours deride us for. Paxman embodies this, for I have never seen anyone else so condescending and brave in interviewing politicians. Another example of this confidence, or arrogance if you prefer, is that we have no need to wear nationalist dress except for Clarksonseque jeans that are two sizes too small as we stumble towards a pensionable age.

Our leading humourists are world-renowned for their eccentric characters, with such examples as the Goons, Basil Fawlty and Jeffery Archer, the last being a parody of himself. We love the ridiculous and if there is a dark side to humour, so much the better. A few years ago the ventriloquist Rod Hull, whose stage act was based solely on having an emu puppet on his arm, fell from his roof to his death. That was while he was trying to get a better reception on his television as he tried to watch his beloved Manchester United. The newsreaders may have announced his death with deadpan faces, but the jokes flew around thick and fast for the rest of the week, based on the premise that if his partnership hadn't involved a flightless bird he might still have been alive.

Even at a national scale, we have the ability to make errors, which we celebrate rather than hide. Our national newspapers, take great pride in reporting our shortcomings as nation. When Eurostar were testing their trains in readiness for the Channel Tunnel, the French found that their trains' windscreens were able to withstand the impact of birds when they impacted at high velocity, but English tests led to frequent calls to Autoglaze. Both sides of the channel used the same technique, which involved firing chickens out of specially built cannons directly at the trains' windscreens, so we requested that one of the French engineers came over to see what we were doing different. He pointed out that in his country they did not use frozen chickens.

I enjoy how our newspapers report outrageous incidents but in a subtle manner, almost in code, so as not to offend the readers' sensibilities. A few years ago, a man was arrested after having been caught in a field having sex with a goat, just as a full passenger train broke down in front of him. The paper reported that, despite being observed by a number of shocked *Financial Times* readers, he pleaded not guilty. The article recorded that a unanimous guilty verdict was delivered by the jury, following forensic evidence that 'goat hairs were discovered in a strategic, though private area.' Even the sanctity of Christmas is a cause for mirth; last year the ambulance service renewed their plea for revellers not to abuse the emergency service line after a drunk telephoned to ask for assistance in finding his trousers, and a woman reported that her snowman had been stolen.

The same eccentricity is to be found in our politics, unlike countries like Italy where politicians are found to be on the payroll of the Mafia, our scandals are usually based on sexual perversion. The late Stephen Milligan, Tory MP, was found dead in his own kitchen, on a chair, after one of its legs gave way. He was found suffocated, naked, except for a rope around his neck, his head was tied up in a plastic bag, he was wearing knickers and suspenders and he had an orange in his mouth. It must be a coroner's nightmare to be given a case involving a Member of Parliament.

Even in the recent MP's' expenses scandal, our politicians abused their claims, not on the purchase of drugs or prostitutes, like any decent French politician, but on cleaning a moat, making repairs to a belfry and my personal favourite, replacing a flagpole. After all, where would democracy be in this glorious country of ours, if one couldn't pipe the Union Jack up one's mast each morning? Mind you, there were arguments put forward by some of the claimants, such as the Right Honourable Member for Gosport, Peter Viggers, who bought a duck house on tax-payers' money. He robustly argued in his defence that, "The ducks didn't like it." An unusual defence, as it was not only an abuse of public taxes, but it was also a waste of money.

Our popular literature also promotes the Great in Britain. Our authors have invented fictional heroes and have ensured that when we created counterparts of whatever nation, they are flawed. Sherlock Holmes and James Bond are superhuman heroes of literature and cinema. The first ironically was created by a Scotsman, Sir Arthur Conan Doyle and the other by an old Etonian, Ian Fleming. While at the same time, we invented the bumbling Inspector Clouseau as an example of French crime detection and even Agatha Christie's Belgian detective Poirot, though no slouch in the little grey cells department, is comically burdened with a ridiculous dead slug of a tash.

This self-belief is expressed from *Henry V* right up to Margaret Thatcher and later, the most dangerous of all, in Tony Blair. Each of these held the belief that they knew what was right for everyone else in the world. Occasionally, we have a crisis of confidence and this insecurity will break out through random football violence in small countries whose names until then simply accumulated a high score in Scrabble (now that names are allowed). A few and it really is only a few, England football fans will attempt to raze these countries to the ground in the belief that they pose a threat to our national sovereignty. They are then struck down with homesickness and police truncheons and, while surrounded by policemen firing teargas, they will run to any TV news camera while wearing a pair of Union Jack shorts to shout, 'I love you, Mum!'

How we cackhandedly reaffirm this stereotypical view of the English was painfully exhibited when I went to visit a girlfriend, Mary, who came from Spain's Basque city of San Sebastian. Again, the Duke of Wellington reappeared in my life. Taking my motorbike, I caught a ferry from Portsmouth to Bilbao and rode through northern Spain until I parked in one of the most beautiful towns I have ever seen.

I said to Mary, "San Sebastian is beautiful; it's so unspoilt, there is not even a McDonald's, or a Starbucks here. The town has certainly kept its heritage; it must have a very strong planning committee."

"No, just the ETA," replied Mary. It seems that the Basque separatist group kept blowing them up, as they were not big fans of globalisation.

That night I was invited to an authentic Basque dinner to meet Mary's family, comprised of her grandmother, her father and mother, three brothers and three sisters, none of whom could speak English. Mary had a task on her hands that night as translator since my Spanish was nonexistent. The evening started well enough; the Rioja was flowing freely and a traditional Basque delicacy of octopus cooked in its own ink was served. Mary smiled at my reaction but was delighted that I dived into the meal with gusto. Everything was going well, until Mary's father asked her something and I could tell by Mary's face that she felt very uncomfortable.

"What is it Mary?" I whispered.

"It does not matter, just a question my father has," she replied.

"Go on, you know me, I can take it."

She took a deep breath, made her apologies in advance and then repeated her dad's question in English.

"My Father wants to know why it is that when English football fans go abroad they drink too much and then they smash up the foreign towns that welcome them as guests?" she said.

I could see why Mary was uncomfortable, but I did not want to make her embarrassment any worse, so I politely replied that I was aware that that occurred on a rare occasion and that I did not condone such violence. Mary was relieved and everyone around the table seemed to welcome my considered response. The Rioja continued to flow at a great rate and Mary and I were delighted that the family seemed to accept the Englishman; everyone started to relax. Then a further question was asked of Mary, this time by her eldest brother Gustav, which resulted in an even more uncomfortable reaction from her. I, along with everyone else was now a little drunk, so I repeated to Mary that it did not matter how embarrassing the question was.

Mary shook her head, but again started to translate, "My brother has heard that the English Channel is so polluted, that your fish have become hermaphrodites?"

I decided to pay the humour card this time, "Yes, I have heard this, but I'm not sure if it's true, however I only eat octopus if the tin has a foreign label."

This went down well once again with everyone including Mary breaking into a smile. By now even granny, who had watch me intensely all evening was smashed and laughing along with the rest of us.

Then the youngest of the party, Mary's little sister Angelica, leant forward and whispered something in Mary's ear. Mary's smile abruptly disappeared to be replaced with one of shock.

"Mary just say it, if your little sister has a question I would be delighted to answer it," I again reassured her. I was confident that I could deal with the Spanish Inquisition, as I was well fortified with wine. Mary looked most uncomfortable, but being drunk and less reserved than usual she repeated her little sister's question in English.

"Please, please don't be offended, but my little sister has read in *Cosmopolitan* that due to poor diet, Englishmen," she could barely finish her sentence, "have the lowest sperm count in the world?"

Even I was taken aback by this question asked on behalf of a twelve-year-old girl. I sat up; I was now totally drunk and in the mood to defend myself and my fellow countrymen.

"Mary, can I return to your father's original question on why the English smash up places when they go abroad? Well it's because when we ask the locals for directions to their football stadiums, we are then asked if our nation is truly surrounded by fish that have an abundance of genitals and if the local girls are truly not at risk of pregnancy, as all Englishman are seedless," I said with a smile and toasted my dinner table companions.

Mary for the first time that night was laughing loudly, as she translated my answer to their youngest daughter's question. Mary told me that she loved my spirited response and was proud of me for standing my ground.

They all liked me, even granny, but even more so for standing up to their onslaught and defending my country, which is the Basque way. After that, more wine flowed, the night merged into a haze of singing, dancing and more laughter at the Englishman's expense.

The next morning Mary woke me to say that her father liked me so much, that he wanted me to join him for a walk up the mountain that overlooked the town.

"Does he want me to carry his gun and a shovel as well?" I asked.

"No, he likes you," Mary replied and added, 'This is a great honour, so I hope you will go."

"OK, but if I don't return by sunset, inform the usual authorities and give the SWAT team Bear's address," I replied.

"Who or what is this Bear character?" asked Mary.

"Both questions are equally valid," I said and added a goodbye to Mary for what I thought might be the last time, thinking that I would be another statistic on the local missing persons list.

Mary's father and I set off and after three hours with him talking away and me dutifully smiling, not knowing what the hell he was talking about, we reached the peak of the mountain. He signalled me to come and stand near him in a clearing at the edge of the mountain, where he pointed to a stone memorial surrounded by freshly cut flowers.

Instantly I recognised the main figure on the memorial; it was the Iron Duke, the Duke of Wellington himself. With my childhood love of Napoleonic history, I knew the stone carving represented the Duke's troops driving out the French from the Spanish peninsula at the turn of the nineteenth century. Having a passion for this period, I felt frustrated that I could not communicate this to Mary's father, who obviously was expressing his thanks to me as an Englishman for driving out the French invaders from his homeland. Just then I realised that I could communicate with him, as I had an English five pound note on me with a picture of our national hero printed on it. I held it up proudly to his face between the thumb and forefinger of each hand, but his reaction was not what I expected, in fact he appeared shocked and then he shook his head in disappointment. We continued back down the hill and he talked, but this time he shook his head disapprovingly.

I had obviously upset him but I could not work out how. I explained what had happened to Mary when we reached the base of the mountain.

"You idiot, the memorial is to the men who were murdered and the woman who were raped by Wellington's army after the French were driven out." Mary said.

"Ah, so when I showed your father the Queen's currency with Wellington on it, I might as well have been a German proudly showing a Jewish person a fifty Deutschmark note with Hitler on it?" I asked. As Bertie would say, I had really put my dick in the paella this time. "Mary, what's Spanish for 'Oh Bollocks'? Can you please tell him that I am sorry and I meant no offence?"

Mary just laughed and said that he would understand, but added "It's best I don't leave you alone from now on."

It just goes to show that the English, even when we are really trying not to offend anyone, have a way of sending our neighbours to their gun cabinet.

Perhaps this is the last aspect of what it is to be English, in that we have the gift to aggravate all nations, even when we do not mean to. Lord Halifax, our Foreign Secretary, on meeting the new Chancellor of Germany Adolf Hitler, supposedly entered the Reichstag and handed him his coat, thinking that he was the butler. More recently, when it was finally agreed that the Channel Tunnel would be built, it was designated that the terminal on our side would be called Waterloo Station. The name of the station commemorated Wellington's great victory or, putting it another way, reminded our neighbours of Napoleon's greatest defeat. Even during the Arab Spring, we sent HMS Cumberland to the Middle East; a battleship named after a pork sausage.

Prince Philip, may have been born a Greek but he's as stiff and unapologetic as our upper lip and a consummate diplomat in the field of how to make friends and influence people. A few years ago, when he met the Indian Ambassador who was wearing his national ceremonial robes, he was quoted as saying, "You look like you have just got out of bed."

"So in my summing up your honour, Bear, Wurzel, Brains and the other rogues are just a symptom of our national identity and that concludes the case for the defence, my Lord."

Chapter 36 – *It's Not the Years, it's the Mileage*

Tommy has now passed the test to be a London a taxi-driver. I think ninety percent of the scenarios given during the training, modelled on the most critical situations a driver might encounter, were probably based on previous encounters involving him. Over the years I started to worry more about Tommy, as he got into a fight with a local gang alongside his friend Nick, whom I did not know that well, but whose picture was regularly published in the tabloids. After an Arsenal game in Copenhagen, the newspapers printed a photo of him appearing to be a contestant in the world international table-throwing championships. Knifes were becoming more prevalent in gang incidents in Trumpton; I feared that unless Tommy changed, I would lose another good friend. Then thankfully, things did change. Nothing can change the direction of a man more than the entrance of a woman.

Tommy met Amy, who takes no nonsense and together they have one of the strongest relationships I have ever witnessed. They have four kids to prove it. They had grown up together and I always knew that Tommy loved her, though he never said anything. Whenever Amy would appear, was the only time I saw Tommy exhibit shyness. When they did finally get together, it was volatile, as will always be the case when two strong characters decide to start a relationship. There were occasions when Tommy would throw Amy's clothes on the doorstep of her mother's house and he would then have a can of coke, unopened, bounced off his head in return. If that isn't love, then I don't know what is. OK, maybe it's not *Oklahoma* or *Doctor Zhivago*, but if they make it into a film, I know the Kids would bunk in to see it. Their passion continues and recently, after an argument, Tommy came home to find Amy, an Arsenal fan, and their four children in front of the TV greeting him with big smiles and all wearing new Arsenal shirts. He has met his equal, and more. He told me this story himself, and he could not help smiling as he told it.

Chelsea Mark has now opened up a diving school in the Arab tourist resort of Sharm el Sheikh on the Egyptian coast. I feel that I have to stress that the diving school is by the water, as with the Kids you never can be sure. Numbnuts, when he heard the news, commented that Chelsea Mark had gone from fairy to Pharaoh. When Chelsea Mark told his mum of his new business venture; a Jewish boy opening up a diving school for tourists during a particularly tense time in the Middle East at the same time as Israel was sending tanks into Palestine, she responded with, "Here is the telephone number of my psychiatrist."

After another terrorist bombing in London, Chelsea Mark sent me an email telling me to get out of England and come out to where he was, as it was safer being in an Arab country. A bomb blew his windows in the following day, covering him in glass, while he was typing a similar email in the nude. When Chelsea Mark informed me of this near-death experience, distressed at the news I quickly wrote back, "What the hell are you doing writing emails to me naked?"

H married Shirley and I was his best man. The wedding was in St Ives and I offered to make all the arrangements, such as organising the hotel for the other guests.

"What guests?" asked H.

It wasn't a big wedding and though Bear was invited, he didn't turn up. I won't judge him too harshly as H did send him in the direction of Aberdeen.

As for the Bear, he moved with his mum down to a pensioners' village on the coast. Their furniture was made of cardboard, as it was adequate to support frail old women, but it was not built for the likes of Bear. When he shortly followed his mum to Folkestone, it wasn't long before they had to make themselves comfortable sitting on the floor. When I asked him how his relocation to the British coast was working out, he informed me that the coast winds could sometimes be gale force, and, that he had to sit on Gladys when the wind gets up to stop her from blowing away. This I would normally find funny, except for the fact that it was probably true. I just hoped he warned the eighty-year-old woman of his arrival.

On Bear's last annual Christmas pilgrimage to London to meet H and me for a drink, he turned up wearing a Chelsea top, which he had got from Oxfam for fifty pence.

"Why the Chelsea top, Bear?" I asked.

"Well, every year I turn up and that fucking psycho punches me in the head as soon as he sees me, so I thought this might provide me with a few minutes' grace," he replied.

At that moment, H walked into the pub and landed a glorious right hook to the side of Bear's head.

"It was never going to work, was it?" said the now wobbling Bear.

"H is old-school remember – if he had a therapist, his skull is would probably be on H's coffee table by now, being used as a bowl for his Werther's Originals," I informed Bear.

"How did your mum's move to Folkestone go?" asked H, carrying on the conversation as if nothing had happened.

"Well, I went through my mum's diary, as you do," said Bear.

H and I looked at each other, acknowledging that the 'as you do' was perhaps normal practice for Bear.

"Then I found her entry for her first week in her new place," said Bear, "and it read: 'I CAN'T BELIEVE IT: THE FAT BASTARD HAS FOLLOWED ME DOWN HERE.'

"I'm worried; who could she be referring to? Do you think we are being stalked? Am I in danger?" asked my anxious friend. Bear was in a pensive mood,

"I'm worried lately about getting old, what happens to old Bears?"

"Their hides are used to cover H's sofa," I replied.

Bear looked even more depressed, "Do you think I will get married?"

"Well, what are you looking for in a woman?" I asked.

"The usual, someone who can hold their own in a fight," he replied.

After a long while leaning on the bar and pondering his future, Bear then said, "I'm fucked aren't I? What a life. All I want is to find someone, so that I can sit at home with my pipe."

"You don't smoke," I added.

"I can start," Bear replied.

Following the warm greeting he had given our friend, H had said little; his mind was elsewhere. After a few minutes he made a comment on the subject of Bear's future.

"I was thinking about this and I have been working on the inscription on your headstone." H said.

Bear perked up, as he liked to be at the centre of any conversation, "What will it say?" he asked.

H described the inscription:

"'The Bear

Born 1961

Achieved nothing.

Died 2011'"

"Bollocks!" responded Bear, who then downed his third pint of mulled wine in one.

Bear downed a few more that afternoon and finally he said that he had better head home as his mum was due to give him his Christmas back wash.

Before he left, I asked my old friend, "What are you doing for Christmas Day Bear?"

"The usual, I'll scoff down twelve custard doughnuts and then take a walk to the park and expose myself to the squirrels," he said.

I wished Bear a Happy New Year; "Are you taking the piss?" replied Bear, just as H gave him one final almighty whack to the back of his head with a tightly rolled-up copy of the Daily Express to help him on his way.

Wurzel had his fortieth birthday recently. When it was discovered that the community hall where his party was to be held had a fruit-machine, the core reason why Wurzel was in debt and why he was barred from his bank, it was quickly arranged that his cash present was made in the form of a cheque. No doubt one with a huge penis printed across it.

I met Wurzel recently having a drink after a game.

"Wurzel how are you? As always your welfare is my greatest concern," I said.

"Caracas," he replied.

I turned to H, "When did he start drinking?" I asked.

"Yesterday," replied H.

Recently, Brains fell off the roof of his house, though his survival instincts kicked in again and thankfully he landed on his head. He has tried to make more of an effort and play a more active part in family life. Recently, to play a more active role in the family, he graciously offered to drive Linda's mother to a dinner she was having with her friends from work. It was going to be a very cold evening and the heater was broken in his car, so somehow he managed to manoeuvre the exhaust pipe through a hole in the floor of the 'Brainsmobile.' Linda's mother never complained of the cold, though she did insist on Brains stopping the car on the hard shoulder of the motorway, so that she could throw up after inhaling carbon monoxide fumes. Her mood wasn't helped by the fact that she was completely covered in black ash; she eventually crawled into the restaurant still gasping for air. However, despite such efforts, Linda finally had enough of his antics; with a heavy heart and for the sake of the kids, she told him to pack his bags and go. At least he had the good sense to leave through the front door this time.

Brains lived with me for a while, as I tried to help him get back on his feet. On the first day, I came home and Brains excitedly informed me that he had stocked up the fridge. I opened it to find twenty-four cans of Stella Artois lager and a pizza margarita for eight. Brains is trying to get in a bit more control of his life and I hope he succeeds, finds someone new, and sets up a new home so that he can descend on to a new patio that he can call his own.

Our Scottish experiment has secured some stability in his life, as he now works for a good friend of ours, John, on a fruit stall on Camden High Street. Surf, who is always the most supportive of us all, went along to assess Jockey's progress after his first week. Surf noticed that Jockey weighed each bag of grapes, but the orders were ninety percent of the time for only a pound of grapes. Surf, always the entrepreneur, then offered some advice to our Scottish experiment on how he could speed up his service and improve his turnover.

"Jockey, why don't you prepare ten bags of grapes, each weighing a pound? Then you don't lose time weighing and bagging them up each time."

"But what if they don't want a pound of grapes?" asked Jockey.

"OK, then just have a few at the side pre-weighed at two pounds," replied Surf, irritated that Jockey couldn't work out such a simple task.

"Ah, but what happens if they want three pounds?" replied Jockey, still questioning Surf's business plan.

"Ah, fuck yer!" screamed Surf, whose patience was exhausted as he stormed off to the pub. Jockey closed the business for a half-day and followed close behind.

Jockey announced recently that he was getting married, so he asked Numbnuts to be his best man. However, his prospective best man did not seem unduly honoured at being asked.

"Why don't you strip bollock-naked and stand in front of the mirror and have a good look to see what a prize prick you really are?" said Numbnuts.

"A simple 'no' would have done," said the loved-up Jockey.

It mattered little, as Jockey's bride to be was deported the following week for drugs offences and prostitution.

Numbnuts has not changed in the slightest and is still raging at the world, though he is now doing his ranting from Bangkok where he is a teacher. When I see Biddy she always asks after the Kids, so I update her in turn on H, Bear, Chelsea Mark, Surf, Jockey and the others,

"Where is Numbnuts these days?" she asked me recently over dinner.

"He's in Bangkok, on a teaching course," I replied.

Biddy carried on eating and just before she sipped another spoonful of soup, the eighty-two-year old woman made a single comment while still focused on her spoon, "Sex-case."

Surf continues to try to find an angle and for that I respect him; he never gives up. His work, indeed his very approach to life is high-risk; as a result, he is a millionaire. Even when his business ventures collapse, nothing saps his spirit. Only once have I ever seen him knocked back, after he was made redundant from a gas company when his company director informed him that its finances had sprung 'a leak'. But, the real low occurred following a heavy drinking session to celebrate the next challenge: he commissioned a street artist in Leicester Square to do a sketch of him. The picture she finally presented him with was a cross between the hidden portrait of Dorian Gray and Edmund Munch's *The Scream*. I think she was a psychologist, registering his inner turmoil rather than his outer, clean-cut, professional business image as he sat on the stool in front of her. Numbnuts laughed so much that he came very close to bursting an artery, while Surf was raging so much that he began to look frighteningly like his portrait. Later when we were in a taxi heading to The Mayflower, Surf, still angry, jumped out of the cab and fell down some street works being dug by another gas company in Tottenham Court Road. Now that's irony.

Bourgeois is still filed under missing-in-action and police are looking for a sixty-year-old woman last seen trying to suck his brain out of his mouth.

My Auntie Julia eventually had to go into a nursing home. I looked after her for two years but her mind was completely gone: she needed better help than Biddy and I could give her. The staff at her home appeared scared of her, and one member of staff asked me to try and get the television for the lobby back out of her room. The rest of the residents seemed unaware of recent events; they remained in their seats looking at the tobacco-stained wall and the white rectangle shape where the TV used to be. The thing was that the TV she had picked up and taken to her room was huge, so I asked her why she went to so much trouble.

"I don't mind people coming around to my house to visit, but sometimes I like to have a bit of time with Bruce Forsyth on my own," explained that lovely old woman. Julia died recently; she was much loved and I do miss her.

Our genes have somehow protected us and, despite the tales I have related, none of my immediate friends are alcoholics, in the sense of having to drink on a daily basis. Drugs have destroyed more of my friends than wars or alcohol. Last year, a memorial service was held for Shep who was found dead in a car in Denver with a needle in his arm. James, his best friend, gave part of the eulogy at his service and related the story of how, while they were on a tourist boat Tanzania, the tour guide warned all of the passengers not to hang their hands over the side as they were in crocodile-infested waters. A splash was heard, as Shep dived in. Perhaps a 'septic' had tried to strike up a conversation with him.

Despite never getting married or having children, I was conscious that my life had provided me with commitments I had to honour. One of those was that Loftus and Suzy left me some money and some possessions, but despite making me their children's guardian, they had made no made provision in their will for what I should do with their children's inheritance. I released some of the money on an annual basis to Monica and her husband Tom, the 'official' guardians of the children, to help with their upkeep. I invested the rest for when the children came of age to pay for their university education, to travel the world, to raise a family or to use as stakes at the tables; it was theirs to do as they wished. When that time had finally arrived, I contacted Monica and sent the money over. My investment turned out to be more than enough to cover their university fees and even to lay down a deposit on a house near the college, which they both shared.

A month later, I received a call from Monica to say the Liam and Mary, Loftus' and Suzy's children wanted to meet me. I said that there was no need to thank me, and just to tell them to get on with their lives and enjoy it.

"You'd better come over, that's not the reason they want to see you," replied Monica in a very anxious voice.

I thought it was strange but Monica would say no more. My thoughts were that maybe they wanted to hear more about their parents from their closest friend. I arrived at Dublin's Shannon airport to be met by Monica but not her husband. As we drove into the city, I confided my thoughts to Monica on why I thought they wanted to see me. I knew from Monica's subdued response that I wasn't right.

"There's something I have not told you and you must swear not to mention it to them," Monica confided.

"Of course, what is it?" I asked. I later regretted making this promise.

"It was very difficult for Tom and me to win the trust of Liam and Mary. We had to break all ties with Jim and Suzy, to allow the children to have a fresh start and as part of that we had to cut off any connection with you." Monica said.

"I understand," I said, and I did. I had not made contact with either child directly but passed everything through Monica to help them build a new life in Dublin.

I had written to Monica and Tom every Christmas and on the children's birthdays to make sure that they were not in need of anything.

"Well, will you understand that we told them you were dead?" asked Monica.

At this point, I paused, but the biggest surprise was yet to come. I continued, "If that was the case, how have you explained my miraculous reappearance?"

"We had to think quickly and we told them we'd had to lie, as we did not want them to know the true reason for your absence."

"Which was?" I persisted.

"You were in jail for twenty years."

"What?" I cried. I should have come out with something more profound, which truly expressed my surprise, but that was all I could muster at the time.

Before I could press further, we were at the rendezvous and there were the sweet little children I remembered, now adults. They were certainly wary of the recently released jailbird. I soon realised that the reason they had wanted to see me was that they thought the money was dishonestly acquired. Like Monica and Tom they had grown up to be very pious individuals. I explained that I had invested all of their parents' money and it was genuinely theirs to keep. I did this without revealing that in a way they were right; the original investment was not due to Loftus being lucky with premium bonds. I think the bit that really hurt about that afternoon, was not what they thought of me, but that they had no interest in their parents whatsoever. When I tried to talk about how Suzy and Loftus had loved them, they both looked away.

When they departed, I turned to Monica.

"The lie did not stop with me did it? You erased the memories of their parents too." I said.

"You don't understand, to make sure that they accepted us, Tom and I had to cut all ties to the past," Monica protested again.

"Bollocks, Loftus and Suzy loved them, you had no right to deny that, and they would still have loved you and Tom. You two bastards did this for you, not for them, you wouldn't share their love. Now I suggest you go and if that husband of yours wants to say any different he will find me drinking across the road, and I guarantee that by midnight he will be explaining himself to Loftus and Suzy in person."

Monica fled and I have never seen her or her husband since.

The two young adults I met after all those years had Suzy's green eyes, her pale complexion and the sharp intrusive chin of their dad. But internally they were clones of their step-parents, both as boring as buggery at Eton, with the same holier-than-thou attitude, suspicious nature and pensive approach to life. I have no doubt that they will be academically successful, continue to be regular churchgoers and each will raise a sensible family of their own. If I had raised them, they would be feckless, mischievous and prolific sinners, but their lives would not be based on lies. That night I got drunk in the bar; I cursed myself for betraying the memory of Loftus and Suzy. It was that day, twenty-five years after the event that I realised that that had been one of the biggest mistakes of my life.

I told Surf my tale when I arrived back in England and he roared out laughing.

"Christ almighty, what crime did you commit to serve twenty years?"

I suppose I had been so busy thinking about Loftus and Suzy; that it hadn't occurred to me to think about Monica's lies about me. Perhaps I was in shock at the time not to question Monica about my life of alleged crime. If I had been a mass murderer with a sideline in necrophilia, I would have been out sooner than that. Whatever crime they concocted for me, I must have been dressed as the Queen Mother at the time to warrant such a lengthy stretch in her daughter's custody.

Often I have been accused, usually by angry ex-girlfriends, of being afraid of commitment. This is nonsense; each year I renew my Chelsea season ticket. I had given up on relationships fairly early on; I was happy to settle for fun with no strings attached. That worked perfectly well and if any women got too close, I called in the marriage police and moved on. So when Lorraine, the beautiful nurse who mended my body on the boat after my battering in Panama, started to talk of marriage, I kissed her on the cheek and gave her my standard lame response, "It's my fault, not yours," speech and moved on. I had done that all my life.

Later that year, at Christmas, Lorraine called me to say that she was in the Royal Free Hospital nearby and she needed my help. When I got there, she was still as beautiful as ever, but there were tears running down her cheeks and fear in her eyes as she supported herself against the corridor wall in the hospital. I had no idea what was going on, but as she started to walk towards me, she stumbled and I grabbed her before she hit the floor. In holding her, I realised that she had lost about a stone in weight and my arm went right around her. At that moment, an equally distressed woman arrived, who turned out to be her mother, Joan.

"What's going on?" I asked, but instead of answering her mum directed me to take Lorraine out of the hospital and away from the crowds.

I lifted Lorraine up in my arms and carried her out of the hospital into a cab, her mother followed and I told the driver to go to my address.

"Thank you so much, but who are you?" asked Lorraine's mother.

"Johnny. I met Lorraine on a boat this summer," I replied.

"Bastard," she retorted.

"No please, no need for thanks, I am happy to help," Christ, I was now getting abuse from people who had never even met me.

By the time we got to my flat, Lorraine was asleep and her mother was in tears. I put Lorraine in my bed and left her to rest. I made Joan some tea in the traditional English manner of dealing with any emotional situation. Joan then told me that Lorraine was receiving treatment in the hospital, but that she had got lost, as she was disorientated following her treatment and that is why her daughter must have called me.

"What is the treatment for? I asked.

"Lorraine has cancer of the pancreas."

I knew that cancer was not the 'Big C' any more, thanks to tremendous advances in medical science, so on that basis I tried to give Joan some words of comfort along the lines of all was not lost.

"Johnny, Lorraine must have loved you for your stupidity. Can't you see that she is past the stage where it's possible to treat her cancer? All that can be done is to delay the inevitable. My daughter is terminal. Thank you for today, but, God, you're an idiot."

When I had ended the relationship with Lorraine, she became ill shortly afterwards. The family thought that she was just upset at our break-up and even her doctor diagnosed her as having severe depression. It was only later, when her condition drastically deteriorated that she was taken into hospital and they discovered that it was terminal cancer. Lorraine could not return to nursing and was so weak that she had had to return home to live in Cambridge with Joan and her stepfather. Joan admitted that Lorraine and her stepfather had an uneasy relationship. On top of that, Lorraine had to make an exhausting trip into London twice a week to visit the hospital for treatment.

After relating her story, Joan too was exhausted, so I put her feet up on the sofa, put a quilt over her and let her sleep. Joan told me that neither of them had had a decent night's sleep for months and that this was exasperated by Lorraine's panic attacks, which were on the increase. I sat there thinking about these two poor women, their nightmare journeys into London, the stepfather who it seemed cared little for Lorraine; I only had one option. So for the next three months, Lorraine stayed with me and on occasion Joan and Lorraine's sister Rachael took turns sleeping in the front room on my sofa bed. At first, I slept on the floor, but Lorraine would join me, as she said having me nearby, for some strange reason, made her feel secure. Even more perversely, her panic attacks subsided.

Occasionally Lorraine would wake up during night, she'd be scared and sweating profusely, but I would hold her tight against me and tell her that she would be alright. If there is a God and he forgives me for repeating lies to a dying woman maybe I'll be allowed into heaven. I'm not sure I would want to go in anyway: I wouldn't know anyone in there apart from my Auntie Julia. I would have a few words for him though.

When Lorraine did wake up following a panic attack, I was usually awake; in fact I hardly slept during those months. I could hear her breathing deeply and feel her hands involuntary gripping the hairs on my chest throughout the night. She was scared, but according to Joan, sleeping close to me finally offered her daughter some peace. Joan was a very pious, churchgoing, middle-class mother, but she was beyond caring what the neighbours thought and just wanted her daughter to have some peace in her final months, even if it was with an idiot.

One day Lorraine's stepfather came to my flat with Joan. He proceeded to scold Lorraine for not spending more time with them to help them with their grief before she died.

"You always only think of yourself, Lorraine. Do you never think of the pain you are putting your poor mother and me through?" he said. Those were the last words he ever spoke to his stepdaughter, as I then threw him out of the house.

Throwing him out on his arse is probably an action you would have expected of me, having read thus far, but Joan's reaction threw me. As her husband started the car, still rubbing the ear by which I had dragged him out of my house, she smiled up at me and then got into the car beside her man without saying a word to him. She seemed to be able to find a compromise in her world; balancing her love for her daughter with her love for her husband. Unfortunately, I see the world largely in black and white, so I think I have the skills to become a leader of a superpower or perhaps the next pontiff. However, for people with a clearer perspective, life is complex and mainly coloured grey.

During her final months, Lorraine fought hard. She laughed more; she even made her way through my DVD collection; however, in hindsight that might have indicated that she was beyond caring. She even wanted me to teach her how to play chess; there was only time to learn the fundamentals, but it mattered not, as she appeared to revel in learning something new. She was determined to fight and we both started to fool ourselves that things were improving, but her cancer was relentlessly eating its way through her.

When we first met, it had been passion; she had been incredibly attractive, sassy and fun. That was then; during her illness, I saw another side to this remarkable woman. I don't think I've ever met anyone as brave as Lorraine was in how she coped with the pain in her final weeks. She never lost her femininity; as she declined in health she grew in grace. She loved her shoes, went straight for the fashion magazine from my Sunday broadsheet and refused to let me help her in the bathroom, unless she was completely drained from being sick all night. Even though I would hear her struggle for half an hour trying to use the toilet, she remained a proud woman refusing my help. Neither death nor I were going to strip her of her dignity. Many a night I lay in bed thinking of how I had tried to help others, in a small way, in various parts of the world, and yet I was unable to help the person who lay, wrapped around my body, scared and wracked with fear.

Lorraine's last days were hell as she could no longer stand. I had to take her to the toilet, wash her and feed her; her anger grew in response to her helplessness and reliance on me. Better writers can come up with more imaginative ways of describing such events, I can't. All I can say is that it was a fucking cruel and painful way to die. On the day before Lorraine died, Joan seemed to know that her daughter's death was imminent, whereas, as usual, I didn't have a clue what was going on. Joan came over to me and hugged me for the first time in all those months.

"I think I preferred it when you just called me an idiot," I said.

"You may be useless, but I know my daughter loves you and you care about her," she said, holding my face tightly in her hands and staring directly into my eyes.

I held firm and did not show how I felt. I would have my time after Lorraine was gone. When the time came, I would grieve in the traditional Irish manner, getting smashed out of my skull and knocking out one of my best friends, even if it cost me a flight to Bangkok.

I said to Joan, "Do you want me to call your personal bastard to come and take you home?"

"No, I will spend my last night here on your sofa," said Joan, again acknowledging something I didn't.

Lorraine would wake up very slowly and try to get her bearings; as she did, she would always give me a tight squeeze, but her hugs were weaker.

On her final morning, she looked up at me and said rather ominously, "When my mum next pops out for a walk, I need to talk to you."

"OK. I will send her out for some condoms." I said.

"I'm serious," she looked stern.

"Ok, let's risk it," I said, as I playfully pulled her towards me.

"God you really are an idiot," she said giggling.

"Yes, but I am consistent."

Later, Joan went for a walk on nearby Hampstead Heath and Lorraine told me to sit on the bed when I brought in some tea.

She wrapped herself around me again but tighter than ever before, which took an extraordinary effort on her part.

"You're a smart, clever man, but you can hurt people without knowing it." Lorraine said.

"Oh, I don't know, when people start throwing chairs at me I usually think something's up." I replied.

"Christ, you're hard work," she then took my face in both her hands, as Joan had done but if I hadn't seen her do it, I would hardly have felt her touch.

Lorraine continued, "You're a loner, I know that, but I don't want you to be alone," she added, "This will mean that you have to try and think before you open your mouth," she paused as talking was draining her and then continued, "Johnny, try and be that one thing that men never want to hear about," Lorraine continued with my face held in her hands, not allowing my eyes to divert from hers.

I paused and then replied, "You mean, 'Hello Mr Floppy'."

Despite how draining this was, Lorraine would not be dissuaded from her task,

"Be nice, Johnny."

She then said the words I wanted to hear for the first time, but being the braver of the two of us, she said them first.

"I love you."

"I love you too," I replied without pausing.

I think it was the first time I had ever said that without weighing up all the ramifications and all the repercussions, I meant it. It was honest, total and incredibly painful. In the traditional family manner, I took Lorraine's face in my hands and gave her the gentlest kiss. I was afraid to break my incredible woman. Lorraine, strangely, for the first time in those last weeks, slept peacefully for most of the day. I remember looking down and seeing a small trickle of blood emerge from her lips, which slid down the side of her mouth and dripped on to a hair on my chest; fortunately hairs have no feelings; then the droplet of blood hit my chest like a hammer. The drips formed a crimson pool on my chest that I gently mopped up; I did the same thing with the rest of the blood from Lorraine's lips during the night. That night, Lorraine could fight no longer and she left me.

Less than a year after I'd met the nurse who helped remove my physical pain, that same nurse, died peacefully in my arms. It was around three in the morning and she was wrapped around me, as she had been every night for her last few months. That night, I was awake as usual, trying to read words off a page that my mind just refused to take in. Lorraine's breath slowly ebbed away; there was no death rattle, no seizure, she just went completely still. Her head was resting on my shoulder and despite the fact that all the energy in her body had gone, her final movement was to grasp the hairs on my chest for the last time, but her fingers did not relax. She had fought long enough. I lay there, kissed her once more and held her for a while. People say it's better when you know someone is going to die; that is a lie, you can try and prepare yourself, but it makes no difference in the end. It fucking hurts.

I went and woke Joan who ran to the bedroom, wrapped her lifeless daughter in her arms and cried softly. You often hear that there is no greater pain than losing the child you brought into the world. It's true: Joan kind of died with her daughter that night. A few weeks later, Joan suffered a mental breakdown and a month after that she took an overdose, which left her severely brain-damaged. That night, I left Joan to mourn her daughter and I went into the living room and called the special number that the hospital had given me. I sat in the dark on my sofa listening to Joan crying next door. I thought of Lorraine and of how she had removed my physical pain that summer, but then left me with a deeper scar. As with all pain, it will heal in time. The scar will remain, but I would never be without it for I am a better man for having known such a wonderful woman.

Chapter 37 – *With Friends Like These Who Needs Enemies?*

Friendships are often based on trust and shared values. This is not the case with mine. I disagree with my friends on mostly everything. This also goes for the relationships that they have with each other. H, a true Chelsea fan to his core, lumps Arsenal fans in with Germans and refuses to speak to Surf and his fellow Trumptonites when they meet. Surf, ever the salesman, ignores this and grabs and shakes H's hand wherever they meet. This can take some time when H's hands are hidden within folded arms. These tensions between one section of my friends can lead to strange alliances amongst the others. Tommy, a Tottenham fan, and H always have a very warm welcome for each other when they meet on Arsenal territory.

"Alright, H?"

"Yer. Alright Tommy?" well that is it really, for them that is as emotional as it gets.

I have taken these tensions between the rogues into account. I have made H and Surf joint executors of my Will, knowing that if I made H my sole executor, Surf, Numbnuts, Brains or any of the Arsenal crowd would not be invited to desecrate my grave. Similarly, I know that if Surf were my sole executor, H, Amelia, Wurzel and the rest of my Chelsea supporting friends, plus the Tottenham contingent, would be absent, as he would only remember to invite them to my funeral after Numbnuts had finished stuffing his face with the last sausage roll at my wake.

With the arrangements in place and all my friends present, with the exception of Bear who will be misdirected by both sides to the British Army's firing range on Salisbury Plain, my funeral will be a day of fun and frolics. The congregation will listen to Elvis's *Return to Sender,* as I make my way down the aisle at a forty-five degree angle. I'm after that pinewood-box dragster look, having Jockey and Churchill, both around four-foot tall, as pall-bearers at the front and Wurzel and Chelsea Mark, both well over six-feet in height, at the back. I am sure that the irony of me finally making my way to the altar will not be lost on the congregation. After Brains' eulogy, which will be short and consist of the words 'I didn't know him,' the ceremony will close with the twenty-two minute version of the John Barry track *Underwater Mayhem* from *Thunderball*. It's a pleasant tune that helps me relax, so I'm sure it will get the congregation in the mood to head to the bar and start the celebrations in earnest, particularly Terry.

Listening to other people's opinions leads to greater understanding, resulting in a consensus of opinion and on the whole a harmonious existence with your fellow man. Of course, I'm talking about reasonable people, not the rogues I know, who are a bit like the Amish in having an area marked out that segregates them from the rest of the world. Amish life is centred around a bright white wooden church, populated with women sitting alongside their well-behaved children who gently swing their legs on hand-carved pews. Outside, men with neatly trimmed beards wearing black wide-brimmed hats and buttoned up white shirts with no ties, secure the horse-drawn carriages that brought them to worship to a white picket fence by the church. The only slight change in my analogy would be that the Kids would have swiped the building materials to construct the church, sold them on the black market and 'reallocated' the money for their personal vices. But if you ignore that, add a fair dollop of sexual activity and mix in some blasphemous language, then you could easily belief you are on a farm in Pennsylvania, provided you have taken enough horse tranquillisers.

'Communities' have rules and to be accepted you have to conform to them. Whether it is a young black man on a Peckham estate or a white 'ned' in a Glasgow tenement, to a gang member, the gang is their community. For them, it's more to do with the gang ethos and gaining respect, having material possessions to impress your friends and to attract women. Even those in the highest echelons of society do not escape the need to play by their communities' rules rather than the laws of the land. Some turn a blind eye to abuses by members of their own community, others go further, proclaiming that traditional values must be upheld, but all the while living in fear of being exposed breaking those same rules and being banished from their social circle. In *The Life of Brian*, Brian pleads with the baying crowd to think for themselves and 'That they are all individuals', which all bar one of them repeats. The lone dissenter who says, 'I'm not', is battered to the ground. It all leads to the same thing. Most of us, me included, succumb to peer pressure as we desperately try to be accepted, perhaps even loved, by our peers.

Not so benevolent rogues: they think for themselves and care nothing for society's boundaries, which means they stand alone. Those like Rory, who watched his regiment head home as he stood fast with the Kurds and was damned as a traitor for it, and Larry, who cared nothing for the edicts of his state department, if it meant abandoning the innocent to a drug-crazed mob. Rory and Larry knew what was right and made a stand. You could accuse Rory and Larry and other benevolent rogues of many things, but never that of simply obeying orders.

Some of the rogues I know are not necessarily trustworthy, brave or even loyal for that matter, but they always remain true to their character, so I never feel betrayed by their actions. My good friend Jockey for example would try to seduce your fiancée as you were walking up the aisle; he would be the first to admit it and he would proudly reply with, "Oh yes, pal." When he texted me recently to ask how my new relationship was going, I replied very well. He responded,

'THAT'S A SHAME. I WAS HOPING IT WASN'T, I'M LOOKING FOR A LODGER.'

The very fact that you know you can't rely on Jockey as well as Brains means that you know where you are with them and that you can depend on them not to be there for you when you need them. It's comforting to have certainty in this ever changing world. I would put Chelsea Mark alongside Jockey, Bear and Brains, for though he is a socialist and a member of EMU, he often quotes Marx (Groucho rather than Karl), "I have principles and if you don't like them I have others."

Although his beliefs are as changeable as the wind, with Chelsea Mark I know where I am. You only get to the truth if you ask him the same question three times. When working in Saudi Arabia, he left the country very abruptly. I asked him why.

"I did it as a protest, making a statement against the abuses committed by the tyrannical ruling Saudi Royal Family," he answered.

"Why?" I asked for a second time.

"My mum wasn't well, so I had to rush home to look after her, as any dutiful son would do."

"Why," I pressed for a third time.

"I owed a thousand pounds in rent. Fuck em!"

Some of the Kids believe that the others are eccentric or just plain mad. Brains more than any of the others, believes that he really is the only sane person in Trumpton. However, he is the only one who has been sectioned recently, having been tasered by the police while he was wearing only a pair of underpants. The Brains family were very upset by this, as he didn't even pay them the common courtesy of being tasered in his own underpants; he was wearing his father Billy's Sunday best.

After a christening of one of Brains' kids, Amy said to me, "You're smart, Baloo, you have a good job," I was working for the Government at the time, "you mix with the rich and the well-educated, so why do you hang around with this lot?"

"Their insanity keeps me sane," I replied.

Amy is an independent spirit, she puts her brain to good use, had a good job and she was raising a great family with Tommy but she knew what I meant. We both knew we needed our benevolent rogues as much as they needed us. We too were rogues after all.

Amelia, Kathleen, Bertie, the Colonel, James, Wayne, Francis, Sam, Maureen and

Eddie and many others helped me get medical supplies to orphanages in Romania and then years later to children's hospitals in Bosnia. Later, Sam, Hannah, H, Surf, Chelsea Mark, Gertrude, Colleen, Larry, Amy, Terry, Smiler, Tracy and even Numbnuts worked in their own ways to help get me across Southern India on an old British Army Enfield *Bullet* to raise money for a children's charity. That was my way of trying to do something positive after Lorraine died. I had witnessed the effect on her family. Lorraine was an adult, so I could only imagine the effect when it was a child.

At the same time, I tried to do more to help women who were in the same position as Biddy had been, trying to protect her family and herself but living in fear of her partner. Some good friends in the police and the military have helped me relocate 'high-risk' families in Britain and the United States.

Perhaps the one person who summed up the philosophy of my rogues, including the benevolent ones, and their approach to life, was my late friend Loftus. Though it is not the most original of comments, I remember, as we sat in a pub in Camden following our release from the police cell, both battered and busted up, he turned to me and said, 'Johnny Boy, well at least we made them sit up and know we were here!' He included some other words, well one six times, but I think you have the gist of it.

Chapter 38 – *Twilight of the Godless*

I made four decades, six years more than my father, so to celebrate I booked the function room of the Celeste Voltaire pub in Camden. I invited the usual 'unusual' suspects to ensure that the experience was scarred into the memory. Simon, the manager of the pub, informed me that if five hundred pounds was taken behind the bar, then he would not charge me for the function room, apart from any breakages. No problem the fee, but the last part would probably involve the sale of my motorbike.

I thought that for such an occasion I would make an effort. I based the cuisine on one of those Scottish celebrity chef books – I think it had the title, "Open the Wrapper and Fucking Eat It!" I threw two carrier bags of French bread, a pound of full-fat butter, half a litre of Branson pickle, ten pounds of English cheddar cheese and twenty packets of processed ham on to the table. I called the catering area the Hougoumont, in honour of the farmhouse where the bitterest hand-to-hand skirmishes took place during the battle of Waterloo. There were some pretty hefty men coming; so, to reduce the inevitable casualties, I had enough forethought to make sure that the cutlery was all plastic.

Eddie had arranged the DJ. He had recently lost three fingers in an accident and had received sympathetic cards along the lines of "How's your darts?" and "I hope that wasn't your wiping hand."

When it came to alcohol, the rogues, benevolent or not, were seasoned drinkers, so they all began the evening at a steady pace; drinking copious amounts of alcohol right from the start. As the host, I decided to balance myself by making sure that I was holding a bottle of Pilsner in each hand, like a tightrope walker with two open umbrellas making his way across Niagara Falls.

As some of the Kids were meeting each other for the first time that night, all sorts of stimulating conversations were taking place. Tommy's friend Nick was on his best behaviour, but then Sam spotted his jewellery and approached him.

"What a face, you're a very angry young man aren't you?" she asked.

Nick turned to me, "Baloo, who the fuck is she?" he said, through gritted teeth.

"Nick, you're on your best behaviour, which means no throwing any of my guests out of the window, and that includes the tables," I reminded him.

Sam then looked at the jewellery on Nick's fingers, "Let me give you a fashion tip love, lose the rings."

Nick kept his promise; he stood on his own away from everyone all evening, but he did provide Sam with a reply, "Let me give you a violence tip, love," he said, showing his clenched fist, which was encrusted with gold rings to form makeshift knuckle-dusters, "these hurt."

"Oh, you're just a confused little boy," said Sam who then tweaked his cheek and roared with laughter as she walked off in search of her next prey.

Nick turned to me afterwards, "What a woman, she's got more balls than anyone I met when I was on the dangerous offender's wing."

Meanwhile, Tommy had been corralled in a corner by Amy. He was bellowing steam through his nose, desperate to drink the place dry and launch a full-scale verbal assault as the first of the Arsenal forces from Trumpton entered the battlefield. Arsenal had beaten Tottenham earlier in the day, so you couldn't see his head through the cloud of steam. Churchill's wife Gillian was as polite and charming as always, but that was all forgotten when the shutter was lifted and the bar declared opened. The first wave of the Chelsea heavy artillery arrived on the battlefield promptly at eight o'clock, with H, Wurzel, and Plank making their way to secure the high ground already claimed by Churchill, that being the right-hand side of the bar nearest the toilets on the landing below. Strategically, that was a major mistake, as Scottie deserted the battlefield before a single shot was fired, disappearing into the toilets with one of the barmaids, which depleted their forces and cut off their access to the latrines.

Five minutes later, the final troops of the French army of Arsenal entered the field and took position. I say French as their manager was French, the team was mostly French and the language in the dressing room was French, according to their only English player. This particular troop was led by Surf, Numbnuts and Snotbox. Even before the outbreak of hostilities the Arsenal had already taken heavy causalities, heavy in the sense that Dale was a big lad, and he had already collapsed blind drunk in The Bobbie. Earlier in the day, he and the others popped in for "just a quick warmer" before the main battle.

"It was terrible. I could barely look at the damage he caused. I think he nearly shat his pants," said Surf, describing the carnage to me.

"I was the one who had to drag him into the TV lounge," said Numbnuts, "and believe me there was no 'nearly' about it."

Tom the Bomb and Don were also lost, as they had remained with Dale in The Bobbie. They were like Napoleon's Old Guard, protecting the exposed rear.

Surf approached H and they shook hands in the way that Wellington and Napoleon might have saluted each other before the first cannons were fired. In reality at Waterloo, the opposing military leaders never met but I would have surmised that there would have been more warmth in their encounter than there was between my two best friends.

Meanwhile, a separate skirmish broke out by the Hougoumont, when Kathleen arrived and Sam went over to introduce her to Angela. Kathleen and Angela had never met, but Sam was able to form an instant and mutual bond between them when she opened with a brutal but effective frontal attack on both.

"I never liked you," Sam said to Kathleen and then added, "But you're better than her," she indicated by directing her thumb at Angela.

Kathleen turned to me raising her eyebrows, "I forgot that not all of your Kids are little boys."

At that point, dissent broke out in the Chelsea ranks, when Gillian belted Anarchy for saying "Wow" when Colleen walked in. In the heat of the battle, lines became blurred. Sam pinched Wurzel's arse and he turned on Plank.

"Cunt off! I'm one hundred percent butter side up ... though I'm still not sure what side that is!" he screamed.

Gillian then tried to win the barmaid's favour to get served before the others. Her advance, though direct, was strategically flawed. The barmaid refused to carry out her orders; quite understandable when someone tries to catch your attention by describing you as having a face like a slapped arse. Meanwhile, alcohol and lips clashed violently.

Like the English and the Prussians at Waterloo, Chelsea and Tottenham were uneasy allies that night, as their mutual hatred for the French took precedence for that one brief, but deadly skirmish. Once the French were defeated the English and the Prussians went back to fighting each other with even greater vehemence for another hundred and fifty years. The animosity between Chelsea and Tottenham will last even longer. Tommy, like the Prussians, was held in reserve during the battle, as he was corralled in a corner without a drink by Amy.

Chelsea had need of reinforcements, so when Scottie reappeared he was sent off to scout for reserves. He returned triumphant and filled the depleted British ranks with women from the downstairs bar. Unfortunately, Scottie soon became a casualty and like Lord Uxbridge, he had to leave the field of battle early. Unlike Uxbridge, his injury was not due to the loss of a leg from cannon grapeshot, but because of an attack to his groin area when Tracy grabbed him there, dragged him downstairs and threw him into a taxi. Scottie did tell me a few weeks later, that although he'd escaped any serious injury, he was still sore and, like Uxbridge, had taken severe injuries to his legs, his knees in particular.

Waffle and Terry had been lured to the engagement not by the allure of rum, as was the case for some mercenaries at Waterloo, but by John Smith's bitter. I say mercenaries in the sense that they had no allegiance to the main protagonists; Waffle was a Hull supporter and Terry was a Liverpool fan. Waffle and Terry went over to say hello to Bourgeois, whom they had not seen since my university days, but Bourgeois had already fallen off his stool drunk. He managed to retain a grip on his *Guardian* newspaper while he slept on the floor. Other mercenaries broke ranks as personal scores needed to be settled. Eddie, a Liverpool fan, picked up a stool with his one-digit hand and attempted to break it over the head of the Everton-supporting Scouse Bob, who put up stout resistance.

"Get to fuck, Nelson!" Eddie shouted, which was confusing, as though it was the right period, Trafalgar was a different battle altogether.

Tommy's Prussian force was firmly penned in by Amy, and they were hidden by steam, so much so that that we lost sight of that corner of the room.

At half past eight Simon ran up to me to say that I would not be charged for the function room, as the bar had taken the agreed amount and surpassed the record set by the police in their fundraising pub-a-thon the year before. They had held the previous record of a very impressive ten minutes past nine.

Churchill, meanwhile, looked leeringly at the barmaids, not noticing that Gillian had appeared on his flank. She delivered a punch to the back of his head. I surveyed the battlefield; bodies were strewn across the ground while other casualties staggered bewildered up and down to the bar. I was reminded of Wellington's words when he commented on his troops, but like 'Nosey,' I beamed with pride at their total commitment to the battle and the drunken debauchery now taking place.

Then, from the mêlée, the DJ staggered into the mid of battle like some bewildered pacifist trying to bring the action to an end. He foolishly informed us that he refused to participate and play any my records. You couldn't have a party selection without a bit of Wham he said, as H looked over at him, menacingly. Before H had a chance to have a quiet word with him, Sam grabbed the DJ by his testicles and threatened to French kiss his girlfriend, unless he took up arms once more. He waved a flag of surrender, playing The Clash's *London Calling,* followed by that quaint little urban tune by The Jam, *A Town Called Malice*. Sam, to the horror of the recipient and her boyfriend, stuck her tongue down the DJ'S girlfriend's throat despite his complete capitulation. It was a bloody affair, no quarter was asked; apart from the pleading by the DJ, none was given.

Tommy was now on the verge of spontaneous combustion at his enforced isolation in the corner without a drink, but just then, his forces broke free, as he circumvented Amy's barricade of a handbag and two bags of shopping. He broke through the French ranks, cutting with the sympathy of a scythe through a cornfield, making his way towards the alcohol-depleted bar.

The Scots Greys, Wellington's heroic cavalrymen, then appeared. Jockey had developed a little snow on his sideburns. Reminiscent of the glorious entrance of his countrymen, who had ridden furiously towards the French cavalry, he entered the battlefield already fortified by two bottles of red wine. He rode fearlessly into the French to get to the bar. Like the Scots Greys, he was met by a ruthless response from the opposing forces.

"You fall into me again and I'll nail your battle-scarred cock to your forehead," said Numbnuts, lancing him with his tongue.

Our little Scots Grey bravely bounced along the French lines without fear, hesitancy or sense until he crashed into the Hougoumont.

As he drunkenly tried to struggle to his feet, Dillon commented, "Fucking drunken kilt-lifer."

"Bastards!" cried our little Scottish experiment; he gave up any further advances and fell back to the ground holding a scotch egg, which he must have brought with him.

At that moment, a tall statuesque blonde walked in and the Chelsea battalions surged forward with a regiment of Tottenham fans in tow, but the Arsenal forces held their position, refusing to budge. Then on closer inspection, I could see why they were keeping their powder dry. Was it the hairy legs or the Arsenal tattoo that indicated that Dancing Spice was the blonde transvestite? H spotted the discrepancy too and he quickly sent a message along the Chelsea line warning that the French were feigning a weakness in their position and leading the

British into an ambush. The Chelsea and Tottenham ranks withdrew to the safety of the bar in horror, as the British troops had done in the last throngs of the battle, forming 'English squares'. That night the 'square' was in the form of vodka shots and lager, piled high on a table between them and Dancing Spice. Unfortunately, the English square nearly collapsed as part of its front line broke off. Wurzel transported his drinks with him to fraternise with the enemy, as he talked to Dancing Spice about the outfit he had bought in Genoa.

That was the turning point in the evening. H and Tommy, leading the allied forces, knew that victory was theirs as they watched the Arsenal troops retreat to the rear of the bar and begin drinking harder to cope with the shame of one of their own entering the fray wearing a blonde wig, suspenders and high heels.

Now that the battle had ended, Brains entered and surveyed the carnage. He was nearly knocked down the stairs by Anarchy as he was chased by Gillian, who was set on landing another blow to her man's head. All combatants put their differences to one side as the Chelsea and Tottenham battalions, along with the Arsenal forces, downed vodka and whisky shots like there was no tomorrow. Many soldiers didn't emerge again until later that week. The price of victory was high in livers and kidneys. A deathly quiet descended on the battlefield and was only interrupted by the sound of an unconscious *Guardian* reader snoring happily away on the floor. After a time it was broken again by repeated and desperate cries from the most able of the wounded, "Ten lagers and the same again in sambucca, love!"

Brains, having only just been released from the sanatorium, surveyed the carnage. "Daloo, everyone seems to be getting on great," he said, stood next to me and viewed the throng of bodies piled up against the bar. He added, in the way that only Brains could, "Yes, you could cut the atmosphere with a cake." Sixty miles away, Bear was sitting on his own wearing his Norwich boob tube in a notorious gay sadist pub in Brighton. He was wondering why I was having my party there, where everyone was and why the huge moustachioed man at the bar with toilet-chains clamped to his nipples was blowing kisses at him.

Printed in Great Britain
by Amazon.co.uk, Ltd.,
Marston Gate.